Environmental Footprints of Drought: Focusing on Emerging Issues and Their Underlying Mechanisms

Environmental Footprints of Drought: Focusing on Emerging Issues and Their Underlying Mechanisms

Editors

Jinping Liu
Quoc Bao Pham
Arfan Arshad
Masoud Jafari Shalamzari

MDPI

Basel • Beijing • Wuhan • Barcelona • Belgrade • Novi Sad • Cluj • Manchester

Editors

Jinping Liu
North China University
of Water Resources and
Electric Power
Zhengzhou
China

Quoc Bao Pham
Institute of Earth Sciences
University of Silesia
in Katowice
Bedzińska
Poland

Arfan Arshad
Biosystems and Agricultural
Engineering
Oklahoma State University
Stillwater
United States

Masoud Jafari Shalamzari
Department of Environment
Tabas Branch
Tabas
Iran

Editorial Office
MDPI
St. Alban-Anlage 66
4052 Basel, Switzerland

This is a reprint of articles from the Special Issue published online in the open access journal *Atmosphere* (ISSN 2073-4433) (available at: www.mdpi.com/journal/atmosphere/special_issues/footprint_drought_mechanisms).

For citation purposes, cite each article independently as indicated on the article page online and as indicated below:

Lastname, A.A.; Lastname, B.B. Article Title. *Journal Name* **Year**, *Volume Number*, Page Range.

ISBN 978-3-0365-9962-5 (Hbk)
ISBN 978-3-0365-9961-8 (PDF)
doi.org/10.3390/books978-3-0365-9961-8

Contents

About the Editors

Jinping Liu

Jinping Liu, Ph.D., works in the North China University of Water Resources and Electric Power and is a visiting scholar at Katholieke Universiteit (KU) Leuven, Belgium. His research activities include cryosphere remote sensing, remote sensing of vegetation, climate change, and hydrological modeling, with particular research experience and a knowledge background in the use of remote sensing and geographic information tools for the identification, impacts, and mechanisms of extreme climatic events, as well as risk assessment. As a supervisor of masters, he has contributed significantly to academia and mentoring students. Also, he has served as a reviewer for several SCI journals and has organized several Special Issues with international scholars as a guest editor.

Quoc Bao Pham

Quoc Bao Pham obtained his Ph.D. from the Department of Hydraulic and Ocean Engineering, National Cheng Kung University, Taiwan. He has more than 10 years of research experience in hydrology, GIS, and Remote Sensing. His research interests are water resources engineering, applied machine learning, climate change, statistical downscaling, hydrology, remote sensing, GIS, and spatial analysis. He has collaborated actively with many international researchers in several other disciplines of engineering. He has published more than 80 SCI/SCIE research papers and is a reviewer of more than 20 SCI/SCIE journals. He is also a guest editor of some SCI/SCIE journals in the fields of water resources, remote sensing, and machine learning. He is also a reviewer of world-leading high-impact journals such as Nature Communications, Journal of Hydrology, Science of The Total Environment, Remote Sensing of Environment, and Journal of Environmental Management.

Arfan Arshad

Arfan Arshad, Ph.D., a Climate and Water Resources specialist working as Research Assistant at Oklahoma State University, United States. His expertise includes integrating geodesy and remote sensing datasets to study local-scale adaptive management and freshwater sustainability in transboundary basins. With extensive international experience, he focuses on hydrological modeling and climate impact assessment, notably in the Indus Basin and high-mountain Asia. He has contributed significantly to academia and mentoring students. He is also a guest editor for Remote Sensing and Atmosphere Journals with MDPI, actively contributing as a reviewer for several MDPI Journals, furthering his commitment to upholding rigorous academic standards and supporting the peer review process across various scientific publications.

Masoud Jafari Shalamzari

Masoud Jafari Shalamzari, PhD, presently lends his expertise to the Iranian Department of Environment. Recognized for several international publications, his focus encompasses remote sensing, GIS, and hydrology, with specific interests in water harvesting for wildlife and drought studies. He is currently the guest editor of the journal of Atmosphere with MDPI along with several other journals. His multifaceted involvement reflects a holistic commitment to fostering both student development and the advancement of research within the international scientific community.

Preface

In recent decades, human activities, along with climatic variabilities, have adversely impacted socioeconomic and environmental conditions, leading to a growing drought susceptibility, mainly in arid and semiarid parts of the world. Droughts are associated with the dynamic nature of human–environment interactions in ecosystems at different scales and dimensions, resulting in a wide range of issues such as soil and air quality deterioration, vegetation degradation, water scarcity, human migration, urban water supply issues, and a reduction in hydropower production. Thus, concerted efforts are required to bring all emergent concerns and their related processes together into a unified framework to serve as a roadmap for research and management. This Special Issue features studies on understanding the climate system's vulnerability, droughts relationships with large-scale climate patterns, underlying effects of droughts, and land–atmosphere feedback using observations or modeling studies, targeted field campaigns, or long-term measurements ranging from local to regional spatial scales. The papers published in the Special Issue offer a comprehensive exploration, ranging from hydro-meteorological, atmospheric feedbacks and socio-economic implications. Notably, publications pivot around forecasting drought events using data-driven approaches and provide applications on how earth observation datasets can used for drought monitoring, comprehensive evaluations of drought characteristics and associated drivers, watershed warmings, drought vulnerability for agricultural systems, and groundwater droughts. We hope this collection of diverse papers provides useful information and encourages additional research that contributes to understanding the land–atmospheric feedback of the earth system.

Jinping Liu, Quoc Bao Pham, Arfan Arshad, and Masoud Jafari Shalamzari
Editors

atmosphere

MDPI

Article

Extremeness Comparison of Regional Drought Events in Yunnan Province, Southwest China: Based on Different Drought Characteristics and Joint Return Periods

Ruxin Zhao [1,2], Siquan Yang [1,2,*], Hongquan Sun [1,2], Lei Zhou [3], Ming Li [1,4], Lisong Xing [1] and Rong Tian [3]

1 National Institute of Natural Hazards, Ministry of Emergency Management of China, Beijing 100085, China; ruxinzhao@ninhm.ac.cn (R.Z.); sunhq@ninhm.ac.cn (H.S.); bqt2200204047@student.cumtb.edu.cn (M.L.); xinglisong22@mails.ucas.ac.cn (L.X.)
2 Key Laboratory of Compound and Chained Natural Hazards Dynamics, Ministry of Emergency Management of China, Beijing 100085, China
3 School of Geomatics and Urban Spatial Informatics, Beijing University of Civil Engineering and Architecture, Beijing 102616, China; zhoulei@bucea.edu.cn (L.Z.); 2108570022093@stu.bucea.edu.cn (R.T.)
4 College of Geoscience and Surveying Engineering, China University of Mining & Technology (Beijing), Beijing 100083, China
* Correspondence: ysq_74@163.com

Abstract: Droughts frequently occur in Yunnan province, the southwest of China, which leads to crop loss, ecosystem degradation, and difficulties in drinking water for people. In order to assess and compare the extremeness for different drought events, this study quantified it by utilizing the joint return period of drought multi-characteristics. Three characteristics at the regional scale: drought duration, severity, and affected areas were obtained by a simple regional drought process methodology, and their relationship was considered based on three types of Archimedean Copulas. Standard Precipitation Evapotranspiration Index at a six-month time scale was selected as the optimal drought index based on actual drought impact data. Results showed that drought events in Yunnan province were mostly short drought duration, low severity, and high drought-affected areas. By comparing the historical reported droughts' loss, the return periods of drought events calculated by the combination of duration and severity and drought-affected area are much more suitable to reflect the real drought situations than those calculated by one- or two-dimensional drought characteristics, especially for extreme drought events. On average, the drought in Yunnan province was almost shown a return period of ~10 yr. The frequency of droughts in Yunnan province has gradually increased due to climate change, and droughts with ~100 yr or even larger return periods occurred in 2009–2010 and 2011–2013.

Keywords: drought duration-severity-affected area; joint return period; Copula; regional drought event; Yunnan province

check for updates

Citation: Zhao, R.; Yang, S.; Sun, H.; Zhou, L.; Li, M.; Xing, L.; Tian, R. Extremeness Comparison of Regional Drought Events in Yunnan Province, Southwest China: Based on Different Drought Characteristics and Joint Return Periods. *Atmosphere* **2023**, *14*, 1153. https://doi.org/10.3390/atmos14071153

Academic Editor: Alexander V. Chernokulsky

Received: 16 June 2023
Revised: 13 July 2023
Accepted: 14 July 2023
Published: 16 July 2023

1. Introduction

Drought is the main natural disaster that causes global grain production reduction. It also causes ecological disasters such as forest fires and vegetation degradation, as well as basic livelihood problems such as water storage for people in ecologically fragile areas. What's more, it has caused millions of deaths throughout human history [1]. Human activities' impact on the global climate has resulted in an increase in severe drought events, and changes in climatic parameters are expected to escalate the severity of droughts. Being influenced by climate change, it is projected that the severity, duration, and affected area of drought may increase continuously in the future [2,3].

Droughts are generalized water deficit phenomenon that can be characterized quantitatively by different indices. Reasonably, the precipitation anomaly percentage (Pa) (Henry, 1906) [4] was initiated and used widely to assess the degree of drought for several decades.

After the Standardized Precipitation Index (SPI) was developed by McKee et al. [5], it was demonstrated to be suitable for drought comparison at different spatial and temporal scales. In addition to precipitation factors, temperature, evapotranspiration, vegetation, soil moisture, etc., are also significant factors reflecting drought. Thus, drought indices considering multiple factors were developed one after another, such as Relative Moisture Index (MI) [6], Palmer Drought Severity Index (PDSI) [7], Standard Precipitation Evapotranspiration Index (SPEI) [8], Multivariate Standard Drought Index (MSDI) [9]. With the development of remote sensing technology, drought indices based on remotely sensed data are mushrooming [10,11]. These drought indices have been developed for drought monitoring throughout the world or regions [12,13]. However, the applicability of the indices varies across regions for reflecting the real drought impact [14]. Therefore, it is necessary to assess drought events for one designated study area based on the optimal drought index, which shows a higher link with actual drought impact [15].

To a certain extent, the drought index only reflects the severity of the drought. To properly evaluate or compare any changes in drought events, additional characteristics of droughts are required. Due to the need to incorporate these characteristics into any modeling, an analysis based on drought frequency alone is not enough if it is not quantitatively related to other information such as duration, severity, and areal extent [16]. Assessing the risk of drought events by integrating different drought characteristics is relevant. The Copula joint probability function can establish a dependence model for multidimensional random variables and is widely used in the study of multidimensional drought characteristics. Due to the dynamic nature of the spatial and temporal evolution of drought impact areas, previous research often used the run theory to identify the drought duration and severity at the site scale and then conducted bivariate drought return period analysis using the Copula [17–19]. Although site-based drought return period analysis can provide useful local information, these results seem to embed with high uncertainty in drought management or drought risk assessment in a wide area. During the development progress for one drought event, the timing and location of the drought both influence its consequences. Droughts are considered regional when the spatial extent exceeds a certain threshold [20], so it is crucial to include the affected area of drought with duration and severity in the drought risk studies [21–23].

Droughts can occur in both arid and humid areas across the world. Despite being located in the humid climate zone, the region of Southwest China has frequently been hit by exceptional and sustained droughts in recent years, with the summer of 2006, the autumn of 2009 to the spring of 2010, the summer of 2011, and the winter of 2019 to the spring of 2020 [24–26]. As one of the five regions that make up Southwest China, Yunnan Province (YP) suffered from frequent and the most severe drought disaster since 2006 [22,27,28]. The drought disaster record in YP indicated that the crop area affected by droughts reached a total of 308,349 km^2 during 1972–2020 [26]. Wang and Yuan [29] indicate that anthropogenic climate change increased the risk of hot and dry extremes in 2019 over YP by 123–157% and 13–23%, respectively. Many researchers have paid attention to the drought evolution in YP. For instance, Li et al. [30] analyzed the drought trend and drought coverage area at various timescales over YP, and the results showed that droughts in YP occurred frequently, and the change point was detected in 2002. Wang et al. [31] used Copulas to analyze the joint return period of drought events by considering two-dimensional drought characteristics: duration and severity. Due to the impact of climate change, extreme events occur frequently in different regions, and it is particularly important to combine multiple characteristics to evaluate the extremity of drought events. However, the joint return period of three-dimensional drought characteristics that include the drought duration, severity, and affected area has been few explorations in YP.

Therefore, this study addresses the spatiotemporal variations of drought duration, severity, and affected areas, using the regional drought characteristics extraction method, and discusses the drought extremeness of YP based on the return periods analysis. The objectives of this study are (1) to select the optimal drought index for the assessment of drought events in YP, (2) to identify the spatiotemporal characteristics of drought events and verify by the actual records of drought events, and (3) to discuss the joint return periods of droughts using Copulas, as well as the differences of return period combining drought characteristics in different dimensions.

2. Materials and Methods

2.1. Study Area

YP is located in Southwestern China, covered with a total area of 3.94×10^5 km^2. The main types of land use in the province are forest land, grassland, and cultivated land, which account for 57%, 23%, and 18%, respectively (Figure 1). This region is the river source or the upstream of many major rivers, such as the Nujiang River, Lancang River, Red River, Yangtze River, and Pearl River. Precipitation is spatially and temporally uneven across the region (ranging from 560 to 2300 mm) due to the influences of the monsoons and complex terrain and is primarily concentrated from May to October [32]. Therefore, droughts occur more frequently and severely in winter, spring, or early summer in YP and have a serious impact on agricultural production and human life [30,33]. For example, the severe drought in northern and central YP from November 2002 to early May 2003 affected 10.53 million people and 0.86 million hectares of crops, which led to a direct economic loss of 1.98 billion yuan [34]. The drought from November 2009 to May 2010 resulted in 24.98 million people affected, with 2.96 million hectares of crops affected and direct economic losses of 27.33 billion yuan [35].

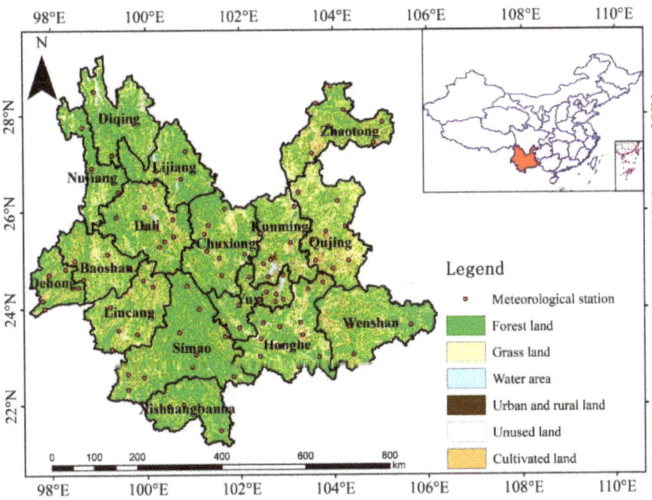

Figure 1. The location of meteorological stations in the YP.

2.2. Data

Three types of data were used in this study: meteorological and remotely sensed data used for the calculation of drought indices, actual drought impact data used for the optimal drought index selection, and drought event records used for the verification of whether the theoretical drought events extracted in this study are consistent with the drought events that actual occurred.

Meteorological data includes daily observations of precipitation and temperature at 101 stations from 1961 to 2020 (Figure 1), which were obtained from the China Meteorological Data Sharing Service System (http://data.cma.cn, accessed on 1 May 2022). Remotely sensed data mainly comes from Moderate Resolution Imaging Spectroradiometer (MODIS) productions to extract Land Surface Temperature (such as LST) and Normalized Difference Vegetation Index (NDVI). The included MODIS11A2 and MODIS13A2 productions ranging from 2002 to 2020 with 1 km spatial resolution, 8 and 16 days respective synthesis periods were used in this study. The data were obtained from the Land Processes Distributed Active Archive Center (LPDAAC) of NASA (https://ladsweb.modaps.eosdis.nasa.gov/, accessed on 20 September 2022).

The actual drought impact data were obtained from the drought statistics reporting system in China. In general, drought impact data are collected after the event has ended; and no data are collected if there is no drought event. Once the drought emergency response levels are triggered, the statistical frequency of drought impact data will increase, such as in the range of 3 weeks or even every day [15]. Six types of drought impact data during the period of 2003–2017 were available and used in this study (listed in Table 1). Considering that these impact data were reported by administrative units, this study will select the optimal drought index by calculating the correlation coefficient based on the sum of drought impact data and mean of drought indices in the sixteen cities (Figure 1) in YP during 2003–2017.

Table 1. Six types of drought impact data used in this study.

Impact Type	Abbreviation	Unit
Crop area affected by light drought	CA_LD	10^3 ha
Crop area affected by severe drought	CA_SD	10^3 ha
Crop area affected by extreme drought	CA_ED	10^3 ha
Water shortage and moisture shortage area of paddy fields	WMA_P	10^3 ha
Water shortage and moisture shortage area of dryland	WMA_D	10^3 ha
Population with drinking water difficulties due to drought	P_D	10^4 persons

The drought event records data were obtained from the China Meteorological Disaster Yearbook (https://data.cnki.net/yearBook/single?id=N2023020114, accessed on 15 August 2022) and the literature of Zhang et al. [36], which were available from 1949 to 2020 and used to verify whether the theoretical drought events extracted in this study are consistent with the drought events that actual occurred.

2.3. *Methods*

2.3.1. Alternative Drought Indices

Drought indices are an effective way to quantify drought phenomena, but there are a variety of drought indices currently available, as mentioned above. In view of this, this study selected six commonly used drought indices (Table 2) as candidates. Then, the rank correlations (Equation (1)) between the values of the indices and the actual drought impact data were calculated to find the optimal one for drought analysis in YP. Until this section; we finished part I of Figure 2.

$$Rc = 1 - \frac{6\sum d_i^2}{n(n^2-1)} \tag{1}$$

where d_i is the rank difference between the drought index and drought impact data series. The drought index series is the mean of the drought index at each station within one city. The drought impact data series is the sum of drought impact data at different counties within one city. n is the length of the used data series.

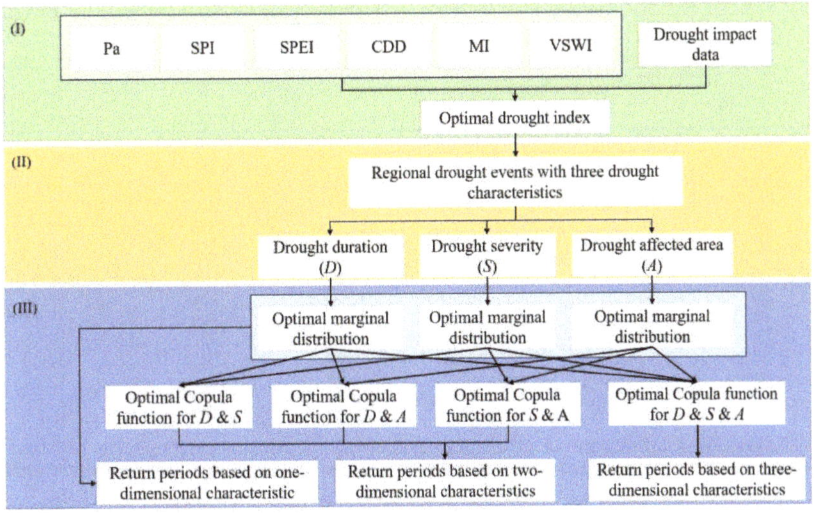

Figure 2. The flow chart of this study (Part I: Optimal drought index selection, Part II: Regional drought and drought characteristics extraction, Part III: Joint return periods based on different drought characteristics).

Table 2. The drought indices candidates used in this study.

Drought Index	Time Scales	References
Percentage of precipitation anomaly (Pa)	monthly	[4]
Standardized precipitation index (SPI)	1, 3, 6, 12-month	[5]
Standardized precipitation and evapotranspiration index (SPEI)	1, 3, 6, 12-month	[8]
Days without continuous rain (CDD)	maximum days without continuous effective rain in one month	[37]
Relative moisture index (MI)	monthly	[6]
Vegetation Supply Water Index (VSWI)	monthly	[38]

2.3.2. Regional Drought and Drought Characteristics Extraction

The occurrence of drought events is not limited to a single point, and the duration (D), severity (S), and affected area (A) are the main characteristics generally used to describe regional drought events from the temporal and spatial perspective. Previous research usually extracted D and S by the run theory method at a single site scale [31,39]. A is difficult to synthesize simultaneously with D and S to describe a drought event at a regional scale.

In order to obtain these three drought characteristics simultaneously, we referenced a regional drought process assessment methodology from Liao et al. [40]. Firstly, we determined the regional average drought index (I_m) and drought impact area (A_m) for each month:

$$I_m = \frac{1}{n}\sum_{i=1}^{k} Index_{opt-i}, if\ Index_{opt-i} \leq drought\ threshold \quad (2)$$

$$A_m = \frac{k}{n} \times 1 \quad (3)$$

where k means the number of stations with optimal drought index under the threshold and n is the total number of stations in the study region. The drought threshold value is given as -0.5 in this study.

Secondly, based on the run theory, the I_m series was used to extract the regional D and S based on a threshold value (-0.5). D is the sum month when I_m less than -0.5 for a drought spell, and S is the cumulative sum of the difference between the I_m value and its threshold value in the drought spell. If a certain two drought events were adjacent to each other for only one month, the two events would be combined as a single event (Figure 3).

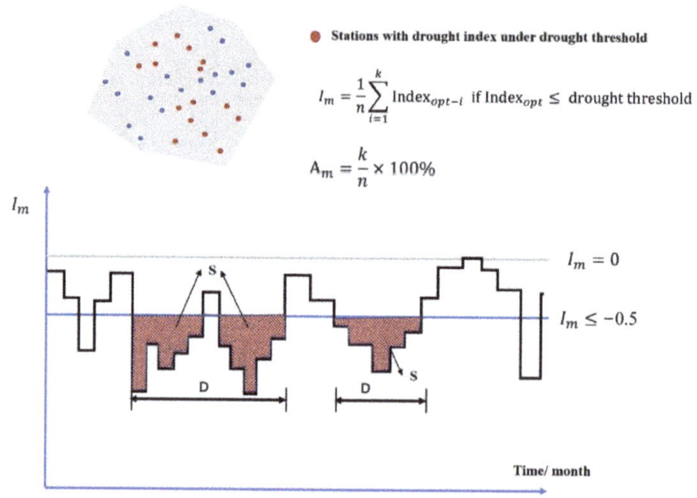

Figure 3. Schematic diagram of the drought event and characteristics extraction process (The bule dots mean the stations with drought index larger than the threshold).

Finally, the drought-affected area (A) for a drought event during the drought duration is calculated as follows, and until this section, we finished part II of Figure 2.

$$A = \frac{\sum_{m=1}^{D} A_m}{D} \tag{4}$$

2.3.3. Framework of Copula-Based Joint Return Period

The joint return period of different drought characteristics is defined as the average time elapsing between two successive realizations of the drought events and can be used to compare the comprehensive risk of drought events. To a certain extent, the larger return period corresponds to the more extreme drought event, the higher the risk of damage it may cause.

The joint return period is calculated as the inverse of exceedance probability of different variables, and the Copulas functions are usually used for this calculation due to their flexibility in modeling the dependence structure among random variables based on their marginal distributions [19,31,41]. After obtaining the appropriate Copula joint distributions, we can proceed with the analysis of drought return periods (as shown in part III of Figure 2).

Copula is defined as the multivariate distribution function of multidimensional random variables $(x_1, x_2, \ldots, x_m)$. It could be served to connect m-dimensional marginal distributions $(F(x_1), F(x_2), \ldots, F(x_m))$ to form a multivariate distribution function on $[0, 1]$:

$$F(x_1, x_2, \ldots, x_m) = C(F1(x_1), F2(x_2), \ldots, Fm(x_m)) = C(u_1, u_2, \ldots, u_m) \tag{5}$$

where C is the cumulative distribution function (CDF) of Copula, and $Fm(x_m)$ is the marginal distribution for random variable x_m, i.e., the marginal distribution.

The formula for the return period of one drought characteristic is expressed as:

$$T_{one-d} = \frac{E(L)}{P(X \geq x)} = \frac{E(L)}{1 - Fx(x)} \tag{6}$$

where $E(L)$ is the expectation of the average drought interval and $P(X \geq x)$ is the CDF of a given value x. F_X is the marginal distribution of variable X.

The joint return period of two or three drought characteristics is expressed as:

$$T_{two-d} = \frac{E(L)}{P(X \geq x \cap Y \geq y)} = \frac{E(L)}{1 - Fx(x) - Fy(y) + C_{XY}(Fx(x), Fy(y))} \tag{7}$$

$$T_{three-d} = \frac{E(L)}{P(X \geq x \cap Y \geq y \cap Z \geq z)} = \frac{E(L)}{1 - Fx(x) - Fy(y) - Fz(z) + C_{XY}(Fx(x), Fy(y)) + C_{XZ}(Fx(x), Fz(z)) + C_{ZY}(Fz(z), Fy(y)) - C_{XYZ}(Fx(x), Fy(y), Fz(z))} \tag{8}$$

where C_{XY} is the optimal Copula describing the joint distribution of random variables X and Y. C_{XYZ} is the optimal Copula describing the joint distribution of random variables X, Y, and Z.

- Marginal distribution

Before the establishment of the joint distribution, the optimal marginal distribution of random variables should first be identified. Six distribution functions (Table 3) were used to fit and optimize the marginal distribution for each of the three drought characteristics (D, S, and A). Kolmogorov–Smirnov (K–S) method [42] was used to test the theoretical distribution and empirical distribution in order to determine the optimal one.

Table 3. Alternative marginal distribution functions for a one-dimensional variable.

Distribution	Formula	Parameters
Exponential (EXP)	$F(x) = 1 - e^{-\frac{x}{\theta}}, x \geq \xi$	θ
Weibull (WBL)	$F(x) = 1 - e^{-(\frac{x}{b})^a}, x > 0$	a, b
Gamma (GAM)	$F(x) = \frac{\beta^{-\alpha}}{\Gamma(\alpha)} \int_0^x t^{\alpha-1} e^{-\frac{t}{\beta}} dt, x > 0$	α, β
Log-normal (LOGN)	$F(x) = \int_0^x \frac{1}{x\sigma_y\sqrt{2\pi}} e^{-\frac{(lnx-\mu_y)^2}{2\sigma_y^2}} dx, x > 0$	μ_y, σ_y
Normal (NOR)	$F(x) = \int_{-\infty}^{x} \frac{1}{\sigma\sqrt{2\pi}} e^{-\frac{(x-\mu)^2}{2\sigma^2}} dx, -\infty < x < \infty$	μ, σ
Logistic (LOGC)	$F(x) = \left(1 + e^{-\frac{x-m}{a}}\right)^{-1}, -\infty < x < \infty$	m, a

- Copula functions

In this study, three common Archimedean Copula functions: Clayton Copula, Gumbel-Hougaard Copula, and Frank Copula, were used. Details are shown in Table 4, where d in the formulas denotes the variable dimensionality. The parameter estimation of the Copula function is performed by the method of great likelihood estimation. The optimal Copula function is tested by using the Akaike information criteria (AIC) and Bayesian Information Criterions (BIC) based on the joint empirical and theoretical probability of the random variables [43]. The smaller the statistic value of AIC and BIC, the better the corresponding Copula function fitting.

Table 4. Basic information of commonly used Archimedean Copula functions.

Copula Type	Formula	Parameter Value Range
Clayton	$C_\theta = (\sum_{i=1}^{d} u_i^{-\theta} - d + 1)^{\frac{-1}{\theta}}$	$\theta > 0$
Gumbel-Hougaard	$C_\theta = exp\left\{ - \left[\sum_{i=1}^{d} \left(-lnu_i \right)^\theta \right]^{\frac{1}{\theta}} \right\}$	$\theta \geq 1$
Frank	$C_\theta = -\frac{1}{\theta} ln[1 + \frac{\prod_{i=1}^{d} (e^{-\theta u_i} - 1)}{(e^{-\theta} - 1)^{d-1}}]$	$\theta \in R$

3. Results

3.1. Optimal Drought Index Selection

The rank correlation coefficients between drought indices and drought impact data in 16 cities of YP are shown in Figure 4. It can be seen that SPI, SPEI, Pa, MI, and VSWI drought indices show a negative correlation with drought impact data, and CDD shows a positive relationship. This phenomenon is consistent in 13 out of 16 cities in YP, while the three cities of Nujiang, Dehong, and Xishuangbanna are not consistent. The topography of YP is complex, with an area of 1000–3500 m accounting for more than 90% of the total area, showing large change in elevation. For example, the area of Nujiang with elevation above 2000–3500 m accounts for ~66% (weaker correlation results shown in Nujiang), and the area of Baoshan with elevation below 2000 m accounts for ~66% (higher correlation results shown in Baoshan); agricultural drought in high altitude areas is not easily assessed by drought index. The proverb says that the weather varies within 10 miles of YP, which is the main result of topographic influence, and coupled with the different drought resistance for different cities, synthetically leading to different drought impacts in adjacent cities even under the influence of the same drought degree.

It is worth noting that for drought indices at longer time scales such as SPI, SPEI indices at 3-, 6-, and 12-month time scales, and VSWI, their negative correlation with drought impact data is more significant with the time scale increasing. While for the drought impact data of CA_ED and WMA_P, the VSWI shows a weaker correlation in the 16 cities compared with SPI and SPEI. For drought indices at shorter time scales, such as SPI and SPEI at a one-month time scale, CDD, Pa, and MI indices, the correlations are weaker, which indicates that using the current month drought index does not fully reflect the actual cumulative damage effect by drought. Although a remotely sensed-based drought index VSWI could indicate a drought situation at a relatively high spatial resolution, sometimes the lower values of VSWI are usually not only affected by drought but may also be affected by a combination of other factors, such as pest disease, forest fire, and human logging. Moreover, the remotely sensed data still have the problem of missing data due to the weather; this will result in drought events not being fully identified.

Based on the coefficient of variation (Cv) of the rank correlation coefficients for 16 cities (Table 5), we could see that the absolute Cv values of SPI and SPEI indices at a 6-month time scale (SPI6 and SPEI6) for different drought impact data are smaller than other drought indices, especially for the drought impact of CA_ED, CA_SD, P_D, and WMA_P (the Cv values are −0.53, −0.54, −0.48, −0.84, and −0.55, −0.55, −0.48, −0.75, respectively), indicating that SPI6 and SPEI6 can generally reflect the drought spatial distribution in YP. We initially suggested to select SPI6 or SPEI6 as the appropriate index to assess the drought situations in YP. Considering that the SPI index has only precipitation data as input, while the SPEI index takes into account the effect of temperature. Therefore, the following study will analyze the drought return period based on SPEI6.

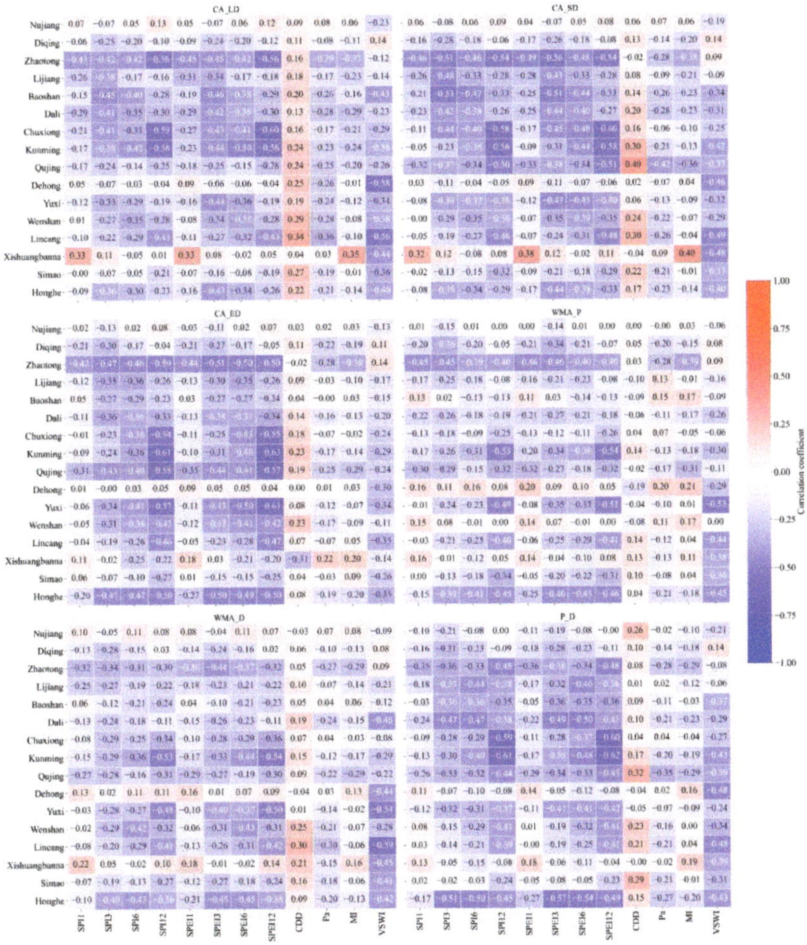

Figure 4. Rank correlation coefficients between drought indices and drought impact data in 16 cities of YP.

Table 5. The coefficient of variation of different drought indices for different drought impact data.

	SPI1	SPI3	SPI6	SPI12	SPEI1	SPEI3	SPEI6	SPEI12	CDD	Pa	MI	VSWI
CA_ED	−1.55	−0.55	−0.53	−0.64	−1.41	−0.59	−0.55	−0.64	1.68	−1.29	−1.94	−0.75
CA_LD	−1.68	−0.60	−0.64	−0.79	−1.35	−0.55	−0.63	−0.79	0.39	−0.60	−1.39	−0.55
CA_SD	−1.71	−0.60	−0.54	−0.69	−1.48	−0.56	−0.55	−0.70	0.80	−0.76	−1.69	−0.64
P_D	−1.44	−0.53	−0.48	−0.53	−1.29	−0.51	−0.48	−0.55	0.92	−0.87	−1.64	−0.57
WMA_D	−2.05	−0.57	−0.79	−0.89	−1.47	−0.57	−0.72	−0.91	0.87	−0.90	−1.86	−0.73
WMA_P	−2.28	−0.86	−0.84	−0.91	−1.85	−0.82	−0.75	−0.91	16.74	−2.64	−4.15	−0.91

3.2. Drought Events and Drought Characteristics in YP

Based on the method in Section 2.3.2, we extracted 41 drought events in YP during 1961–2020 (Table A1). Averagely, a drought event in YP generally lasts 5 months; cumulative S is almost 4.2, A is about 60% of the total stations, and drought starts mostly in February or December and ceases in March (Figure 5d). Since the turning point of climate change in China is thought to have occurred around 1990 [44], we compared the difference of drought characteristics before and after No. 20 drought event (~the year of 1990). Before 1990, the average *D* of drought in YP was 4 months, with an average *S* of 2.7 and an average *A* of 59%; while after 1990, the average *D* increased to 6 months, the average *S* increased to 5.6, and the average *A* was 61% (Figure 5a–c), this is main caused by the drought No. 35 and No. 36, which shows higher *D* and *S* than other events. However, the affected areas of these two droughts do not stand out among other events. This well illustrates the multidimensional spatiotemporal nature of drought events. Therefore, in order to facilitate comparison and assessment of the extremeness of a drought event among multiple events, it is necessary to analyze a drought event by jointing different drought characteristics.

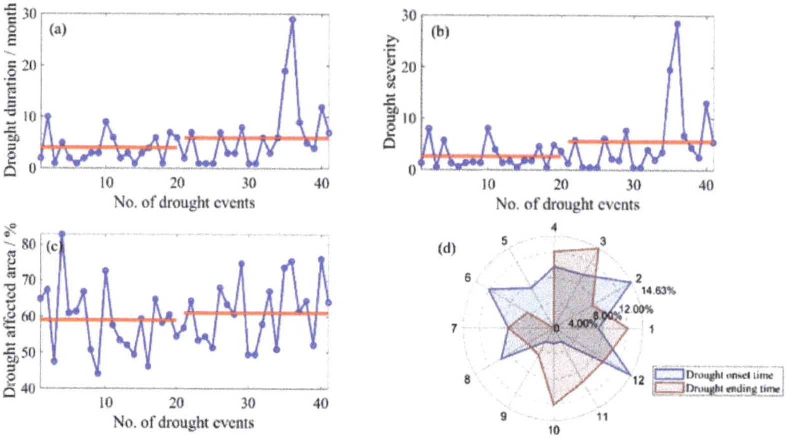

Figure 5. Changes of different drought characteristics in 41 drought events ((**a**): Drought duration, (**b**): drought severity, (**c**): drought affected area, (**d**): drought onset and ending time).

3.3. Optimal Marginal Distribution and Joint Distribution for Different Drought Characteristics

The statistical results of the K–S test of marginal distributions for drought characteristics *D*, *S*, and *A* are shown in Table 6. We can see that the marginal distribution to fit the one-dimensional drought characteristic was not unique, so the optimal distribution based on the minimum statistical value of K–S (D) was selected. The optimal marginal distributions for *D*, *S*, and *A* in YP were WBL, LOGN, and LOGN, respectively (Table 6).

Table 6. The optimal marginal distributions for one-dimensional drought characteristics.

Distribution	Parameter	*D*	*S*	*A*
EXP	θ	0.193	0.237	0.016
	K–S(D)	0.147 *	0.132 *	0.521
	Pass or not	Yes	Yes	No
WBL	*a*	1.149	0.947	6.932
	b	5.467	4.093	64.098
	K–S(D)	0.105 *	0.128 *	0.094 *
	Pass or not	Yes	Yes	Yes
GAM	α	1.433	1.016	45.05
	β	0.277	0.241	0.749
	K–S(D)	0.112 *	0.135 *	0.086 *
	Pass or not	Yes	Yes	Yes

Table 6. *Cont.*

Distribution	Parameter	*D*	*S*	*A*
LOGN	μ_y	1.255	0.871	4.085
	σ_y	0.882	1.061	0.148
	K–S(D)	0.114 *	0.107 *	0.082 *
	Pass or not	Yes	Yes	Yes
NOR	μ	5.171	4.212	60.132
	σ	5.240	5.349	9.035
	K–S(D)	0.204 *	0.244	0.096 *
	Pass or not	Yes	No	Yes
LOGC	m	4.346	3.237	59.754
	a	2.334	2.236	5.276
	K–S(D)	0.185 *	0.227	0.091 *
	Pass or not	Yes	No	Yes

Note: * means that *p*-values of K–S test larger than the significant level of 0.05 and the variable is obeyed the corresponding distribution. The optimal marginal distribution is selected when the corresponding value of K–S(D) is minimal.

For the joint distribution of multidimensional drought characteristics, the statistical results are shown in Table 7. Based on the minimum statistical values of AIC and BIC, the optimal joint distributions for *D* and *S*, *D* and *A*, *S* and *A*, and *D* and **S** and *A* in YP were Frank, Frank, Gumbel-Hougaard, and Frank Copula, respectively (Table 7).

Table 7. The optimal joint distributions for multidimensional drought characteristics.

Copulas	Drought Characteristics' Combination	Parameter	AIC	BIC	Accepted
Clayton	*D* and *S*	14.067	−52.444	−50.730	
	D and *A*	1.734	−8.546	−6.832	
	S and *A*	2.357	−4.118	−2.405	
	D and *S* and *A*	1.600	−56.428	−54.715	
Frank	*D* and *S*	29.904	−120.319	−118.605	Yes
	D and *A*	4.513	−16.367	−14.653	Yes
	S and *A*	5.226	−23.823	−22.109	
	D and *S* and *A*	6.636	−73.887	−72.173	Yes
Gumbel-Hougaard	*D* and *S*	5.546	−99.423	−97.710	
	D and *A*	1.600	−14.207	−12.493	
	S and *A*	1.897	−26.404	−24.690	Yes
	D and *S* and *A*	2.069	−72.625	−70.911	

3.4. *Return Period of Drought Events in YP*

According to the optimal distribution for different dimensional drought characteristics (Tables 6 and 7) and Equations (6)–(8), we calculated the return periods (RP) for the 41 drought events. We can see that from Figure 6, although the fluctuation of RP assessed by one-dimensional drought characteristics is consistent with those assessed by three-dimensional characteristics, the results based on *D* are different from those based on *S* or *A*, and the assessment results for specific events are significantly different. For drought events with relatively small RPs, the assessment results based on three-dimensional drought characteristics are similar to those considering only one-dimensional characteristics. For example, the four drought events (No. 12–15) that occurred during 1982–1985, which lasted 1–3 months, with severity between 0.52 and 1.82, and affected areas less than 60%, all of which were lower than the average drought characteristics of YP.

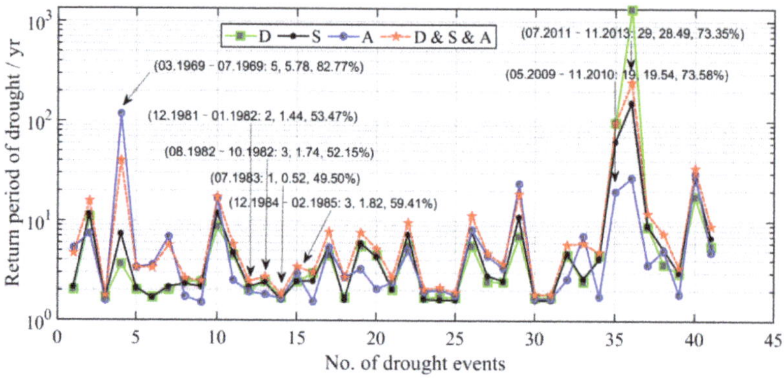

Figure 6. The comparison of return period based on one- and three-dimensional drought characteristics.

Since the significant correlation between *D* and *S*, the drought RPs assessed based on *D* or *S* are very close, but the results of RPs based on A differ significantly from *D* or *S*. For example, the No. 4 (March 1969–July 1969) drought event's RP assessed only by *A* was 117 yr, while the RPs of these two drought events of No. 35 (May 2009–November 2010) and No. 36 (July 2011–November 2013) were only 19 yr and 27 yr. However, the actual impacts of the latter two drought events were more severe according to the relevant records of the China Water and Drought Disaster Bulletin. As for extreme drought events, only using one-dimensional drought characteristics to assess the RP has a large discrepancy for each other; and for the common drought events, the RP based on three-dimensional characteristics was larger than those based on one-dimensional. This indicates that there is an overestimation or underestimation of the extremeness of a drought event if the characteristics are not fully considered.

When considering a combination of two drought characteristics to assess the RP for a given drought, we found that the assessment results were more consistent with the results by considering three-dimensional drought characteristics for the drought events where the magnitude of drought characteristics was small (Figure 7). For example, the RPs assessed by two-dimensional drought characteristics for the No. 11 to No. 25 drought events were more consistent with the results assessed by three-dimensional characteristics, but there were still differences for certain drought events with larger drought features. The No. 4 drought event lasted for 5 months, with a severity of 5.78 and a drought-affected area of 82.77%. The RP of this event was 7 yr assessed based on the combination of *D* and *S*, but after considering the drought-affected area, the RPs based on *D* and *A*, and *S* and *A* were assessed to be 119 yr.

Similarly, the *D*, *S*, and *A* of the No. 40 drought event (April 2019–March 2020) were 12 months, 13.06, and 75.91%. Although *A* was smaller than the No. 4 drought event, it was larger in *D*, and *S*. The RP assessed by *D* and *S* was 32 yr, while the RPs assessed by *D* and *A*, and *S* and *A* was 101 yr and 49 yr, respectively, which were larger than the RP assessed by *D* and *S* and *A*. This suggested that drought characteristic *A* was an important factor for assessing the extremeness of one drought event since it is associated with the impact degree of drought disaster, such as cropland areas affected by different severity of drought, population with drinking water difficulties due to drought. As we can see that when we combined other drought characteristics with variable *A*, the RP assessed by two-dimensional characteristics for the 41 drought events was most consistent with RP assessed by three-dimensional characteristics, especially for the common drought events (No. 11–25). The RP assessed by *D* and *A* were all higher than RP assessed by three-dimensional characteristics. If RP assessed by *D* and *S* or *D* and *A* was used to assess the extremeness of the No. 36 drought event, the result was higher than 1000 yr; it might be overestimated since the *A* was the same as the event of No. 35.

The above joint RP analysis indicates that extreme drought events do not imply that all features are large in magnitude; it is possible that they are the combined result of multiple features with different magnitudes. The identified drought events also show that it is rare for different drought characteristics to reach the maximum magnitude at the same time. In a word, a comprehensive assessment of the extremeness of a drought event using the joint RP based on spatial and temporal multidimensional characteristics is more reasonable.

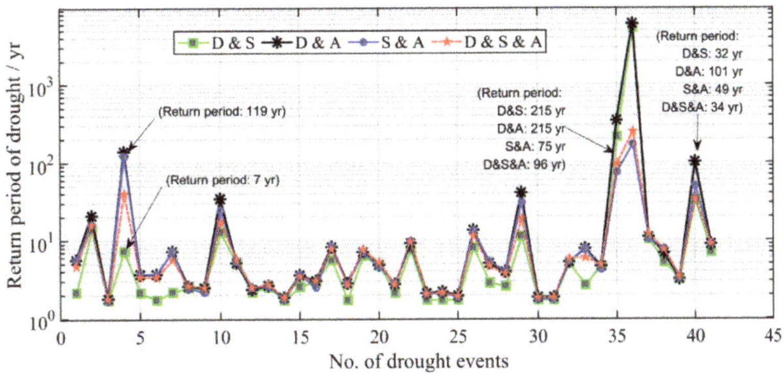

Figure 7. The comparison of return period based on two- and three-dimensional drought characteristics.

3.5. Case Validation

In order to prove the credibility of the RP assessed by three-dimensional drought characteristics, we verified the results based on the actual drought-affected area, grain loss rate, and the textual description of the drought events in YP.

The change of RPs was consistent with the grain loss rate, and the correlation coefficient was 0.74, which indicates that RP could reflect the real damage caused by droughts, especially by the extreme drought in 2009–2010, 2011–2013, and 2019–2020 (Figure A1). There were night years the crop area affected by drought exceeded 1 million hectares since 1961, which were all reflected in the theoretical drought events extracted in this study (Table 8). With the exception of the drought event in 2005, for which the RP was less than 10 yr, the RPs of all other high-impact drought events were greater than 10 yr, and the actual textual description of drought events is also consistent with the expression of RP size.

For the three drought events (No. 4, 35, 36) with more controversial RP results (in Figures 6 and 7), No. 4 was recorded as few major drought spells since 1949; it was more reasonable to assess this event with RP of 40 yr rather than 10 yr or 100 yr (RPs assessed by one- or two-dimensional characters).

The No. 35 drought occurred during 2009–2010 and was recorded as the most harmful extreme drought since meteorological records began. Not only did the textual material show how serious this drought was at that time, but the research published afterward may also indicate the severity of the drought's impact on society. The ratio of actual drought-affected area to total sown area was 44%, which were highest during 1972–2020. Therefore, the RP of 96 yr was reasonable to describe the extremeness of this drought event.

The No. 36 drought extracted from this study showed that the drought event lasted for nearly three years (July 2011–November 2013). The actual drought process had interruptions during the three years, so the final impact extent was not as severe as the drought in 2010, but the actual drought-affected area exceeded 1 million hectares each year, which indicates the cumulative damage effect formed by the three-year drought. Similar to the drought in 2009–2010, the RP of No. 36 drought in 2011–2013 should also be on the century scale rather than the millennium scale.

Table 8. Comparison of drought record information and theoretical drought event in this study.

No.	Year	Drought-Affected Crop Area/10^4 ha	Records of Actual Drought in Statistical Yearbook of YP	Theoretical Drought Event Extracted in This Study (RP)
1	1963	102.67	The droughts in YP in 1963 and 1969 were described as few major drought years since 1949 [45]	December 1962–September 1963 (16 yr)
2	1969	—		March 1969–July 1969 (40 yr)
3	1979	121.27	In 1979, following the winter drought of the previous year, YP continued to receive low precipitation and experienced a rare drought in its history [45]	December 1978–August 1979 (17 yr)
4	2005	205.33	This year was the worst drought disaster year in nearly 50 years	May 2005–October 2005 (6 yr)
5	2009	103.67	This drought was the longest-lasting, most extensive, and most harmful mega-drought since meteorological records began	May 2009–November 2010 (96 yr)
6	2010	283.85		
7	2011	123.05	YP suffered from persistent drought, with 25 counties breaking historical records of least precipitation	July 2011–November 2013 (248 yr)
8	2012	107.27	Drought in YP continued to worsen and finally affected tens of millions of people	
9	2013	117.74		
10	2019	138.12	The impacted scope, intensity, and duration of drought disasters in Yunnan in 2019 ranked the second severity of drought disaster since 2010	April 2019–March 2020 (32 yr)

4. Discussion

4.1. Uncertainty of Regional Drought Event Identification

It is necessary and challenging to effectively and accurately identify regional drought events. The development of regional drought identification methods has been ongoing for nearly 20 years. Ren et al. [20] provide a systematic review of methods for identifying regional extreme drought events, with a representative example of Andreadis et al. [46] and Lloyd-Hughes [21], who used spatial clustering methods to extract drought patches and propose a severity-area-duration method (abbreviated as SAD). Zhang et al. [47] used the precipitation data from 22 meteorological stations and the SAD methods to identify drought events with D, S, and A in YP during 1961–2018. This study took 101 meteorological stations as the basis; we referenced the method proposed by Liao et al. [40], simplified and combined it with the run theory to identify 41 regional drought events in YP from 1961 to 2020. And the extremeness of theoretical drought events is basically consistent with the result of Zhang et al. [47] and the historical actual drought records (Table 8). In order to reflect the three-dimensional dynamic evolution process of drought, quantitative analysis of the whole process of drought events from the time–longitude–latitude has become a hot topic [48].

The results of regional drought event identification have uncertainties due to the optimal drought index selected, drought threshold, spatial and temporal resolutions of the data used, the weight for each station when to calculate the regional drought index, and the methods adopted, etc. In future work, we will use grid data with higher resolution and set up tests with different drought thresholds to systematically analyze the joint RP risk of drought in YP.

4.2. Comparison of Joint Return Period Based on SPI and SPEI

The correlation results in Section 3.1 show that the SPI6 could also reflect the change of drought impact; this study wanted to detect the differences between SPI6 and SPEI6 in the assessment of drought RPs in YP.

For one-dimensional drought characteristics, it was found that the RPs of drought events with shorter duration, less severe, or less affected areas (such as events of E1, E3, and E5 in Figure 8) were similar for SPI6 and SPEI6; however, with the magnitude of drought characteristics increase, the SPI6-based RPs were higher than those of SPEI6 (such as the events of E2, E4, and E6 in Figure 8).

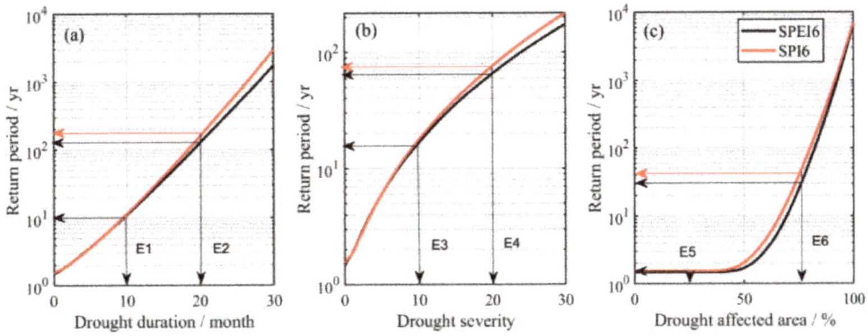

Figure 8. Comparison of the return period of one-dimensional drought characteristics based on SPI and SPEI ((**a**): return period based on *D*, (**b**): return period based on *S*, (**c**): return period based on *A*).

The RPs' difference is more obvious when considering two-dimensional drought characters. As the drought characteristics magnitude increases, the SPI6-based RP is higher than that of SPEI6 (the solid line is always away from the dotted line with the same color). For example, for drought events with $D \geq 7$ and $S \geq 7$, the RP is about 10 yr for both SPI- and SPEI-based evaluations (Figure 9a). E (20,15) in Figure 8a means a drought event with $D \geq 20$ and $S \geq 15$. The SPEI6-based RP of this event is shown to be 200 yr, while it is greater than 200 years based on SPI6. The same phenomenon is also observed in the joint RPs of *D* and *A* (Figure 9b) and *S* and *A* (Figure 9c).

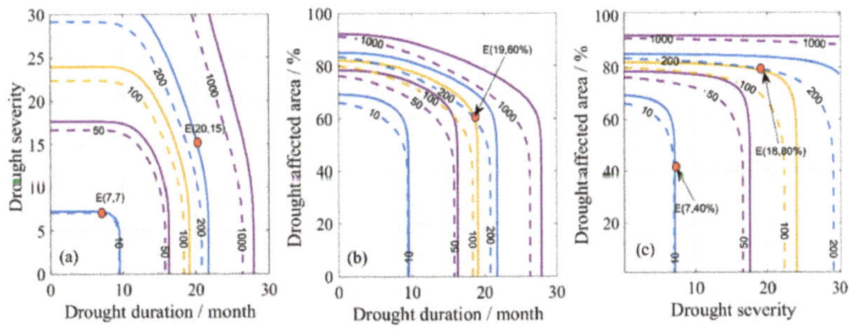

Figure 9. The return periods (with different colors) of two-dimensional drought characteristics based on SPI6 (the dotted line) and SPEI6 (the solid line) ((**a**): the return period based on *D* and *S*, (**b**): the return period based on *D* and *A*, (**c**): the return period based on *S* and *A*).

Figure 10 shows the RP surfaces (5 yr, 10 yr, 50 yr, and 100 yr) with three-dimensional drought characters. Similarly, we can see that if a drought event is evaluated by SPI6, the RP is higher than that by SPEI6. For example, the RP of the drought event ($D > 4 \cap S > 2.74 \cap A > 57\%$) in Figure 10a is on the 5 yr surface of SPI6, while it is inside the 5 yr surface of SPEI6, which indicates that this event is less than 5 yr return period assessed by SPEI6. The same result is also reflected in the comparison of the 10 yr return period surface (Figure 10b). However, for the cases with larger RPs (Figure 10c,d), instead of showing a complete surface like 5 yr and 10 yr return periods, 50 yr and 100 yr RPs show depressions on the sides, and top of the surface, but the red surface is still surrounded by the blue. This means that the magnitude of drought characteristics of drought events with 50 yr or 100 yr RP assessed by SPI6 are less than SPEI6. For the occurrence of depressions in RP surface, it also reflects that the correlation structure between drought characteristics may change under extreme conditions: drought events with high RPs showing longer *D*, less *S*, and larger *A*, or higher *S*, shorter *D*, and larger *A* is recorded in YP. However, the drought events with large magnitude for *D*, *S*, and *A* have a more irregular RP surface, which indirectly indicates the rare frequency of such events.

4.3. Effect of Global Warming on Drought in YP

The frequency of extreme events has changed under the background of global warming, with events previously considered to be extreme becoming more frequent and will probably no longer be classified as extreme events [49–51]. Southwestern China has also been particularly affected by serious droughts; this was especially true in 2009 and 2010. The regional circulation patterns are strongly affected by the high-elevation Himalayas Mountains, which block the flows of air masses, and the large variation in landform conditions combine to increase the drought frequency [52]. Researchers have also analyzed that anthropogenic climate change increased the drought risk in the region of Southwestern China [53]. Global warming is the main feature of climate change. Since 1961, YP has shown a significant increasing warming tendency ($p < 0.01$) either at seasonal or annual scale [30]. Meanwhile, the interval time between the end of one drought event and the beginning of the next drought event in YP showed a decreasing trend (Figure 11a). This suggests that YP is more and more prone to occur drought, and it is confirmed by the decreasing trend change in the cumulative anomaly of RP of the 41 drought events (Figure 11b).

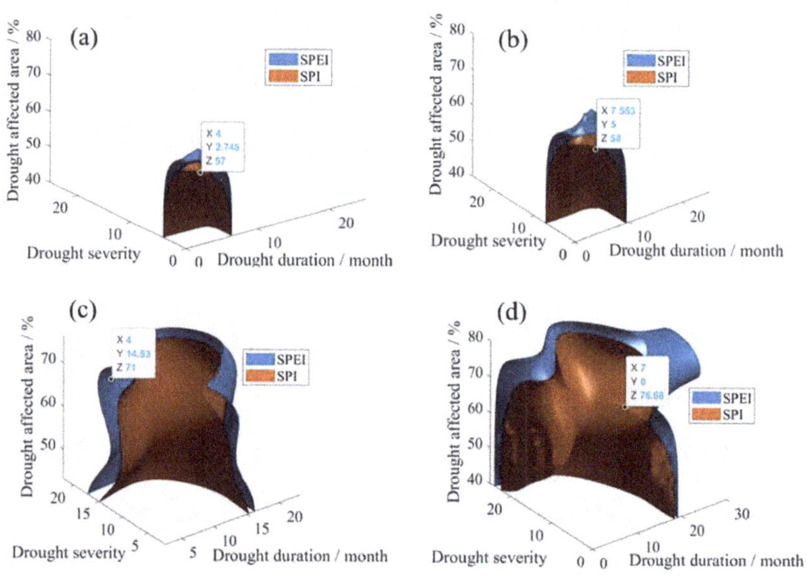

Figure 10. Comparison of the return period of three-dimensional drought characteristics based on SPI and SPEI. ((**a**): 5 yr RP, (**b**): 10 yr RP, (**c**): 50 yr RP, (**d**): 100 yr RP; x means *D*, y means *S*, z means *A*).

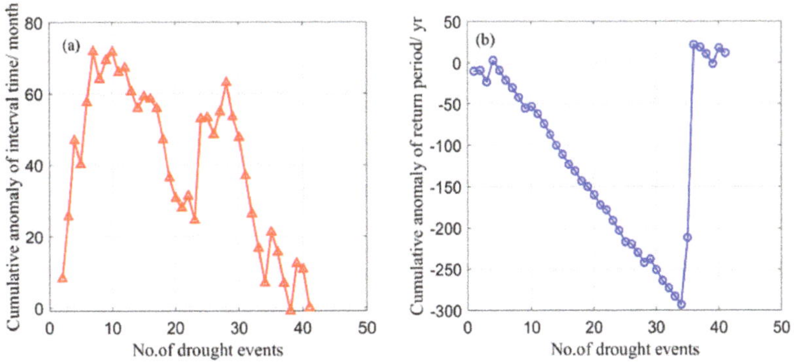

Figure 11. Trends of the cumulative anomaly of drought interval (**a**) and return period (**b**) in YP.

5. Conclusions

In this study, several drought indices were first compared with the drought impact dataset, and the optimal one was selected. Then we explored the spatial and temporal dynamics risk of drought events in YP by considering *D*, *S*, and *A*. There were 41 drought events identified during 1961–2020 in YP. Different types of RP for droughts were estimated based on Copulas and compared with each other. The main findings can be summarized as follows.

(1) The drought indices on a shorter time scale can only characterize the degree of water shortage in the current month; the longer time scale drought indices can better reflect the actual drought-impacted situation in YP, especially SPEI and SPI at more than six-month time scale.

(2) Drought is more likely to occur in YP during the winter and spring. The fluctuations of drought characteristics magnitude have been relatively stable over the study periods, except for the years 2009–2013. For the 41 drought events, the optimal marginal distribution functions for *D*, *S*, and *A* were WBL, LOGN, and LOGN distribution, respectively. Frank Copula function is suitably used to construct the joint distribution of *D* and *S*, *D* and *A*, and *D* and *S* and *A*, and Gumbel Copula is the optimal joint distribution function for *S* and *A*.

(3) The joint RP analysis showed that the 41 events during 1961–2020 differed in different combinations of *D*, *S*, and *A*. When assessing the RP based on one-dimensional drought characteristics only, the *D*-based and *S*-based RPs' fluctuation was similar and differed with *A*-based RP. When assessing the joint RP based on two-dimensional drought characteristics, the *D* and *A*-based and *S* and *A*-based assessment results show consistent fluctuations. For drought events with large differences in drought characteristics, it is more reasonable to utilize joint RP based on *D* and *S* and *A* for their extremeness assessment.

(4) The interval between drought events in YP has gradually decreased, and the risk of drought has gradually increased since 1961. Compared with other drought events, the extremeness of two droughts during 2009.05–2010.11 and 2011.07–2013.11 was highest with RP larger than 100 yr.

Author Contributions: R.Z.: Manuscript writing, Data analysis. S.Y.: Conceptualization, Content review. H.S.: Content review, Software. L.Z.: Content review. M.L.: Software. L.X. and R.T.: Data collection and calculation. All authors have read and agreed to the published version of the manuscript.

Funding: This study was supported by the National Key Research and Development Project (Grant Number: 2021YFB3901203).

Institutional Review Board Statement: Not applicable.

Informed Consent Statement: Not applicable.

Data Availability Statement: The data used in this study can be obtained from the Website listed in Section 2.2. The dataset is available from ruxinzhao@ninhm.ac.cn upon reasonable request.

Acknowledgments: We gratefully acknowledge the China Meteorological Administration for providing the dataset.

Conflicts of Interest: The authors declare no conflict of interest.

Appendix A

Table A1. Theoretical drought events identified in this study.

No.	Drought Events	*D*/Month	*S*	*A*/%
1	01/1961~02/1961	2	1.41	64.85
2	12/1962~09/1963	10	8.01	67.33
3	04/1966~04/1966	1	0.54	47.52
4	03/1969~07/1969	5	5.78	82.77
5	02/1970~03/1970	2	1.36	60.89
6	10/1972~10/1972	1	0.66	61.39
7	02/1975~03/1975	2	1.42	66.83
8	09/1975~11/1975	3	1.62	50.83
9	06/1977~08/1977	3	1.43	44.22
10	12/1978~08/1979	9	8.12	72.61
11	04/1980~09/1980	6	4.02	57.59
12	12/1981~01/1982	2	1.44	53.47
13	08/1982~10/1982	3	1.74	52.15
14	07/1983~07/1983	1	0.52	49.50
15	12/1984~02/1985	3	1.82	59.41
16	03/1986~06/1986	4	1.83	46.29
17	05/1987~10/1987	6	4.59	64.85
18	03/1988-03/1988	1	0.53	58.42
19	06/1988~12/1988	7	4.89	60.54
20	08/1989~01/1990	6	3.68	54.62
21	12/1990~01/1991	2	1.29	56.93
22	06/1992-12/1992	7	5.79	64.36
23	07/1993~07/1993	1	0.57	53.47
24	01/1997~01/1997	1	0.53	54.46
25	03/1998~03/1998	1	0.53	51.49
26	12/1998~06/1999	7	6.17	68.03
27	02/2001~04/2001	3	2.22	63.37
28	02/2003~04/2003	3	1.87	60.73
29	08/2003~30/2004	8	7.69	74.63
30	11/2004~11/2004	1	0.52	49.50
31	02/2005~02/2005	1	0.50	49.50
32	05/2005~10/2005	6	3.98	57.92
33	02/2006~04/2006	3	2.01	67.00
34	08/2006~01/2007	6	3.49	50.99
35	05/2009~11/2010	19	19.54	73.58
36	07/2011~11/2013	29	28.49	75.35
37	04/2014~12/2014	9	6.74	61.50
38	06/2015~10/2015	5	4.37	64.36
39	01/2018~04/2018	4	2.48	52.23
40	04/2019~03/2020	12	13.06	75.91
41	06/2020~12/2020	7	5.44	64.07

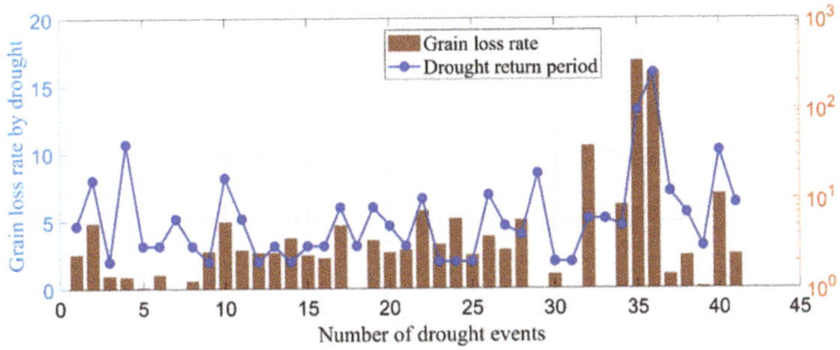

Figure A1. Variations in the extent of grain loss rate and the joint return period calculated by three-dimensional drought characteristics for 41 drought events in Yunnan province.

References

1. Eklund, L.; Theisen, O.M.; Baumann, M.; Tollefsen, A.F.; Kuemmerle, T.; Nielsen, J. Societal drought vulnerability and the Syrian climate-conflict nexus are better explained by agriculture than meteorology. *Commun. Earth Environ.* **2022**, *3*, 85. [CrossRef]
2. Intergovernmental Panel on Climate Change (IPCC). Food Security and Food Production Systems. In *Climate Change 2014: Impacts, Adaptation, and Vulnerability*; Porter, J.R., Xie, L., Challinor, A.J., Cochrane, K., Howden, S.M., Iqbal, M.M., Lobell, D.B., Travasso, M.I.., Eds.; Cambridge University Press: London, UK, 2014. [CrossRef]
3. Huang, J.; Yu, H.; Guan, X.; Wang, G.; Guo, R. Accelerated dryland expansion under climate change. *Nat. Clim. Chang.* **2016**, *6*, 166–171. [CrossRef]
4. Henry, A.J. *The Climatology of the United States*; Weather Bureau Bulletin Q: Washington, DC, USA, 1906; pp. 51–58.
5. McKee, T.B.; Doesken, N.J.; Kleist, J. The relationship of drought frequency and duration to time scales. In Proceedings of the 8th Conference on Applied Climatology: American Meteorological Society, Boston, MA, USA, 17–22 January 1993; pp. 179–183.
6. China Meteorological Administration. *GB/T 20481-2017*; Grades of meteorological drought (TEXT OF DOCUMENT IS IN CHINESE). Grades of Meteorological Drought: Beijing, China, 2017.
7. Palmer, W.C. *Meteorologic Drought*; US Department of Commerce, Weather Bureau: Silver Spring, MD, USA, 1965; p. 45.
8. Vicente-Serrano, S.M.; Beguería, S.; López-Moreno, J.I. A Multiscalar Drought Index Sensitive to Global Warming: The Standardized Precipitation Evapotranspiration Index. *J. Clim.* **2010**, *23*, 1696–1718. [CrossRef]
9. Hao, Z.C.; AghaKouchak, A. Multivariate Standardized Drought Index: A parametric multi-index model. *Adv. Water Resour.* **2013**, *57*, 12–18. [CrossRef]
10. Sandholt, I.; Rasmussen, K.; Andersen, J. A simple interpretation of the surface temperature/vegetation index space for assessment of surface moisture status. *Remote Sens. Environ.* **2002**, *79*, 213–224. [CrossRef]
11. Brown, J.F.; Wardlow, B.D.; Tadesse, T.; Hayes, M.J.; Reed, B.C. The Vegetation Drought Response Index (VegDRI): A new integrated approach for monitoring drought stress in vegetation. *GIScience Remote Sens.* **2008**, *45*, 16–46. [CrossRef]
12. Mishra, A.K.; Singh, V.P. A review of drought concepts. *J. Hydrol.* **2010**, *391*, 202–216. [CrossRef]
13. Zargar, A.; Sadiq, R.; Naser, B.; Khan, F.I. A review of drought indices. *Environ. Rev.* **2011**, *19*, 333–349. [CrossRef]
14. Parsons, D.J.; Rey, D.; Tanguy, M.; Holman, I.P. Regional variations in the link between drought indices and reported agricultural impacts of drought. *Agric. Syst.* **2019**, *173*, 119–129. [CrossRef]
15. Wang, Y.; Lv, J.; Hannaford, J.; Wang, Y.; Sun, H.; Barker, L.J.; Ma, M.; Su, Z. Linking drought indices to impacts to support drought risk assessment in Liaoning province, China. *Nat. Hazards Earth Syst. Sci.* **2020**, *20*, 889–906. [CrossRef]
16. Shao, D.; Chen, S.; Tan, X.; Gu, W. Drought characteristics over China during 1980–2015. *Int. J. Climatol.* **2018**, *38*, 3532–3545. [CrossRef]
17. Chamorro, A.; Ivanov, M.; Tölle, M.H.; Luterbacher, J.; Breuer, L. Analysis of future changes in meteorological drought patterns in Fulda, Germany. *Int. J. Clim.* **2020**, *40*, 5515–5526. [CrossRef]
18. Vergni, L.; Todisco, F.; Di Lena, B.; Mannocchi, F. Bivariate analysis of drought duration and severity for irrigation planning. *Agric. Water Manag.* **2020**, *229*, 105926. [CrossRef]
19. Samantaray, A.K.; Ramadas, M.; Panda, R.K. Assessment of impacts of potential climate change on meteorological drought characteristics at regional scales. *Int. J. Climatol.* **2021**, *41*, E319–E341. [CrossRef]
20. Ren, F.-M.; Trewin, B.; Brunet, M.; Dushmanta, P.; Walter, A.; Baddour, O.; Korber, M. A research progress review on regional extreme events. *Adv. Clim. Chang. Res.* **2018**, *9*, 161–169. [CrossRef]
21. Lloyd-Hughes, B. A spatio-temporal structure-based approach to drought characterisation. *Int. J. Clim.* **2012**, *32*, 406–418. [CrossRef]
22. Ren, J.; Zhang, W.; Wan, Y.; Chen, Y. Advances in the Research of Yunnan's Arid Climate and Extreme Drought. *Atmos. Clim. Sci.* **2017**, *7*, 23–35. [CrossRef]
23. Zhu, Y.; Liu, Y.; Wang, W.; Singh, V.P.; Ma, X.; Yu, Z. Three dimensional characterization of meteorological and hydrological droughts and their probabilistic links. *J. Hydrol.* **2019**, *578*, 124016. [CrossRef]

24. Sun, C.; Yang, S. Persistent severe drought in southern China during winter–spring 2011: Large-scale circulation patterns and possible impacting factors. *J. Geophys. Res. Atmos.* **2012**, *117*, D10112. [CrossRef]
25. Wang, L.; Chen, W.; Zhou, W. Assessment of future drought in Southwest China based on CMIP5 multimodel projections. *Adv. Atmos. Sci.* **2014**, *31*, 1035–1050. [CrossRef]
26. Compilation Group of China Flood and Drought Disaster Prevention Bulletin. Summary of China Flood and Drought Disaster Prevention Bulletin 2020. *China Flood Drought Manag.* **2021**, *31*, 26–32. (In Chinese)
27. Qiu, J. China drought highlights future climate threats. *Nature* **2010**, *465*, 142–143. [CrossRef] [PubMed]
28. Wang, L.; Chen, W.; Zhou, W.; Huang, G. Drought in Southwest China: A review. *Atmos. Ocean. Sci. Lett.* **2015**, *8*, 339–344. [CrossRef]
29. Wang, Y.; Yuan, X. Anthropogenic Speeding Up of South China Flash Droughts as Exemplified by the 2019 Summer-Autumn Transition Season. *Geophys. Res. Lett.* **2021**, *48*, e2020GL091901. [CrossRef]
30. Li, Y.; Wang, Z.; Zhang, Y.; Li, X.; Huang, W. Drought variability at various timescales over Yunnan Province, China: 1961–2015. *Theor. Appl. Climatol.* **2019**, *138*, 743–757. [CrossRef]
31. Wang, L.; Zhang, X.; Wang, S.; Salahou, M.K.; Fang, Y. Analysis and Application of Drought Characteristics Based on Theory of Runs and Copulas in Yunnan, Southwest China. *Int. J. Environ. Res. Public Health* **2020**, *17*, 4654. [CrossRef]
32. Duan, X.; Gu, Z.; Li, Y.; Xu, H. The spatiotemporal patterns of rainfall erosivity in Yunnan Province, southwest China: An analysis of empirical orthogonal functions. *Glob. Planet. Chang.* **2016**, *144*, 82–93. [CrossRef]
33. Zhang, W.; Zheng, J.; Ren, J. Climate characteristics of extreme drought events in Yunnan. *J. Catastrophol.* **2013**, *28*, 59–64, (In Chinese with English abstract).
34. China Meteorological Administration. *China Meteorological Disaster Yearbook (2004)*; China Meteorological Press: Beijing, China, 2004; pp. 117–119.
35. China Meteorological Administration. *China Meteorological Disaster Yearbook (2011)*; China Meteorological Press: Beijing, China, 2011; pp. 144–145.
36. Zhang, S.F.; Su, Y.S.; Song, D.D.; Zhang, Y.Y.; Song, H.Z.; Gu, Y. *The Drought History in China (1949–2000)*; Hohai University Press: Nanjing, China, 2008; pp. 672–676. (In Chinese)
37. Duan, Y.; Ma, Z.; Yang, Q. Characteristics of consecutive dry days variations in China. *Theor. Appl. Climatol.* **2017**, *130*, 701–709. [CrossRef]
38. Yan, H.; Zhou, G.; Lu, X. Comparative analysis of surface soil moisture retrieval using VSWI and TVDI in karst areas. *Intell. Earth Obs. Syst.* **2015**, *9808*, 37–49. [CrossRef]
39. Zhou, Y.; Zhou, P.; Jin, J.; Wu, C.; Cui, Y.; Zhang, Y.; Tong, F. Drought identification based on Palmer drought severity index and return period analysis of drought characteristics in Huaibei Plain China. *Environ. Res.* **2022**, *212*, 113163. [CrossRef]
40. Liao, Y.; Zhang, C.; Zou, X.; Ye, D.; Wang, X.; Li, W.; Cheng, J.; Duan, J. *Regional Drought Process Monitoring and Assessment Method (QX/T 597-2021)*; Meteorological Press: Beijing, China, 2021.
41. Dai, M.; Huang, S.; Huang, Q.; Leng, G.; Guo, Y.; Wang, L.; Fang, W.; Li, P.; Zheng, X. Assessing agricultural drought risk and its dynamic evolution characteristics. *Agric. Water Manag.* **2020**, *231*, 106003. [CrossRef]
42. Mosaedi, A.; Abyaneh, H.Z.; Sough, M.G.; Samadi, S.Z. Quantifying Changes in Reconnaissance Drought Index using Equiprobability Transformation Function. *Water Resour. Manag.* **2015**, *29*, 2451–2469. [CrossRef]
43. Amirataee, B.; Montaseri, M.; Rezaie, H. Regional analysis and derivation of copula-based drought Severity-Area-Frequency curve in Lake Urmia basin, Iran. *J. Environ. Manag.* **2018**, *206*, 134–144. [CrossRef]
44. Zhao, R.; Wang, H.; Zhan, C.; Hu, S.; Ma, M.; Dong, Y. Comparative analysis of probability distributions for the Standardized Precipitation Index and drought evolution in China during 1961–2015. *Theor. Appl. Climatol.* **2020**, *139*, 1363–1377. [CrossRef]
45. Wen, K.G.; Liu, J.H. *China Meteorological Disaster Dictionary—Yunnan Volume*; Meteorological Publishing House Press: Beijing, China, 2006; pp. 47–96. (In Chinese)
46. Andreadis, K.M.; Clark, E.A.; Wood, A.W.; Hamlet, A.F.; Lettenmaier, D.P. Twentieth-Century Drought in the Conterminous United States. *J. Hydrometeorol.* **2005**, *6*, 985–1001. [CrossRef]
47. Zhang, L.; Yang, X.; Ren, L.; Sheffield, J.; Zhang, L.; Yuan, S.; Zhang, M. Dynamic multi-dimensional identification of Yunnan droughts and its seasonal scale linkages to the El Niño-Southern Oscillation. *J. Hydrol. Reg. Stud.* **2022**, *42*, 101128. [CrossRef]
48. Liu, Y.; Zhu, Y.; Ren, L.; Singh, V.P.; Yong, B.; Jiang, S.; Yuan, F.; Yang, X.-L. Understanding the Spatiotemporal Links Between Meteorological and Hydrological Droughts From a Three-Dimensional Perspective. *J. Geophys. Res. Atmos.* **2019**, *124*, 3090–3109. [CrossRef]
49. Dai, A. Increasing drought under global warming in observations and models. *Nat. Clim. Chang.* **2013**, *3*, 52–58. [CrossRef]
50. Carrão, H.; Naumann, G.; Barbosa, P. Global projections of drought hazard in a warming climate: A prime for disaster risk management. *Clim. Dyn.* **2018**, *50*, 2137–2155. [CrossRef]
51. Samaniego, L.; Thober, S.; Kumar, R.; Wanders, N.; Rakovec, O.; Pan, M.; Zink, M.; Sheffield, J.; Wood, E.F.; Marx, A. Anthropogenic warming exacerbates European soil moisture droughts. *Nat. Clim. Chang.* **2018**, *8*, 421–426. [CrossRef]

52. Han, L.; Zhang, Q.; Ma, P.; Jia, J.; Wang, J. The spatial distribution characteristics of a comprehensive drought risk index in southwestern China and underlying causes. *Theor. Appl. Climatol.* **2015**, *124*, 517–528. [CrossRef]
53. Zhao, R.; Sun, H.; Xing, L.; Li, R.; Li, M. Effects of anthropogenic climate change on the drought characteristics in China: From frequency, duration, intensity, and affected area. *J. Hydrol.* **2023**, *617*, 129008. [CrossRef]

atmosphere

MDPI

Article

Identification and Spatiotemporal Migration Analysis of Groundwater Drought Events in the North China Plain

Jia Huang, Lianhai Cao *, Lei Wang, Liwei Liu, Baobao Yu and Long Han

The College of Surveying and Geo-informatics, North China University of Water Resources and Electric Power, Zhengzhou 450046, China; b2019081812@stu.ncwu.edu.cn (J.H.)
* Correspondence: caolianhai@ncwu.edu.cn

Abstract: Groundwater droughts can explain developments and changes in groundwater from a climatological perspective. The North China Plain (NCP) is a typical underground funnel area. Therefore, groundwater drought studies in the NCP can provide better understanding of the local hydrogeological characteristics from new perspectives. In this paper, the GRACE groundwater drought index (GGDI) was used to evaluate groundwater drought events in the NCP. Additionally, a new method was proposed in this study for investigating groundwater drought events at the spatiotemporal scale. On this basis, the centroid theory was used to construct an appropriate groundwater drought migration model for the NCP. The results showed that (1) the groundwater drought frequency in the NCP was 24.54%. In addition, the most severe groundwater drought events in the study occurred in March 2020. (2) In total, 49 groundwater drought events occurred in the NCP over the 2003–2020 period. The most intense groundwater drought event occurred over the June 2018–December 2020 period (DE.49), covering the entire study area. DE.29 was the second most intense groundwater drought event over the August 2012–September 2013 period (14 months), resulting in a maximum arid area of 75.57% of the entire study area. (3) The migration of the groundwater drought events was in the southwest–northeast and northeast–southwest directions, which was consistent with the terrain inclination, while most of the groundwater drought centroids were concentrated in Area II. The groundwater drought event identification method and the groundwater drought migration model were effective and reliable for assessing groundwater drought events in the NCP and provided a better understanding of developments and changes in groundwater droughts, which is of great practical significance and theoretical value for the rational development and use of groundwater resources, as well as for guiding industrial and agricultural activities.

Keywords: groundwater drought; GGDI; drought event; centroid migration; North China Plain

Citation: Huang, J.; Cao, L.; Wang, L.; Liu, L.; Yu, B.; Han, L. Identification and Spatiotemporal Migration Analysis of Groundwater Drought Events in the North China Plain. *Atmosphere* **2023**, *14*, 961. https://doi.org/10.3390/atmos14060961

Academic Editors: Jinping Liu, Quoc Bao Pham, Arfan Arshad and Masoud Jafari Shalamzari

Received: 20 April 2023
Revised: 24 May 2023
Accepted: 26 May 2023
Published: 31 May 2023

1. Introduction

Groundwater is an important water source for human beings and has accompanied the development of human civilization. In arid and semi-arid regions, the development of human society depends on depleted groundwater resources [1]. However, the development of human society has disrupted the natural cycle of groundwater, causing the depletion of aquifers around the world at an alarming rate [2–4]. The UN World Water Development Report 2022 focused on groundwater issues, with the conference theme of "Groundwater: making the invisible visible" [5], the same theme as that of World Water Day 2022, highlighting groundwater problems. Climate change is a new challenge facing human civilization. The fifth assessment report of the Intergovernmental Panel on Climate Change (IPCC) highlighted a potential increase in the frequency of extreme climate events [6]. In addition, several studies have demonstrated significant increases in the temperatures of arid regions worldwide over the past 100 years, resulting in drought areas almost doubling, which may potentially continue to expand and cause the further intensification of drought degree [7–9]. Measures to combat climate warming are required urgently [10,11].

There is still no clear and unified definition of the groundwater drought concept. Indeed, scholars have interpreted groundwater droughts differently according to different assessment methods. The "groundwater drought" term was first proposed by Rutulis [12] to distinguish between droughts that occur in groundwater aquifers and other drought types (e.g., meteorological, hydrological, and agricultural droughts). He defined a groundwater drought as a significant natural drop in groundwater levels, causing substantial aquifer dewatering, thus resulting in severe water supply problems and emphasizing natural conditions, non-human interference, and the transmission of meteorological droughts to groundwater droughts. Groundwater droughts under natural conditions are often cyclical and caused by reduced groundwater recharge and storage over periods of time, while other drought types are due to insufficient precipitation [13]. However, groundwater droughts can also be caused by anthropogenic factors, such as groundwater extraction rates that are greater than recharge and groundwater storage amounts, which thus intensifies groundwater droughts to a certain extent. In either case, groundwater droughts are characterized by low groundwater levels and low groundwater abstraction rates (or even dry wells). Therefore, groundwater droughts may be defined as a lack of groundwater recharge or the lack of groundwater in terms of reserves or groundwater heads over a specific period of time [13]. However, groundwater droughts cannot only be evaluated from the perspectives of natural and human factors. In fact, Marchant and Bloomfield [14] defined a groundwater drought as a decrease in groundwater level below the normal level or a reduction in spring flow. Van Lanen and Peter [13] believed that groundwater droughts occur in three consecutive phases: temporal decreases in groundwater recharge, followed by decreases in groundwater levels and decreases in discharge rates. Indeed, the emergence of groundwater droughts can first affect groundwater ecosystems and then spread to other ecosystems. In addition, a groundwater drought is a type of drought that occurs following a meteorological drought that has caused hydrological and agricultural drought events [6]. Drought propagates into groundwater systems through mechanisms such as pooling in river catchments, hysteresis, and the prolongation of drought signals [15]. In the absence of long-term precipitation deficiency, variations in groundwater drought characteristics could be affected by anthropogenic warming activities [16]. Therefore, the occurrence of groundwater droughts is an indicator of the significant impacts of natural and/or anthropogenic factors on the water cycle (e.g., severe climatic influences and groundwater overuse).

At present, the most commonly used methods for evaluating groundwater droughts are the standardized precipitation index (SPI) [17], the standardized groundwater level index (SGI) [18,19], the standardized water level index (SWI) [20], and the GRACE groundwater drought index (GGDI) [21,22]. In addition, other methods have also been used in some studies to assess groundwater droughts, such as the groundwater drought index (GWI) [23], the groundwater resource index (GRI) [24], and the groundwater recharge drought index (GRDI) [25]. Among them, the GGDI has been extensively used to assess groundwater droughts because it is based on remote sensing data, which are highly available, thereby overcoming the limitations of spatiotemporal groundwater drought monitoring [26,27].

Some previous studies have assessed the temporal and spatial evolution of drought events separately, resulting in inaccurate descriptions of the evolution process of drought events [28]. Therefore, the evolution of drought dynamics at the spatiotemporal scale has attracted considerable attention from some scholars. Zhou et al. [29] proposed a new method for studying meteorological drought migration based on topological spatial relationships to accurately describe the spatiotemporal dynamic behavior of droughts. Guo et al. [30] used an improved three-dimensional clustering algorithm (longitude–latitude–time and space) to identify drought episodes and described the characteristics of drought events according to several indicators, including drought duration, severity, intensity, area, centroid, and trajectory. Herrera-Estrada et al. [31] used Lagrangian methods to monitor worldwide spatiotemporal drought events over the 1979–2009 period, analyzing their characteristics and behaviors. Wen et al. [28] proposed a new 3D drought structure extraction method to assess the long-term spatiotemporal distribution patterns of droughts during their development. Han et al. [32] constructed an agricultural drought migration model using the migration of drought centroids to describe the spatiotemporal evolution of agricultural droughts. Understanding the spatiotemporal variations in groundwater droughts is of great significance for the comprehensive assessment of groundwater droughts. However, other studies on groundwater drought events have been carried out from the perspective of time [12,33,34]. Therefore, it is necessary to assess groundwater drought events at both the temporal and spatial scales.

The North China Plain (NCP) is one of three plains in China. Indeed, it has attracted much attention because of its relevant political and economic characteristics, its importance in food production, and its severe water shortage. The large groundwater funnel in this area makes it an interesting area for hydrogeological studies, thus contributing to sustainable groundwater development [35]. However, in recent years, relevant departments have implemented certain measures in the NCP, such as groundwater overexploitation controls, the South-to-North Water Diversion, and ecological water replenishment, to prevent the continuous decrease in the groundwater level in the NCP, especially in urban areas [36]. However, although these measures have resulted in slight increases in groundwater levels, the magnitude of the groundwater level drop created by long-term groundwater depletion is still huge [37]. In the context of climate change, groundwater drought research in the NCP needs to consider climate change aspects and discuss groundwater changes from a climatological perspective. Although groundwater management policies have been effective in slightly increasing groundwater levels in the NCP, the effects of climate warming on groundwater droughts are still unclear. In addition, the spatiotemporal variations in groundwater drought events in the NCP need to be assessed. Indeed, Wang et al. [34] used GGDI to evaluate groundwater drought events in the NCP over the 2003–2015 period, identifying the time of the most severe groundwater drought events in the region and analyzing the influence of teleconnection factors on groundwater drought events. The results demonstrated the significant impact of ENSO on groundwater drought events in the NCP. Meanwhile, Zhang et al. [38] studied groundwater drought trends in the NCP using the GWI and singular spectrum analysis. However, they assessed the spatial heterogeneity of groundwater drought events in a fixed time without analyzing temporal variations in the groundwater drought events. Therefore, further studies on the spatiotemporal changes in groundwater drought events in the NCP may provide a better understanding of the evolution characteristics of groundwater droughts, which is crucial for ensuring the sustainable use of groundwater resources.

The main purpose of this study was (1) to quantitatively assess groundwater drought events in the NCP using the GGDI, (2) propose a new method for investigating the spatiotemporal patterns of groundwater drought events in the NCP from 2003 to 2020, and (3) construct a specific groundwater drought migration model to analyze the characteristics of groundwater drought migration in the NCP. Therefore, this study interpreted developments and changes in groundwater from the perspective of climatology and examined the spatial and temporal evolution of groundwater drought events from a new perspective, which is of great practical significance and theoretical value for the rational exploitation of groundwater resources and the guidance of industrial and agricultural production.

2. Materials and Methods

2.1. Study Area

The NCP (113°10′–119°25′ E; 34°52′–40°29′ N) is one of the three major plains in China (Figure 1). It is located in the southern part of the Taihang Mountains and the northern part of the Yellow River, covering a total area of about 13.9 × 104 km^2 [39]. The geomorphological features of the NCP can be divided into three main areas, namely, the piedmont alluvial–pluvial inclined plain (Area I), the central and alluvial lacustrine plain (Area II), and the eastern alluvial marine plain (Area III). According to the type of groundwater, the pre-mountain plain in North China can be divided into carbonate rock karst water, bedrock fracture water, and loose rock pore water. The available groundwater is mainly stored in loose pore space, and the regional flow direction is northwest to southeast. Locally, due to the formation of landing funnels by groundwater overexploitation, the groundwater dynamic field has changed and groundwater no longer flows in one direction, forming a seepage field from the edge of the landing funnels to the center of each funnel [40]. Groundwater in the NCP circulates mainly in the quaternary pores of underground aquifer rock series. According to burial conditions, circulation characteristics, and retention time, groundwater in the NCP can be divided into two types, namely, shallow and deep groundwater. The shallow groundwater level varies from 50 to 210 m, while the depth of the deep groundwater level ranges from 100 m in Area I to 600 m in Area III. On the other hand, the aquifer lithology in the study area consisted of gravel, medium-coarse sand, and medium-fine sand in Area I, medium-fine sand, fine sand, and silty sand in Area II, and silty fine sand and silty sand in Area III [41]. The NCP has a semi-arid and semi-humid continental monsoon climate, with an average annual rainfall of about 500–600 mm, occurring mainly between July and September [42]. In addition, the evaporation rates increase with air temperature and decrease with latitude. Indeed, the uneven spatiotemporal distributions of precipitation and evaporation have directly impacted groundwater resources in the NCP [37].

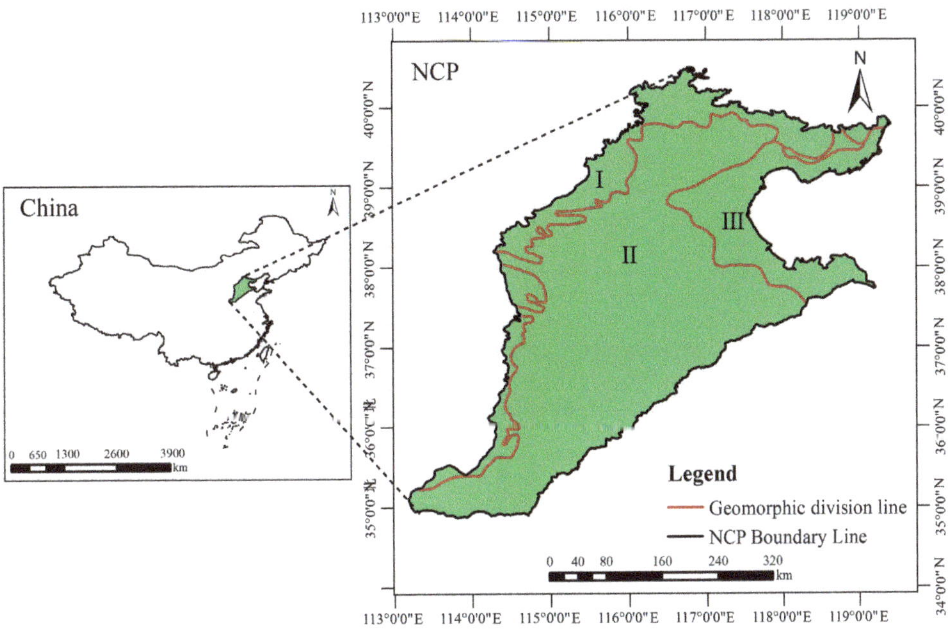

Figure 1. The geographical location of the North China Plain.

The exploitation of groundwater in the North China Plain began in the mid-20th century, and the addition of human activities has led to significant variability in groundwater systems. Humans have influenced the evolution of groundwater systems through groundwater extraction, river management, sewage discharge and management, and the construction of water conservancy. In recent decades, groundwater extraction in the North China Plain has intensified, and the form of the groundwater cycle has been greatly altered. According to statistics, until the 1960s, the smaller scale of groundwater extraction left the groundwater cycle in its natural state, with precipitation as its main source of recharge and evaporation as its main mode of discharge. After the turn of the century, precipitation remained the main factor in maintaining the dynamic balance of groundwater, but groundwater was subjected to higher intensity extraction, resulting in the thickening of the bale zone in the pre-mountain plain area, an increase in the depth of groundwater burial, and a serious lag in the infiltration of precipitation for recharge. In parts of the eastern plain, although the infiltration of precipitation for recharge has increased, changes in the substrata due to urbanization have hindered the infiltration of precipitation and the main mode of groundwater discharge has changed from evaporation to artificial extraction. Deep groundwater in the North China Plain is buried deeper, circulation is slower, and it is a non-renewable resource. In its natural state, the recharge of the deep groundwater mainly comes from runoff and vertical infiltration from alluvial fans and hidden carbonate karst water in front of the mountains.

2.2. *Data Collection*

2.2.1. Gravity Recovery and Climate Experiment (GRACE)

According to their different inversion principles, the GRACE gravity satellite inversion methods can be divided into two categories, namely, the spherical harmonic coefficient and Mascon methods. Terrestrial water storage change data, provided by the latest RL06Mascon data product released by the University of Texas Space Research Center (http://www2.csr.utexas.edu/grace/RL06_mascons.html, accessed 28 October 2021), were used in this study to invert terrestrial water storage anomalies. The data covered the April 2002-September 2021 period, with a monthly scale and a spatial resolution of 0.25° × 0.25°. Some missing data were handled using an efficient and widely used linear interpolation method [29,43]. The data from January 2003 to December 2020 were selected as the basic data after inverting the groundwater storage change data to calculate the GWI.

2.2.2. Global Land Data Assimilation System (GLDAS)

The global land data assimilation system (GLDAS), jointly developed by the National Aeronautics and Space Administration and the National Oceanic and Atmospheric Administration, can use data assimilation technology to merge satellite data and ground observation data and generate surface state quantities and flux [44]. In total, four data products were launched based on the four land surface models, namely, VIC, CLM, CLSM, and Noah. Indeed, the Noah10_M 2.1 (https://disc.sci.gsfc.nasa.gov/data-holdings, accessed on 19 November 2021.) data product provides consistent data with the same spatial resolution as the GRACE RL06Mascon. Therefore, the Noah10_M 2.1 data product, covering the April 2003–September 2021 period, was used to invert the GRACE groundwater storage change data.

2.3. *Methods*

2.3.1. GGDI

The GRACE satellite data were first used to invert the terrestrial water storage anomalies (TWSAs) and then the GLDAS hydrological model was used to invert the surface water storage anomalies (SWSAs), soil moisture storage anomalies (SMSAs), snow water equivalent anomalies (SWEAs), and canopy water storage anomalies (CWSAs). Afterward, the GGDI was calculated according to Thomas et al. [21], as follows:

$$GWSA = TWSA - SWSA - SMSA - SWEA - CWSA, \tag{1}$$

$$C_i = \frac{1}{n_i}\sum\nolimits_1^{n_i} GWSA_i, \tag{2}$$

where Ci is the mean $GWSA$ of the ith month (i = 1, 2, 3, ..., 12) and n is the number of the ith month in the data column.

To remove any seasonal effects on the final results, Ci was deducted from each month's $GWSA$ using the following formula:

$$GSD = GWSA - C_i, \tag{3}$$

where *GSD* is the net groundwater storage deviation.

In addition, to reflect drought conditions, the normalized net groundwater storage deviation (GGDI) was calculated using the following formula:

$$GGDI = \frac{GSD - \overline{x}_{GSD}}{S_{GSD}}, \tag{4}$$

where $\overline{x}_{GSD}$ is the mean value of *GSD* and S_{GSD} is the standard deviation of *GSD*.

2.3.2. Groundwater Drought Event Identification

In previous studies, most groundwater drought events have been determined from a temporal perspective. This study proposed a new method for identifying groundwater drought events from spatial and temporal perspectives, using the following steps:

(1) Drought grid identification

The study area was divided into 221 grids, with grid cells of 0.25 × 0.25. When the center of gravity of the grid was within the study area, the grid was considered to belong to the study area (Figure 2a). When the GGDI of the grid was less than −0.5, the grid was considered to belong to the drought area [29].

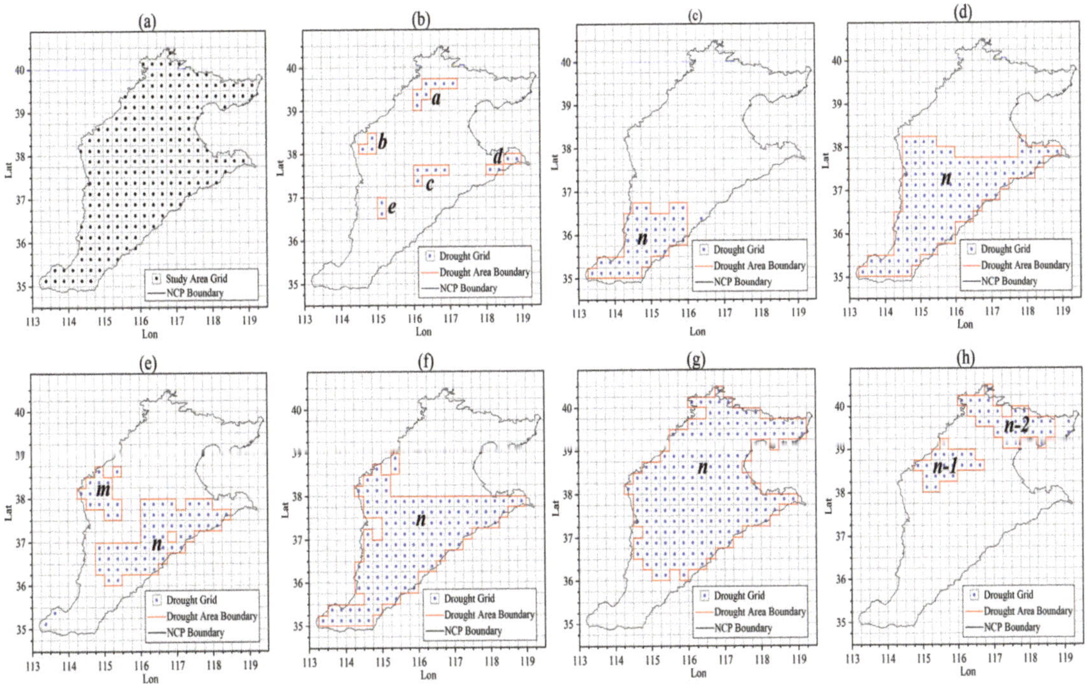

Figure 2. The groundwater drought event identification process (**a**) shows the drought grid identification, (**b**) shows the drought event spatial continuity identification, (**c**,**d**) shows the drought event temporal continuity identification, and (**e**–**h**) shows the drought event spatial-temporal continuity identification.)

(2) Identification of the spatial continuity of drought events

Overlapping drought grid points were observed in certain months, which showed the same drought event (a, c, and d in Figure 2b). Therefore, a drought event was considered absent when the drought area was smaller than the minimum drought area (A_0) (b and e in Figure 2b) [45]. Indeed, A_0 was equal to 1.6% of the total surface of the study area [46], representing four grids in this study.

(3) Identification of the temporal continuity of drought events

After identifying the arid areas, the temporal continuity of the drought events was determined. Indeed, a temporal drought event was considered continuous, i.e., belonging to the same drought event, when the overlapping area between the arid areas of two adjacent months was greater than or equal to A_0 (Figure 2c,d).

(4) Identification of the spatiotemporal continuity of drought events

The drought events m (DE-m) and n (DE-n) in month t were merged into a single drought event in month t + 1. The merged event was called DE-n since the DE-m grid in the merged drought event was smaller than that of the DE-n grid. The original drought events m and n were two sub-events of the merged drought event n, denoted by (n-1) and (n-2), respectively. (Figure 2e,f).

As shown in Figure 2g, DE-(n-1) and DE-(n-2) belonged to DE-n, even though DE-n was divided into several spatially disconnected drought events in the next month (Figure 2h) and the overlapping DE-(n-1), DE-(n-2), and DE-n grids were not less than A_0.

2.3.3. Construction of the Drought Migration Model (DMM)

A centroid refers to a hypothetical point where mass is considered to be concentrated within a material system. Centroids were considered in this study to analyze the spatiotemporal migration characteristics of groundwater drought events. The specific steps were as follows:

(1) Determination of groundwater drought events

Previous studies have shown that only drought events that last for longer than 3 months can reflect the characteristics of drought migration [32]. Therefore, only drought events with a duration greater than or equal to 3 months were considered for the construction of the DMM.

(2) Determination of the locations of groundwater drought centroids

The coordinates of the groundwater drought centroids were determined using the following equations [32]:

$$(X, Y) = \begin{cases} X = \sum_{i=1}^{n} P_i X_i / \sum_{i=1}^{n} P_i \\ Y = \sum_{i=1}^{n} P_i Y_i / \sum_{i=1}^{n} P_i \end{cases}, \quad (5)$$

where X and Y are the longitude and latitude of the groundwater drought centroid, respectively, i represents the ith groundwater drought grid, n denotes the total number of grids contained in the groundwater drought area, X_i and Y_i are the longitude and latitude of the ith groundwater drought grid, respectively, and P_i denotes the GGDI of the ith groundwater drought grid.

(3) Connecting the groundwater drought centroids

The trajectory, direction, speed, and other characteristics of groundwater drought migration were determined by connecting the groundwater drought centroids in chronological order.

(4) Calculation of the groundwater drought intensities

Centroids can not only represent the spatiotemporal characteristics of points but they can also be used as indicators to measure monthly drought intensity. According to previous studies, the centroid of the GGDI (S_{am}), representing a drought intensity of DE. (S_a) in month m, can be calculated using the following equations [32]:

$$S_{am} = -\frac{1}{k}\sum_{i=1}^{k} (GGDI_i - S_0), \quad (6)$$

where k denotes the drought grid number of DE.a in month m and S_0 denotes the groundwater drought threshold (−0.5), and

$$S_a = \sum_{m=1}^{n} S_{am}, \tag{7}$$

where n denotes the total number of groundwater drought months in DE.a.

3. Results

3.1. GRACE Data Validation

The reliability and authenticity of the GLDAS-based SWSA, SMSA, SWEA, and CWSA data have been confirmed in previous studies; thus, they could be combined with the GRACE data to invert the groundwater storage data [47–49]. The GLDAS data were used in this study to validate the GRACE data. The GRACE- and GLDAS-based storage changes showed significant positive correlation coefficients ($p < 0.01$) in each geomorphic area of the NCP ($p < 0.01$). The obtained Pearson correlation coefficients ranged from 0.322 to 0.552 (Table 1).

Table 1. The correlations between the GRACE and GLDAS data in the geomorphic areas of the NCP.

Area	Pearson Correlation Coefficient
Area I	0.322 **
Area II	0.552 **
Area III	0.487 **
NCP	0.527 **

** Significant correlation at the $p < 0.01$ level.

The GLDAS- and GRACE-based water storage in the NCP ranged from 13.06 to 6.81 mm and 13.76 to 41.18 mm, respectively, indicating relatively large values. These results could be explained by the fact that the GRACE-based land water storage included both surface water and groundwater, while that of the GLDAS excluded groundwater storage. The GRACE uncertainties in the NCP and its sub-areas were calculated according to the method proposed by Landerer and Swenson [50] (Table 2). The results showed decreasing trends in the GRACE-based water storage in the NCP and its sub-areas (Table 3). According to the specific situation of the NCP, these decreasing trends could have been due to the ecological restoration projects implemented by the government increasing the consumption of soil water in the study area, as well as groundwater overexploitation. The temporal trends in the GRACE- and GLDAS-water storage in sub-areas of the NCP are shown in Figure 3, indicating consistent changes in water storage from 2003 to 2013. However, gradual increases in the differences between the GRACE- and GLDAS-water storage trends were observed after 2013. Indeed, Area II exhibited the largest mean difference (−19.36), followed by Area I and Area III, with mean differences of −14.80 and −9.30, respectively. These findings suggested increased rates of decline in groundwater storage in the NCP after 2013. According to the obtained results, Area II exhibited the highest decline rate, followed by Area I and Area III. This finding was consistent with those reported by Zhang et al. [38]. In summary, the obtained GRACE data for the NCP were reliable and could be used to effectively investigate groundwater drought events in this study area.

Table 2. The GRACE uncertainties in the NCP and its sub-areas.

Area	Uncertainty (mm/month)
Area I	11.51
Area II	15.11
Area III	7.96
NCP	13.44

Table 3. A comparison of the GRACE and GLDAS data from the North China Plain and its various geomorphic areas.

Area	Decline Rate (mm/Month)
Area I	0.0578
Area II	0.0620
Area III	0.0452
NCP	0.0590

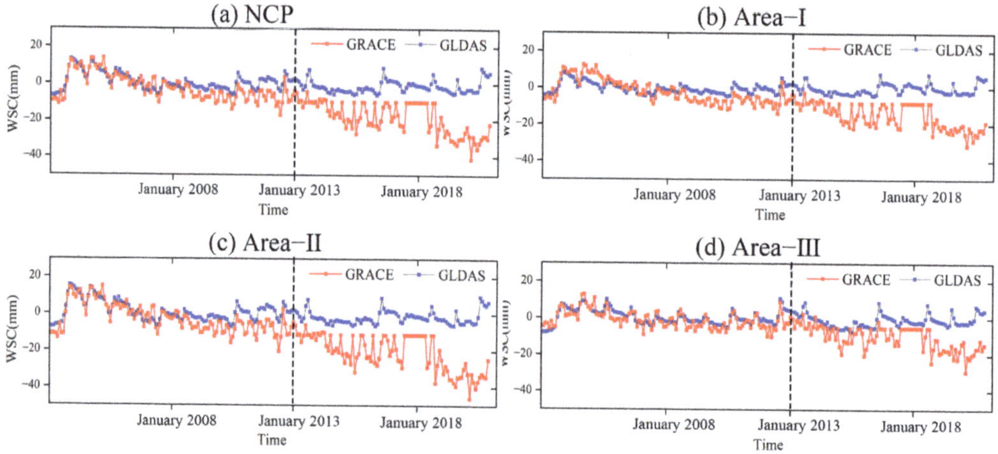

Figure 3. A comparison of the GRACE and GLDAS data from the North China Plain and its various geomorphic areas ((**a–d**) indicates different partitions).

3.2. *Quantitative Assessment of Groundwater Drought Events*

The obtained quantitative evaluation results of groundwater drought events in the NCP and its geomorphic sub-areas using the GGDI are shown in Figure 4. The groundwater drought trends in the geomorphic sub-areas were consistent with those observed across the entire NCP, showing gradually decreasing trends. In addition, the most severe groundwater drought event in all areas was observed in March 2020. Indeed, the GGDI values for the NCP and its geomorphic sub-areas in March 2020 were −3.02, −2.76, −3.07, and −3.10, respectively. The groundwater drought frequency in the entire NCP was 24.54%, with mild, moderate, severe, and extreme groundwater drought frequencies of 8.33, 8.80, 3.70, and 3.70%, respectively. The groundwater drought frequency in Area I was 26.39%, with mild, moderate, severe, and extreme groundwater drought frequencies of 10.65, 9.72, 3.24, and 2.78%, respectively. The groundwater drought frequency in Area II was 24.54%, with mild, moderate, severe, and extreme groundwater drought frequencies of 7.87, 7.41, 5.09, and 4.17%, respectively. The groundwater drought frequency in Area III was 23.61%, with mild, moderate, severe, and extreme groundwater drought frequencies of 9.72, 6.02, 4.63, and 3.24%, respectively. The highest frequency of groundwater drought events was observed in Area I, followed by Area II and Area III. However, unlike Area I and Area III, which exhibited mainly mild and moderate drought events, Area II mainly experienced severe and extreme groundwater drought events.

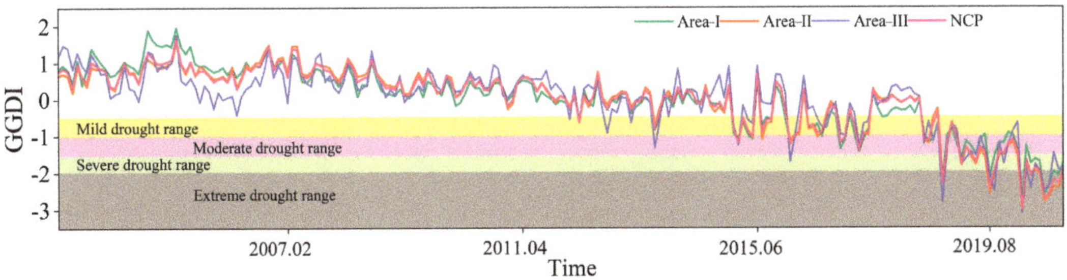

Figure 4. The GGDI times series of the North China Plain and its sub-areas.

3.3. *Groundwater Drought Event Identification*

According to the identification principles reported in Section 2.3, groundwater drought events in the NCP from 2003 to 2021 were identified using GGDI. The results showed that a total of 49 drought events occurred in this region. The duration of each drought event is shown in Figure 5.

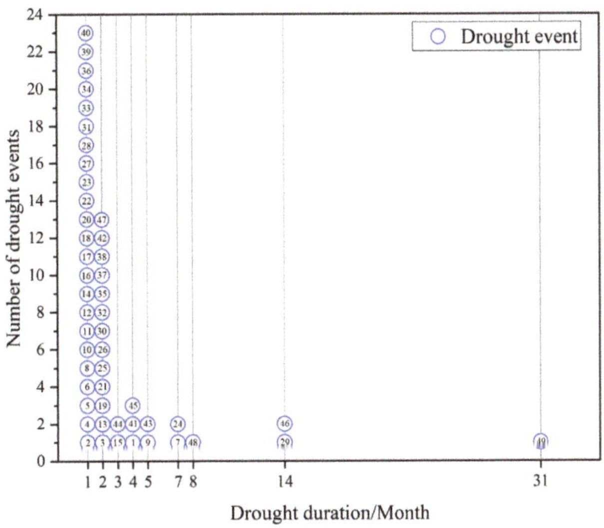

Figure 5. The groundwater drought events in the North China Plain.

In total, 49 groundwater drought events occurred in the NCP, with the shortest and longest duration of 1 month and 31 months, respectively. In addition, there were 23, 13, and 2 drought events with a duration of 1, 2, and 3 months, accounting for 46.94, 26.53, and 4.08% of the total groundwater drought events, respectively. Whereas 3, 2, 2, and 1 groundwater drought events lasted 4, 5, 7, and 8 months, accounting for 6.12, 4.08, 4.08, and 2.04% of the total groundwater drought events, respectively. In addition, 2 and 1 drought events lasted 14 and 31 months, accounting for 4.08 and 2.04% of the total groundwater drought events, respectively.

3.4. *Characteristics of the Groundwater Drought Centroid Migration*

Based on the DMM construction principles, 11 groundwater drought events from 2003 to 2020 were selected in this study to investigate the groundwater drought centroid migration characteristics in the NCP. The characteristics of these drought events are reported in Table 4.

Table 4. Characteristics of 11 groundwater drought events.

Drought Event	Occurrence Time	Drought Duration (Months)	Monthly Average Minimum GGDI	Monthly Maximum Drought Grid Number	Drought Intensity
1	2003.01–2003.04	4	−1.14	14	1.84
7	2005.06–2005.12	7	−1.57	8	4.41
9	2006.02–2006.06	5	−1.55	8	3.08
24	2011.10–2012.04	7	−0.87	155	0.88
29	2012.08–2013.09	14	−2.47	167	11.53
41	2014.08–2014.11	4	−1.03	67	1.24
43	2015.01–2015.05	5	−1.25	203	2.50
45	2015.12–2016.03	4	−1.29	213	2.17
46	2016.05–2017.06	14	−1.42	216	6.67
48	2017.09–2018.04	8	−0.96	9	2.36
49	2018.06–2020.12	31	−3.12	221	38.17

According to Table 4, DE.49 exhibited the highest intensity, longest duration, lowest monthly average minimum GGDI, and highest drought grid number, with 38.17, 31 months, −3.12, and 221, respectively. Therefore, this event was the most serious groundwater drought event over the considered period, followed by DE.29, with drought duration, monthly minimum GGDI, and drought intensity of 14 months, −247, and 11.53, respectively. DE.46 showed the third highest groundwater drought intensity, with the same groundwater drought duration as DE.29 (14 months) and a maximum number of groundwater drought grids of 216, which was second only to DE.49. In addition, it can be seen from Table 4 that DE.24, DE.41, and DE.1 were the groundwater drought events with the lowest drought intensities among the 11 drought events. The drought duration, monthly average minimum GGDI, monthly maximum drought grid number, and groundwater drought intensity of DE.24 were 7 months, −0.87, 155, and 0.88, respectively, making it the least severe event among the 11 groundwater drought events. DE.1 and DE.41 lasted for 4 months in the study area, showing monthly average minimum GGDI values of −1.14 and −1.03, respectively. In addition, DE.1 and DE.41 exhibited monthly maximum grid numbers of 14 and 67 and drought intensities of 1.84 and 1.24, respectively.

In this study, the drought centroid of each drought event was obtained based on the calculation method of the above-mentioned drought centroid. In addition, the migration trajectory of each groundwater drought event was determined by connecting them in chronological order (Figure 6). According to the obtained results, the migration direction of the 11 drought events was mainly southwest –northeast. In contrast, only DE.41 exhibited a southeast–northwest direction. Because the drought migration in the DMM was on a monthly scale, the longer the distance between the drought centroids of two adjacent months, the higher the drought migration rate. Figure 6a shows that the maximum drought migration rate (1.62×10^8 km/month) of all drought events occurred in the fourth to fifth months of DE.49, resulting in the largest overall migration distance. The mid-month drought centroid was more concentrated. In addition, DE.1 exhibited the smallest drought migration distance, followed by DE.48. It can also be seen from Figure 6 that the centroid migration of the 11 drought events was mostly concentrated in Area II. The drought centroids of DE.7, DE.9, DE.29, and DE.49 were in Area II and Area III, while that of DE.48 was concentrated in Area I.

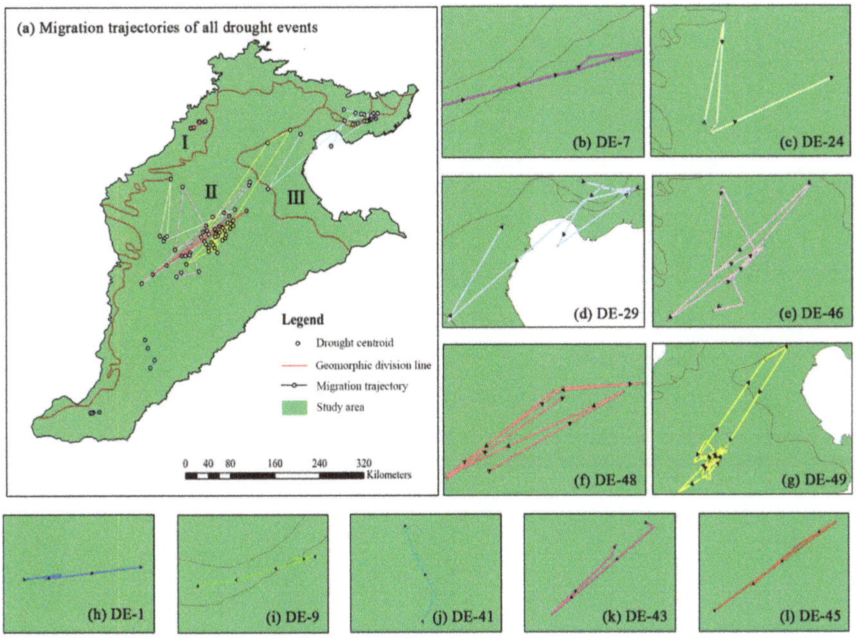

Figure 6. The spatial distributions of the groundwater drought event paths ((**a**) is the migration trajectory of all drought events, and (**b–l**) is the zoomed-in migration trajectory of each drought event).

In order to reveal the developments and changes in the drought events more intuitively, the three drought events with the highest drought intensities among the 11 drought events, namely DE.49, DE.29, and DE.46, were visualized. The drought centroids of these three drought events were first connected in a monthly series, then their drought intensities were represented by spheres. The larger the sphere, the more serious the groundwater drought event.

(1) DE.49

DE.49 occurred from June 2018 to December 2020, lasting 31 months. It was the longest drought event with the highest monthly drought intensity. According to Figures 6 and 7, the highest drought intensities of DE.49 were observed mainly in the central part of the NCP (Area II) over the last 10 months (from March 2020 to December 2020) of the entire drought event. The intensity of the entire drought event showed four obvious peaks in the 5th, 15th, 22nd, and 27th months. The highest drought intensity was observed in the 22nd month (2.62), followed by those in the 27th (2.31), 15th (1.91), and 5th (1.81) months. In addition, the lowest DE.49 drought intensity (0.31) was observed in the 7th month, followed by that in the 4th month (0.51). The entire DE.49 period showed slightly different drought grid numbers and drought intensities, without exhibiting any apparent correlations between them. The number of drought grids fluctuated greatly in the first 9 months, then remained basically at a high value before substantially decreasing in the 21st month. The results showed three minimum drought grid numbers over the DE.49 drought event, which were 71, 159, and 196 in the 4th, 7th, and 21st months, respectively.

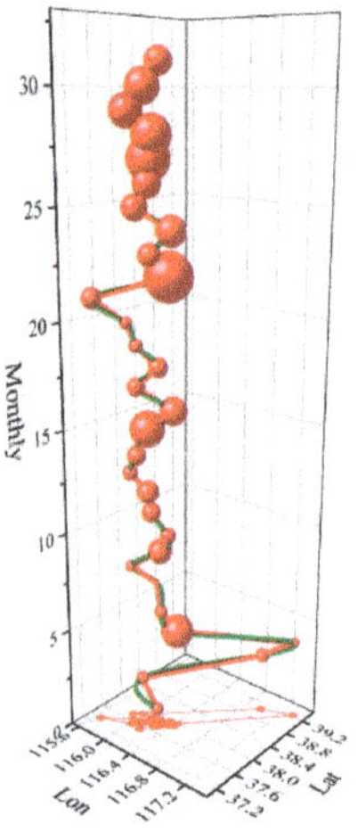

(a)

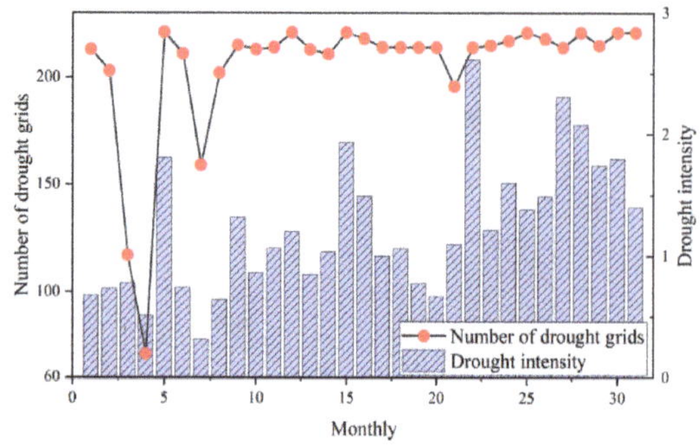

(b)

Figure 7. Changes in the DE.49 groundwater drought event characteristics: (**a**) migration processes and changes in the intensity of arid plasmas; (**b**) changes in drought areas and drought intensities during drought development.

(2) DE.29

DE.29 occurred from August 2012 to September 2013, lasting 14 months. According to Figures 6 and 8, the highest monthly groundwater drought intensities were observed mainly in the Bohai Bay area of the NCP in the 4–9th months. The drought intensity of the entire drought event experienced two obvious peaks of 1.97 and 0.75 in the 4th and 12th months, respectively. However, the lowest groundwater drought intensity (0.33) was observed in the 14th month, followed by that in the 10th month (0.37). In addition, the results showed that DE.29 had slightly different monthly groundwater drought grid numbers and intensities. The highest drought grid numbers were observed in the 3rd and 13th months, without presenting high drought intensities. The groundwater drought grid number in the month with the highest groundwater drought intensity was 10, while that in the month with the second highest groundwater drought intensity was 18. It can be seen that small and large groundwater drought areas exhibited high and low drought intensities, respectively.

(3) DE.46

DE.46 occurred from May 2016 to June 2017, lasting 14 months. It can be seen from Figures 6 and 9 that the highest monthly drought intensities were observed in the central part of the NCP (Area II). In addition, the highest groundwater drought intensities were 0.62, 0.72, and 0.92 in the 4th, 8th, and 12th months, respectively. Meanwhile, the lowest drought intensities were 0.17 and 0.20 in the 6 and 10th months, respectively. On the other hand, the highest DE.46 groundwater drought grid number was 216 in the 9th and 12th months, while the lowest DE.46 groundwater drought grid number was 16 in the 6th month. The obtained results showed that the lowest drought intensity was associated with the lowest drought grid number.

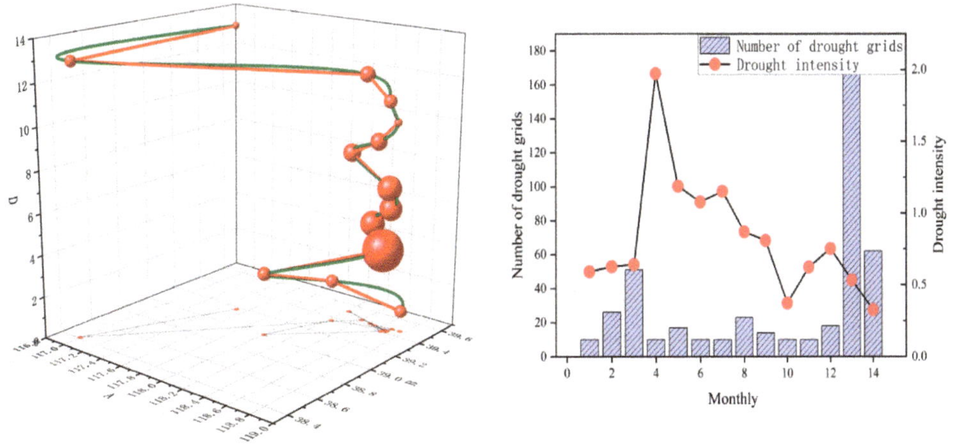

(a) (b)

Figure 8. Changes in the DE.29 groundwater drought event characteristics: (**a**) migration processes and changes in the intensity of arid plasmas; (**b**) changes in drought areas and drought intensities during drought development.

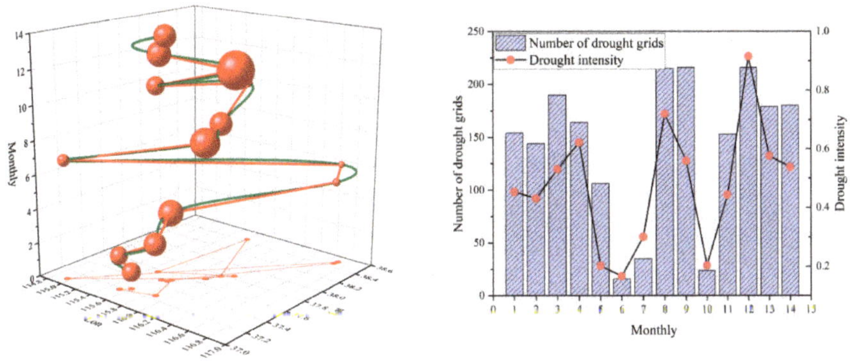

(a) (b)

Figure 9. Changes in the DE.46 groundwater drought event characteristics: (**a**) migration processes and changes in the intensity of arid plasmas; (**b**) changes in drought areas and drought intensities during drought development.

4. Discussion

According to the results of our drought event assessment of the NCP over the 2003–2020 period, the most intense groundwater drought event occurred in the June 2018–December 2020 period (DE.49), followed by that in the August 2012–September 2013 period (DE.29). Wang et al. (2020) assessed drought events in the NCP from 2003 to 2015 and their results showed that the most severe groundwater drought event in the NCP occurred over the August 2013–September 2014 period, while the results of this study showed that the most severe groundwater drought event occurred over the July 2003–July 2006 period. In addition, the occurrence times of DE.29 were inconsistent with those reported by Wang et al. [27], which could have been due to two main reasons. The first reason was related to the difference in the scales of the study areas. Indeed, Wang et al. [27] considered the entire Henan Province and Shandong Province in their groundwater drought assessment of the NCP, while only some parts of Henan Province and Shandong Province were considered in the current study. According to the results of Wang et al. [27], the groundwater drought events in the Henan and Shandong regions were more severe than those in other regions during the August 2013–September 2014 period. However, these regions were not considered in the present study. In fact, Wang et al. [27] assessed the severity of GGDI-based groundwater drought events using the average intensity observed across the entire study area, which also explained the different results obtained. The second reason was related to the identification scale of the groundwater drought events. Indeed, the groundwater drought events in the NCP were assessed from a spatiotemporal perspective in this study, while Wang et al. [27] only assessed temporal variations in the groundwater drought events, which explained the different conclusions obtained.

The NCP is one of the largest groundwater funnels in the world. The Chinese government is committed to improving the local hydrogeological conditions by implementing several appropriate policies. The South-to-North Water Diversion Project is a large-scale water conservancy project established by the Chinese government to alleviate pressure on water resources in North China. Indeed, the Central South–North Water Diversion Project opened to the North China Plain in December 2014 [51]. In addition, relevant departments have implemented water-saving pressure mining policies and carried out ecological water replenishment, which has played an important role in raising the groundwater level in the NCP [38]. However, the results of this study revealed that 7 of the 49 groundwater drought events in the NCP occurred after the implementation of the South-to-North Water Diversion Project and 5 of the 11 droughts that lasted over 3 months occurred after the implementation of the Central South–North Water Diversion Project. Although these measures have raised the groundwater level in the NCP [34,38], the groundwater drought events caused by long-term groundwater overexploitation cannot be mitigated in the short term. Indeed, continuous improvements to water replenishment and water-saving pressure mining measures can enhance groundwater resilience to climate warming. A study by Yang Huifeng et al. showed that groundwater levels in the urban areas of the North China Plain rose significantly after the South–North Water Diversion was opened but water levels in agricultural areas still showed a continuous decline [37] and those agricultural areas in the North China Plain accounted for a large proportion of the GGDI and dominated the regional average. Therefore, although a small recovery in groundwater levels has occurred in some areas of the North China Plain, according to reported studies, no immediate effects on groundwater droughts caused by long-term groundwater overexploitation have been observed, and only by continuously improving water replenishment measures and continuing to carry out water conservation and suppression work will the groundwater systems be able to withstand the test of climate warming.

In this study, a new method for identifying groundwater drought events was proposed from a spatiotemporal perspective to improve our understanding of groundwater droughts. Indeed, the constructed drought migration model may have a beneficial effect on our understanding of the developments and changes in groundwater droughts. The results of our drought migration assessment showed that the centroids of the drought events were mostly concentrated in the deep funnel area [32], indicating that the center of gravity of a groundwater drought is in the deep funnel area. The groundwater drought migration was in the southwest–northeast and northeast–southwest directions, which was the same direction as the slope of the terrain in the study area. There is still an academic gap in our knowledge of the mechanisms of drought migration characteristics. Therefore, we could not explain this phenomenon. Indeed, migration mechanisms may play a crucial role in the prediction and early warning of groundwater drought events. Therefore, future studies should assess the migration mechanisms of groundwater drought events.

5. Conclusions

This paper used GGDI to quantitatively assess groundwater drought events in the North China Plain, proposed a spatiotemporal groundwater drought event identification method, introduced the center of mass theory, constructed a groundwater drought center of gravity migration model, analyzed the migration characteristics of groundwater drought events in the North China Plain, and conducted a visual mapping study of the three groundwater drought events with the greatest drought intensities. The following conclusions were drawn from this study.

(1) The validation results of the GRACE data showed significant Pearson correlation coefficients ($p < 0.01$) between the changes in the GRACE- and GLDAS-based water storage in the NCP and its geomorphological areas, ranging from 0.322 to 0.552. Therefore, the GRACE-based results were reliable and could be used to effectively investigate groundwater drought events in the NCP.

(2) The groundwater drought frequencies in the NCP, Area I, Area II, and Area III over the 2003–2020 period were 24.54, 26.39, 24.54, and 23.61%, respectively. Although Area I showed a higher groundwater drought frequency than the other sub-areas, mild and moderate groundwater drought events were the most prevalent. In addition, Area II showed a high frequency of moderate groundwater drought events, but its severe and extreme groundwater drought frequencies were relatively higher than those in the other sub-areas.

(3) According to the new identification principle for groundwater drought events, 49 groundwater drought events were identified in the NCP over the 2003–2020 period. The maximum duration of drought was 31 months and the minimum was 1 month. Drought events with a drought duration of 1 month were the most frequent, accounting for 46.94% of the total number of drought events, followed by those with a drought duration of 2 months, accounting for 26.53% of the total. Drought events with a drought duration of 8 or 31 months were the least frequent, both accounting for 2.04% of the total. The obtained results indicated that DE.49 was the most severe groundwater drought event, with a drought intensity, duration, and grid number of 38.17, 31 months, and 221, respectively. Meanwhile, DE.29 was the second most intense groundwater drought event, with a drought intensity, duration, and grid number of 11.53, 14 months, and 167, respectively.

(4) A total of 11 groundwater drought events were selected from the 49 drought events to construct a drought migration model. The migration direction of 10 of the groundwater drought events was southwest–northeast, which was in line with the slope of the NCP. However, only DE.41 exhibited a southeast–northwest migration direction. The centroids of the groundwater drought events were mostly concentrated in Area II. The three groundwater drought events with the highest drought intensities were DE.49, DE.29, and DE.46. According to the obtained results, the highest drought intensities of DE.49 were observed mainly in the March 2020–December 2020 period, in which the drought center of gravity was concentrated in Area II, whereas the highest groundwater intensities of DE.29 were concentrated over the 4th–9th month period, in which the drought center of gravity was concentrated in the northeastern coastal area of the NCP. The lowest drought intensities of DE.46 were in the September 2016-November 2016 period, showing a drought center of gravity in Area II. On the other hand, the results indicated a lack of correlation between the drought intensities and drought grid numbers of DE.49, DE.29, and DE.46.

In summary, the principles for identifying groundwater drought events proposed in this study could reveal regional groundwater drought changes from both temporal and spatial perspectives. In addition, the groundwater drought migration model proposed based on these principles could use the center of gravity and migration direction of groundwater drought events to capture their migration characteristics. The principles for defining groundwater drought events, as well as the groundwater drought migration model, were not only applicable to the NCP but could also be used in other regions, thus providing an in-depth understanding of groundwater development laws and promoting the sustainable use of groundwater resources. However, due to the constraints of research time and data, this paper did not conduct a more in-depth study of groundwater drought migration mechanisms or the causes of such migration characteristics. However, the study of groundwater drought migration mechanisms is important to fully grasp groundwater drought development patterns. Therefore, this would be a good direction for future research.

Author Contributions: Conceptualization, J.H. and L.C.; Data curation, J.H.; Formal analysis, J.H. and L.L.; Investigation, J.H. and L.C.; Methodology, J.H. and L.C.; Resources, L.C.; Software, J.H., L.W., and L.H.; Supervision, J.H. and L.C.; Validation, B.Y.; Visualization, J.H. and L.H.; Writing—original draft, J.H.; Writing—review and editing, J.H. and L.C. All authors have read and agreed to the published version of the manuscript.

Funding: This research received no external funding.

Institutional Review Board Statement: Not applicable.

Informed Consent Statement: Not applicable.

Data Availability Statement: The data are available in a publicly accessible repository that does not issue DOIs. Publicly available datasets were analyzed in this study. These data can be found here: http://www2.csr.utexas.edu/grace/RL06_mascons.html, accessed on 28 October 2021; https://disc.sci.gsfc.nasa.gov/data-holdings, accessed on 19 November 2021.

Conflicts of Interest: The authors declare no conflict of interest.

References

1. Sanginabadi, H.; Saghafian, B.; Delavar, M. Coupled Groundwater Drought and Water Scarcity Index for Intensively Over-drafted Aquifers. *J. Hydrol. Eng.* **2019**, *11*, 04019003. [CrossRef]
2. Mohamed, A.; Al Deep, M.; Othman, A.; Taha, A.I.; Alshehri, F.; Abdelrady, A. Integrated Geophysical Assessment of Groundwater Potential in Southwestern Saudi Arabia. *Front. Earth Sci.* **2022**, *10*, 937402. [CrossRef]
3. Othman, A.; Abdelrady, A.; Mohamed, A. Monitoring Mass Variations in Iraq Using Time-Variable Gravity Data. *Remote Sens.* **2022**, *14*, 3346. [CrossRef]
4. Famiglietti, J.S.; Ferguson, G. The hidden crisis beneath our feet. *Science* **2021**, *372*, 344–345. [CrossRef]
5. United Nations Educational, Scientific and Cultural Organization (UNESCO). *UN World Water Development Report*; UNESCO: Dakar, Senegal, 2022.
6. Han, Z.M.; Huang, S.Z.; Huang, Q.; Leng, G.Y.; Wang, H.; Bai, Q.J.; Zhao, J.; Ma, L.; Wang, L.; Du, M. Propagation dynamics from meteorological to groundwater drought and their possible influence factors. *J. Hydrol.* **2019**, *578*, 124102. [CrossRef]
7. Huang, J.P.; Yu, H.P.; Guan, X.D.; Wang, G.Y.; Guo, R. Accelerated dryland expansion under climate change. *Nat. Clim. Chang.* **2015**, *6*, 166–171. [CrossRef]
8. Ren, Y.; Liu, J.; Shalamzari, M.J.; Arshad, A.; Liu, S.; Liu, T.; Tao, H. Monitoring Recent Changes in Drought and Wetness in the Source Region of the Yellow River Basin, China. *Water* **2022**, *14*, 861. [CrossRef]
9. Mohamed, A.; Faye, C.; Othman, A.; Abdelrady, A. Hydro-Geophysical Evaluation of the Regional Variability of Senegal's Terrestrial Water Storage Using Time-Variable Gravity Data. *Remote Sens.* **2022**, *14*, 4059. [CrossRef]
10. Ren, Y.; Liu, J.; Zhang, T.; Shalamzari, M.J.; Arshad, A.; Liu, T.; Willems, P.; Gao, H.; Tao, H.; Wang, T. Identification and Analysis of Heatwave Events Considering Temporal Continuity and Spatial Dynamics. *Remote Sens.* **2023**, *15*, 1369. [CrossRef]
11. Kinouchi, T.; Sayama, T. A comprehensive assessment of water storage dynamics and hydroclimatic extremes in the Chao Phraya River Basin during 2002–2020. *J. Hydrol.* **2021**, *603*, 126868. [CrossRef]
12. Rutulis, M. Groundwater drought sensitivity of Southern Manitoba. *Can. Water Resour. J.* **1989**, *14*, 18–33. [CrossRef]
13. Van Lanen, H.A.J.; Peters, E. Definition, effects and assessment of groundwater droughts. In *Drought and Drought Mitigation in Europe*; Springer: New York, NY, USA, 2000; pp. 49–61.
14. Marchant, B.; Bloomfield, J. Spatio-temporal modelling of the status of groundwater droughts. *J. Hydrol.* **2018**, *564*, 397–413. [CrossRef]
15. Brauns, B.; Cuba, D.; Bloomfield, J.P.; Hannah, D.M.; Jackson, C.; Marchant, B.P.; Heudorfer, B.; Van Loon, A.F.; Bessière, H.; Thunholm, B.; et al. The Groundwater Drought Initiative (GDI): Analysing and understanding groundwater drought across Europe. *Proc. Int. Assoc. Hydrol. Sci.* **2020**, *383*, 297–305. [CrossRef]
16. Bloomfield, J.P.; Marchant, B.P.; McKenzie, A.A. Changes in groundwater drought associated with anthropogenic warming. *Hydrol. Earth Syst. Sci.* **2019**, *23*, 1393–1408. [CrossRef]
17. Fiorillo, F.; Guadagno, F.M. Long karst spring discharge time series and droughts occurrence in Southern Italy. *Environ. Earth Sci.* **2011**, *65*, 2273–2283. [CrossRef]
18. Bloomfield, J.P.; Marchant, B.P. Analysis of groundwater drought building on the standardised precipitation index approach. *Hydrol. Earth Syst. Sci.* **2013**, *17*, 4769–4787. [CrossRef]
19. Saghafian, B.; Sanginabadi, H. Multivariate groundwater drought analysis using copulas. *Hydrol. Res.* **2020**, *51*, 666–685. [CrossRef]

20. Bhuiyan, C. Various drought indices for monitoring drought condition in Aravalli terrain of India. In Proceedings of the 20th ISPRS Conference, Istanbul, Turkey, 12–23 July 2004.
21. Thomas, B.F.; Famiglietti, J.S.; Landerer, F.W.; Wiese, D.N.; Molotch, N.P.; Argus, D.F. GRACE Groundwater Drought Index: Evaluation of California Central Valley groundwater drought. *Remote Sens. Environ.* **2017**, *198*, 384–392. [CrossRef]
22. Kumar, K.S.; AnandRaj, P.; Sreelatha, K.; Bisht, D.S.; Sridhar, V. Monthly and Seasonal Drought Characterization Using GRACE-Based Groundwater Drought Index and Its Link to Teleconnections across South Indian River Basins. *Climate* **2021**, *9*, 56. [CrossRef]
23. Li, B.; Rodell, M. Evaluation of a model-based groundwater drought indicator in the conterminous U.S. *J. Hydrol.* **2015**, *526*, 78–88. [CrossRef]
24. Mendicino, G.; Senatore, A.; Versace, P. A Groundwater Resource Index (GRI) for drought monitoring and forecasting in a mediterranean climate. *J. Hydrol.* **2008**, *357*, 282–302. [CrossRef]
25. Goodarzi, M.; Abedi-Koupai, J.; Heidarpour, M.; Safavi, H.R. Development of a New Drought Index for Groundwater and Its Application in Sustainable Groundwater Extraction. *J. Water Resour. Plan. Manag.* **2016**, *142*, 04016032. [CrossRef]
26. Zhu, Q.; Zhang, H. Groundwater drought characteristics and its influencing factors with corresponding quantitative contri-bution over the two largest catchments in China. *J. Hydrol.* **2022**, *609*, 127759. [CrossRef]
27. Wang, F.; Lai, H.; Li, Y.; Feng, K.; Zhang, Z.; Tian, Q.; Zhu, X.; Yang, H. Identifying the status of groundwater drought from a GRACE mascon model perspective across China during 2003–2018. *Agric. Water Manag.* **2021**, *260*, 107251. [CrossRef]
28. Wen, X.; Tu, Y.H.; Tan, Q.F.; Li, W.Y.; Fang, G.H.; Ding, Z.Y.; Wang, Z.N. Construction of 3D drought structures of meteorological drought events and their spa-tio-temporal evolution characteristics. *J. Hydrol.* **2020**, *590*, 125539. [CrossRef]
29. Zhou, H.; Liu, Y.B.; Liu, Y.W. An approach to tracking meteorological drought migration. *Water Resour. Res* **2019**, *55*, 3266–3284. [CrossRef]
30. Guo, H.; Bao, A.M.; Ndayisaba, F.; Liu, T.; Jiapaer, G.; El-Tantawi, A.M.; De Maeyer, P. Space-time characterization of drought events and their impacts on vegetation in Central Asia. *J. Hydrol.* **2018**, *564*, 1165–1178. [CrossRef]
31. Herrera-Estrada, J.E.; Satoh, Y.; Sheffield, J. Spatiotemporal dynamics of global drought. *Geophys. Res. Lett.* **2017**, *44*, 2254–2263. [CrossRef]
32. Han, Z.M.; Huang, Q.; Huang, S.Z.; Leng, G.Y.; Bai, Q.J.; Liang, H.; Wang, L.; Zhao, J.; Fang, W. Spatial-temporal dynamics of agricultural drought in the Loess Plateau under a changing environment: Characteristics and potential influencing factors. *Agric. Water Manag.* **2020**, *244*, 106540. [CrossRef]
33. Bloomfield, J.P.; Marchant, B.P.; Bricker, S.H.; Morgan, R.B. Regional analysis of groundwater droughts using hydrograph classification. *Hydrol. Earth Syst. Sci.* **2015**, *19*, 4327–4344. [CrossRef]
34. Wang, F.; Wang, Z.; Yang, H.b.; Di, D.Y.; Zhao, Y.; Liang, Q.H. Utilizing GRACE-based groundwater drought index for drought characterization and teleconnection factors analysis in the North China Plain. *J. Hydrol.* **2020**, *585*, 124849. [CrossRef]
35. Shi, J.; Zhang, Y. North China Plain: Strengthening the rational development of groundwater is the key. *China Nat. Resour. News* **2020**, *3*, 7.
36. Zhang, C.; Duan, Q.; Yeh, P.J.; Pan, Y.; Gong, H.; Gong, W.; Di, Z.; Lei, X.; Liao, W.; Huang, Z.; et al. The Effectiveness of the South-to-North Water Diversion Middle Route Project on Water Delivery and Groundwater Recovery in North China Plain. *Water Resour. Res.* **2020**, *56*, e2019WR026759. [CrossRef]
37. Yang, H.F.; Cao, W.G.; Zhi, C.S.; Li, Z.Y.; Bao, X.L.; Ren, Y.; Liu, F.T.; Fan, C.L.; Wang, S.F.; Wang, Y.B. Evolution of groundwater level in the North China Plain in the past 40 years and suggestions on its overexploitation treatment. *Geol. China* **2021**, *48*, 1142–1155.
38. Zhang, H.; Ding, J.; Zhu, Q.; Wang, G.Q. Analysis on spatial-temporal characteristic of groundwater drought based on GRACE in North China Plain. *Yangtze River* **2021**, *52*, 107–114.
39. Wenpeng, L.; Longfeng, W.; Huifeng, Y.; Yuejun, Z.; Wengeng, C.; Ke, L. The groundwater overexploitation status and counter-measure suggestions of the North China Plain. *China Water Resour.* **2020**, *13*, 26–30.
40. Zhang, G.L.; Liu, H.Y.; Guo, H.M.; Sun, Z.X.; Wang, Z.; Wu, T.H. Occurrences and health risks of high nitrate groundwater in the typical piedmont areas of the North China Plain. *Earth Sci. Front.* **2023**, 1–21. [CrossRef]
41. Dou, T.T.; Cheng, H.H.; Zhou, Y.Z.; Shi, Y.L. The influence of groundwater mining on regional seismicity in the North China Plain. *Chin. J. Geophys.* **2022**, *65*, 2931–2944. (In Chinese)
42. Zhang, D. *Study of the Allowable Groundwater Withdrawal in the North China Plain Based on Environmental Limited Groundwater Level*; University of Geosciences: Beijing, China, 2017.
43. Lin, M.; Biswas, A.; Bennett, E.M. Spatio-temporal dynamics of groundwater storage changes in the Yellow River Basin. *J. Environ. Manag.* **2019**, *235*, 84–95. [CrossRef]
44. Rodell, M.; House, P.R.; Jambor, U.; Gottschalck, J.; Mitchell, K.; Meng, C.J.; Arsenault, K.; Cosgrove, B.; Radakovich, J.; Bosilo-vich, M.; et al. The Global Land Data Assimilation System. *Bull. Am. Meteorol. Soc.* **2004**, *85*, 381–394. [CrossRef]
45. Sheffield, J.; Andreadis, K.M.; Wood, E.F.; Lettenmaier, D.P. Global and Continental Drought in the Second Half of the Twentieth Century: Severity–Area–Duration Analysis and Temporal Variability of Large-Scale Events. *J. Clim.* **2009**, *22*, 1962–1981. [CrossRef]
46. Wang, A.; Lettenmaier, D.P.; Sheffield, J. Soil Moisture Drought in China, 1950–2006. *J. Clim.* **2011**, *24*, 3257–3271. [CrossRef]
47. Han, Z.; Huang, S.; Huang, Q.; Bai, Q.; Leng, G.; Wang, H.; Zhao, J.; Wei, X.; Zheng, X. Effects of vegetation restoration on groundwater drought in the Loess Plateau, China. *J. Hydrol.* **2020**, *591*, 125566. [CrossRef]

48. Ganapuram, S.; Nagarajan, R.; Sekhar, G.C. Identification of groundwater drought prone zones in Pedda vagu and Ookachetti vagu watersheds, tributaries of the Krishna River, India. *Geocarto Int.* **2015**, *31*, 385–407. [CrossRef]
49. Mustafa, S.M.T.; Abdollahi, K.; Verbeiren, B.; Huysmans, M. Identification of the influencing factors on groundwater drought and depletion in north-western Bangladesh. *Hydrogeol. J.* **2017**, *25*, 1357–1375. [CrossRef]
50. Landerer, F.W.; Swenson, S.C. Accuracy of scaled GRACE terrestrial water storage estimates. *Water Resour. Res.* **2012**, *48*, 1–11. [CrossRef]
51. Yao, Y.Y.; Zheng, C.M.; Andrews, C.; He, X.; Zhang, A.J.; Liu, J. Integration of groundwater into China's south-north water transfer strategy. *Sci. Total. Environ.* **2018**, *658*, 550–557. [CrossRef]

atmosphere

MDPI

Article

Application of Informer Model Based on SPEI for Drought Forecasting

Jiandong Shang [1,2,†], Bei Zhao [2,†], Haobo Hua [1,3], Jieru Wei [1,*,†], Guoyong Qin [1,*] and Gongji Chen [1]

1 National Supercomputing Center in Zhengzhou, Zhengzhou University, Zhengzhou 450000, China; sjd@zzu.edu.cn (J.S.); hhbcfd@163.com (H.H.); chengj@zzu.edu.cn (G.C.)
2 The School of Computer and Artificial Intelligence, Zhengzhou University, Zhengzhou 450000, China; bayzhao@163.com
3 Department of Mathematics, Zhengzhou University of Aeronautics, Zhengzhou 450046, China
* Correspondence: jieruwei@zzu.edu.cn (J.W.); qinguoyong@zzu.edu.cn (G.Q.)
† These authors contributed equally to this work.

Abstract: To increase the accuracy of drought prediction, this study proposes a drought forecasting method based on the Informer model. Taking the Yellow River Basin as an example, the forecasting accuracies of the Autoregressive Integrated Moving Average (ARIMA), Long Short-Term Memory (LSTM), and Informer models on multiple timescales of the Standardized Precipitation Evapotranspiration Index (SPEI) were compared and analyzed. The results indicate that, with an increasing timescale, the forecasting accuracies of the ARIMA, LSTM, and Informer models improved gradually, reaching the best accuracy on the 24-month timescale. However, the predicted values of ARIMA, as well as those of LSTM, were significantly different from the true SPEI values on the 1-month timescale. The Informer model was more accurate than the ARIMA and LSTM models on all timescales, indicating that Informer can widely capture the information of the input series over time and is more effective in long-term prediction problems. Furthermore, Informer can significantly enhance the precision of SPEI prediction. The predicted values of the Informer model were closer to the true SPEI values, and the forecasted SPEI trends complied with the actual trends. The Informer model can model different timescales adaptively and, therefore, better capture relevance on different timecales. The NSE values of the Informer model for the four meteorological stations on SPEI24 were 0.968, 0.974, 0.972, and 0.986.

Keywords: drought forecasting; SPEI; Informer; Yellow River Basin; multi-timescale

Citation: Shang, J.; Zhao, B.; Hua, H.; Wei, J.; Qin, G.; Chen, G. Application of Informer Model Based on SPEI for Drought Forecasting. *Atmosphere* **2023**, *14*, 951. https://doi.org/10.3390/atmos14060951

Academic Editors: Jinping Liu, Quoc Bao Pham, Arfan Arshad and Masoud Jafari Shalamzari

Received: 28 April 2023
Revised: 22 May 2023
Accepted: 24 May 2023
Published: 29 May 2023

1. Introduction

From a global perspective, anthropogenic climate change, carbon emissions, deforestation, and urbanization have increased the frequency of drought [1]. The World Meteorological Organization (WMO) classifies drought according to the affected domain as meteorological, agricultural, hydrological, and socio-economic. In the world, few natural hazards are as devastating as drought [2]. The frequent and persistent occurrence of drought can lead to substantial losses in the socio-economic sphere, particularly in agriculture, and it can cause various detrimental ecological and environmental impacts, such as water scarcity, desertification, and frequent occurrences of sand and dust storms [3]. Drought prediction is a crucial field in addressing climate change and effectively managing water resources. Drought, characterized by prolonged water scarcity, has severe impacts on global ecosystems, agriculture, economies, and societies. The ability to accurately forecast drought events and their spatiotemporal patterns is of paramount importance for taking proactive measures and minimizing adverse impacts [4–7]. Conducting a series of studies on drought monitoring, assessment, and prediction has become a hot issue of great global concern and is of great practical significance [8]. Monitoring drought and issuing timely warnings are essential precursors for disaster mitigation and prevention. Accurately predicting the occurrence of drought offers useful resources for risk management and pre-warning, helping to reduce disaster damage to the greatest extent possible [3].

The use of a drought index is crucial for the quantitative assessment of drought severity and impacts [9]. Several meteorological drought indexes have been developed over the last few decades, such as SPI [10], SPEI [11], PDSI [12], and SMDI [13], which are utilized extensively at distinct spatial scales on global, regional, national, and different basins [14]. Vicente Serrano et al. [15] proposed SPEI, which builds on the algorithms used in both SPI and PDSI [16] and incorporates multi-scale features to evaluate the effects of temperature variations on drought conditions [3]. At present, there are two potential evapotranspiration models commonly used in the SPEI calculation process in China, which are Thornthwaite and Penman–Monteith. The occurrence and evolution of drought usually form a multi-timescale process, and the selection of different scales of SPEI is important for drought research. So, this study selected the timescales of SPEI at 1 month (SPEI1), 3 months (SPEI3), 6 months (SPEI6), 9 months (SPEI9), 12 months (SPEI12), and 24 months (SPEI24).

At present, drought prediction methods can be classified into two types: numerical prediction and statistical prediction. Numerical prediction [17,18] builds on meteorological principles to predict drought conditions by solving atmospheric dynamics equations. The effectiveness of the numerical prediction method relies on the precision of model parameters, the stability of driving variables, and the support of a lot of meteorological statistics [2,3]. Statistical prediction uses mathematical modeling techniques, such as regression prediction and grey system prediction, to model meteorological data [18]. However, the statistical prediction method has difficulty in accurately predicting future drought conditions during meteorological leaps and bounds [18]. With the rapid development of artificial intelligence [19], some new intelligent drought prediction models have emerged and become the mainstream methods for drought prediction. Hu et al. [20] adopted the LSTM model for SPEI spatiotemporal prediction on multiple timescales, and the results suggested that the forecasting efficiency of LSTM gradually improved as the SPEI timescale increased. Xu et al. [3] introduced a hybrid model that combines ARIMA and LSTM for drought prediction based on the deep learning method, and the results suggested that the hybrid model predicts SPEI with high precision on long timescales and with lower precision on short timescales. Zhang et al. [21] utilized two integration methods, Bagging and Boosting, which integrate multiple single models into a more powerful model with predictions on different timescales. Through a comparison of the forecasting results of various models with actual observations, the study found that the models based on the integration methods have higher accuracy and stability relative to the single models. Xu et al. [22] combined Complementary Ensemble Empirical Mode Decomposition (CEEMD) and ARIMA, and they showed that the CEEMD-ARIMA model was applicable to drought prediction; the model could also identify multiple modalities of drought variability on diverse timescales [23], improving the comprehensiveness and accuracy of the prediction.

Currently, most of the machine learning methods widely applied for drought prediction on multiple timescales are mostly based on recurrent neural networks, which can solve the sequence prediction problem better than other deterministic and traditional models [24]. For these problems, some new methods have been proposed, such as the Transformer [25] model and the Informer [26] model, which can handle long series data and increase the precision of prediction.

The Informer model used in this paper is an effective improvement to the Transformer model. A sequence-to-sequence model proposed by a Google team in 2017, Transformer adopts a self-attentive mechanism to handle sequential information as a whole and can avoid the recursion of information while enabling attention to be paid to local information with strong relevance [27]. Informer is essentially an improvement on Transformer. By modifying the structure of Transformer and the probabilistic sparsification of the original self-attentive mechanism, Informer speeds up the computation speed of Transformer and effectively improves the precision of sequence prediction.

The accuracy of SPEI prediction on short timescales is still low in existing studies. Thus, this paper adopts the multi-layer Transformer structure of the Informer model; adopts a novel position encoding method introduced to capture the long-term and short-term dependencies in time series; and incorporates an attention mechanism, which effectively improves the accuracy of short-timescale SPEI prediction. In this article, a drought prediction model is constructed using the Informer algorithm, it is validated with four meteorological stations in the Yellow River Basin, and it is verified with the LSTM and ARIMA models to demonstrate the higher precision of the model's prediction.

2. Materials and Methods

2.1. Study Area

The Yellow River Basin is a major watershed in China and is known as the Mother River (Figure 1). It is a major agricultural and economic region of China. The Yellow River Basin is located at 90°33′ − 122°25′ E and 24°30′ − 35°45′ N, with a mainly temperate monsoon climate [8]. The temperature difference throughout the year is extremely large [19]. However, environmental problems, such as severe land sanding and water shortage, also exist in the basin.

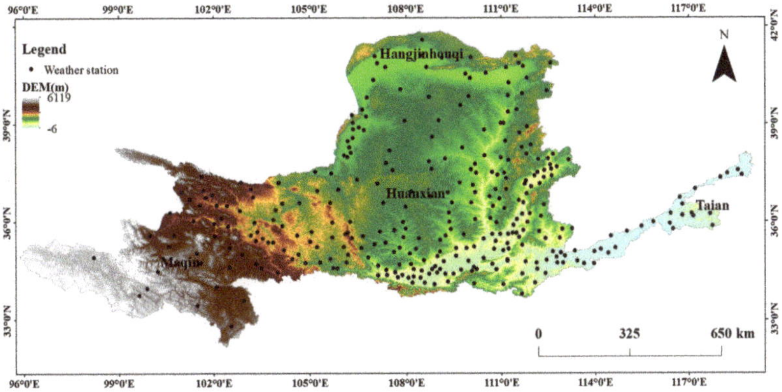

Figure 1. Study area.

2.2. Data Source

The meteorological data in this paper were obtained from the monthly value dataset of the terrestrial climate information from the China Meteorological Data Network (https://www.data.cma.cn/ accessed on 6 April 2022), and they include precipitation (mm), maximum temperature (°C), minimum temperature (°C), average temperature (°C), wind speed (m·s^{-1}), sunshine hours (h), latitude (°), longitude (°), and altitude (m) for the period of 1960–2019. This study selected 4 meteorological stations in the Yellow River Basin to apply validation. Table 1 shows the information of the 4 representative stations.

Table 1. Profile about representative meteorological stations.

Station ID	Station Name	Longitude (°E)	Latitude (°N)	Altitude (m)
53420	Hangjinhouqi	107.12	40.85	1024
53821	Huanxian	107.3	36.57	1255.6
54827	Taian	117.15	36.17	129.8
56043	Maqin	100.23	34.48	3719

2.3. Methods

2.3.1. Standardized Precipitation Evapotranspiration Index

This study uses the Penman–Monteith model to estimate potential evapotranspiration by calculating multi-scale SPEI values for four meteorological stations located within the study area for the period of 1960 to 2019 [15], which allows for the determination of the influence of precipitation, temperature, and evapotranspiration on drought in an integrated manner and has the advantages of multiple timescales and clarity of the mechanism. The procedures for calculating SPEI_PM are as follows [16]:

(1) The Penman–Monteith model is utilized to generate the reference crop evapotranspiration ET_0, which is determined using the following equation:

$$ET_0 = \frac{0.408\,\Delta\,(R_n - G) + \gamma\,\frac{900}{T+273}U_2\,(e_s - e_a)}{\Delta + \gamma\,(1 + 0.34\,U_2)} \tag{1}$$

where ET_0 indicates the evaporation from the reference crop (mm/d); Δ is the saturated hydraulic pressure curve slope (kPa/°C) [6]; γ is the moisture constant (kPa/°C); R_n means solar net radiation (MJ·m^{-2}·d^{-1}); G is the thermal flux of the soil (MJ·m^{-2}·d^{-1}) [11]; T is the mean temperature for the calculation period (°C); U_2 is the mean speed of the wind at a height of 2 m above the ground; e_s is the pressure of saturated water (kPa); and e_a is the real water pressure (kPa) [15].

(2) The monthly values of the difference between precipitation and evaporation is calculated.

$$D_i = P_i - E\,T_0 \tag{2}$$

where D_i indicates the difference between precipitation and evapotranspiration; P_i represents the precipitation amount per month; ET_0 is the monthly actual evaporation volume [15].

(3) The data series of D_i is normalized. D_i is fit with the cumulative probability distribution function $F(x)$, and the corresponding SPEI value for each D_i [15] is calculated, making the data fit the probability distribution.

$$F(x) = \left[1 + \left(\frac{\alpha}{x-\gamma}\right)^{\beta}\right]^{-1} \tag{3}$$

where $F(x)$ is the probability distribution function, and the other parameters are as follows:

$$\alpha = \frac{(a_0 - 2\,a_1)\beta}{\tau(1+1/\beta)\tau(1-1/\beta)} \tag{4}$$

$$\beta = \frac{2\,a_1 - a_0}{6\,a_1 - a_0 - 6\,a_2} \tag{5}$$

$$\gamma = a_0 - \alpha(1+1/\beta)\tau(1-1/\beta) \tag{6}$$

where τ is the factorial function; a_0, a_1, and a_2 are the weighted moment of the probability of data series D_i [15].

The probability of exceeding a certain value of D_i can be written as $P = 1 - F(x)$. Then, SPEI can be written as a function of P as follows:

$$\text{SPEI} = \begin{cases} w - \dfrac{g_0 + g_1 w + g_2 w^2}{1 + e_2 w + d_1 w^2 + e_3 w^3}, & \text{with } w = \sqrt{-2\ln P}, \quad \text{for } P \le 0.5 \\ -w + \dfrac{g_0 + g_1 w + g_2 w^2}{1 + e_2 w + e_1 w^2 + e_3 w^3}, & \text{with } w = \sqrt{-2\ln(1-P)}, \quad \text{for } P > 0.5 \end{cases} \tag{7}$$

where $w = \sqrt{-2\ln(1-P)}$. The other parameters in Equation (7) are $e_1 = 1.432788$, $e_2 = 0.189269$, $e_3 = 0.001308$, $g_0 = 2.515517$, $g_1 = 0.802853$, and $g_2 = 0.010328$. Referring to the national standard meteorological drought grade (GB/T20481-2017) stipulated by the drought grading standard, the drought categories classified according to the SPEI values are shown in Table 2.

Table 2. Drought classification based on SPEI.

Level	Type	SPEI
1	No drought	SPEI ≥ -0.5
2	Mild drought	$-1.0 \leq$ SPEI < -0.5
3	Moderate drought	$-1.5 \leq$ SPEI < -1.0
4	Severe drought	$-2.0 \leq$ SPEI < -1.5
5	Extreme drought	SPEI ≤ -2.0

2.3.2. Informer

Informer is considered a supervised learning model built on the attention mechanism, which, as a whole, consists of two components: an encoder and a decoder [26]. Informer is a Transformer-based time series prediction model that better captures the long-term dependencies of time series by adding processing steps, such as position encoding, the block attention mechanism, and adaptive length sequence sampling, where the encoder is used to obtain a long-term dependence on the robustness of the original input sequence and the decoder can further implement sequence prediction. The structure of the Informer model is illustrated in Figure 2. The left encoder primarily receives longer sequence inputs and incorporates sparse self-attention [27], an alternative to the conventional self-attention mechanism [28]. The trapezoidal component refers to the extracted operation of self-attention, which can dramatically reduce the size of the network, while the stacking of multiple layers further enhances the model's robustness again [28]. The right decoder takes the input of the long-term sequence, padding the target elements to zero, by which an attention-weighted constituent of the feature graph is measured; then, these elements are output in a rapidly generated format [29].

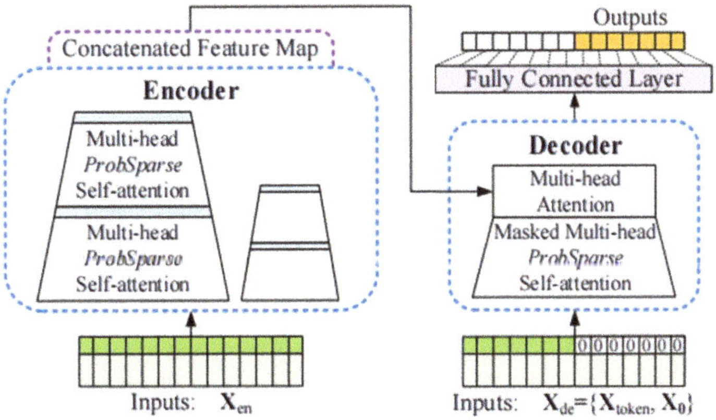

Figure 2. Informer model structure.

Informer Model Inputs

The input data at time t are as follows:

$$x^t = \{x_1^t, \cdots, x_{L_x}^t \mid x_i^t \in R^{d_x}\} \tag{8}$$

and the output is the corresponding sequence of predictions.

$$y^t = \{y_1^t, \cdots, y_{L_y}^t \mid y_i^t \in R^{d_y}\} \tag{9}$$

where L_x and L_y are the input length and output length, respectively; d_x and d_y are the feature dimensions.

For time series prediction problems, the sequence of the data is particularly important. To keep the order structure of the series data from being lost after they are input to the model, Informer encodes the location information $PE_{(pos,2j)}$ and $PE_{(pos,2j+1)}$ for each set of input data, and the specific formulae are implemented as follows:

$$PE_{(pos,2j)} = \sin \frac{pos}{(2L)^{2j/d_{model}}} \quad (10)$$

$$PE_{(pos,2j+1)} = \cos \frac{pos}{(2L)^{2j/d_{model}}} \quad (11)$$

where pos is the position (sequence order). The index j =1, 2, $\cdots$, $d_{model}/2$, indicates the dimension. d_{model} represents the dimensionality of the characteristics represented by the input, and L is the input sequence.

Self-Attention Mechanism of Informer Model

In probability form, the $A(q_i, K, V)$ of the attention coefficient for the i-th Query is as follows:

$$A(q_i, K, V) = \sum_j \frac{k(q_i, k_j)}{\sum_l k(q_i, k_l)} V_j = E_{p(k_j|q_i)}[V_j] \quad (12)$$

where $p(k_j \mid q_i) = (k(q_i, k_j))/(\sum_l k(q_i, k_l))$, and $k(q_i, k_j)$ selects the asymmetric exponential kernel $\exp\left((q_i k_j^T)/\sqrt{d}\right)$ [23].

To measure the sparsity of $Query$, Informer uses Kullback–Leibler divergence. Ignoring the constant, the sparsity measure formula for the i-th $Query$ is equated as follows:

$$M(Q_i, K) = \ln \sum_{j=1}^{L_K} \exp\left(\frac{q_i k^T}{\sqrt{d}}\right) - \frac{1}{L_K} \sum_{j=1}^{L_K} \frac{q_i k_j^T}{\sqrt{d}} \quad (13)$$

where the first is the logarithmic sum expansion (LSE) of q_i on all the keys, and the second is their arithmetic average [26].

According to the proposed measurement, the formula of ProbSparse self-attention can be written as follows:

$$A(Q, K, V) = \mathrm{Softmax}\left(\frac{\bar{Q}K_T}{\sqrt{d}}\right)V \quad (14)$$

where $\bar{Q}$ is a sparse matrix of the same size as q, which only contains $Top - u$ queries under the sparsity measurement $M(q, M)$ [26].

Encoder for Informer Model

The aim of the encoder is to capture the long-range dependency of the robustness of the long sequence of inputs [26]. A sketch of the encoder is shown in Figure 3. The procedure of the distillation operation from layer j-th to layer $(j+1)$-th is as follows:

$$X_{j+1}^t = \mathrm{MaxPool}\left(\mathrm{ELU}\left(\mathrm{Convld}\left(\left[X_j^t\right]_{AB}\right)\right)\right) \quad (15)$$

where $[X_j^t]_{AB}$ represents the attention module, which includes the multi-head ProbSparse self-attention and basic operations. Concld represents one-dimensional convolution operations on a time series, which is performed by using ELU as the activity functions [30].

The self-attention distillation mechanism proposed by Informer enables each decoder layer to reduce the input sequence length by half, which dramatically saves the memory spending and computational time of the encoder [26].

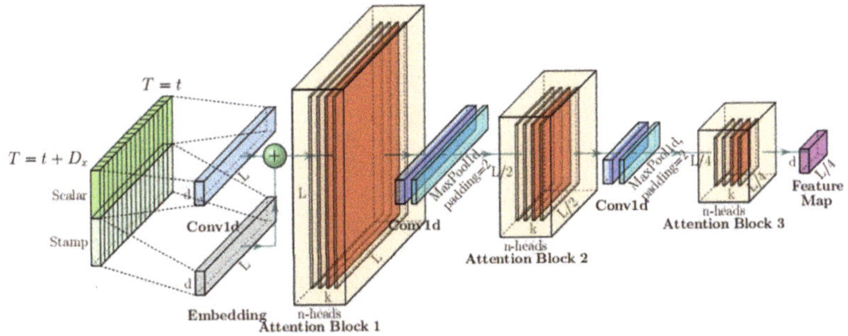

Figure 3. The single stack in Informer's encoder.

Decoder for Informer Model

The standard decoder structure is used in part of the decoder, proposed by VASWANI in 2017 [27], which is composed of two identical multi-head attention layers. The decoder X^t_{de} is supplied with the following vectors:

$$X^t_{de} = \text{Concat}(X^t_{token}, X^t_0) \in R^{(L_{token}+L_y)\times d_{model}} \tag{16}$$

where $X^t_{token} \in R^{L_{token}\times d_{model}}$ is the start token; $X^t_0 \in R^{L_y \times d_{model}}$ is a placeholder for the target sequence [30].

ProbSparse self-attention adopts blocked multi-headed attention, fully connected layer output dimensions to determine uni/multivariate predictions, and a generative structure to shorten the prediction decoding time.

2.3.3. Long Short-Term Memory

Long Short-Term Memory (LSTM) is a recurrent neural network (RNN) architecture for handling sequential data; it was developed as an improvement over traditional RNN [31], and it effectively resolves the problem of prolonged dependence by using three gating mechanisms and a memory unit. By contrast with the ordinary RNN, LSTM incorporates a memory cell to determine whether the information is available [32]. The cell state is the key of LSTM. To protect and control the state of a memory cell, three control gates are placed in a memory cell, called the input gate, forget gate, and output gate [33]. Each control gate consists of a neural network layer containing a sigmoid function and a dot product operation [34]. The LSTM memory cell structure is illustrated in Figure 4.

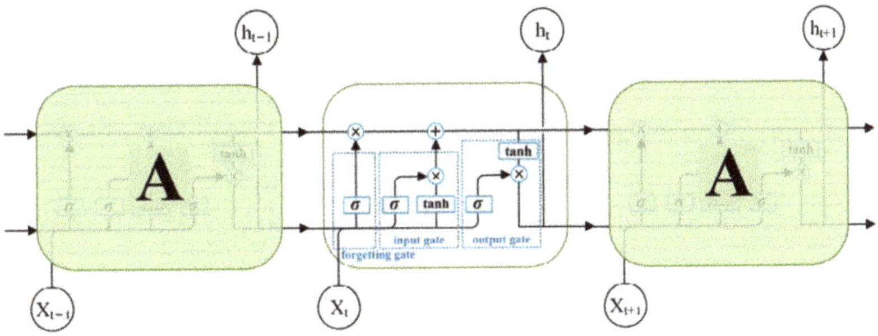

Figure 4. The structure of the LSTM memory cell.

2.3.4. Autoregressive Integrated Moving Average

ARIMA is the combination of AR, MA, and Difference (Diff), which converts unsteady time series into a steady-state series by performing one or more differences and then fitting it with ARIMA [35]. Its composition is as follows:

$$\mathrm{ARIMA}(p,d,q) = \mathrm{AR}(p) + \mathrm{Diff}(d) + \mathrm{MA}(q) \tag{17}$$

where $\mathrm{AR}(p)$ represents the autoregressive model; $\mathrm{Diff}(d)$ indicates the difference model; $\mathrm{MA}(q)$ indicates the moving-average model; p, d, and q are the parameters corresponding to the three models. The ARIMA model prediction equation for $C(t)$ is as follows:

$$C(t) = \varphi_0 + \sum_{i=1}^{p} \varphi_i C_{t-i} + \varepsilon_t + \sum_{i=1}^{q} \gamma_i \varepsilon_{t-i} \tag{18}$$

where $C(t)$ represents the reconstructed component time series formed after the SE algorithm; ε_t represents the current period random error disturbance; φ_i and γ_i represent model parameters; p denotes the quantity of autoregressive terms; d indicates the variance number in a steady time series; q denotes the amount of terms in the moving average [36].

2.3.5. Evaluation Metrics

To estimate the efficiency of the contrasting model more reasonably, NSE, RMSE, and MAE were used in this paper to perform an evaluation. The formula used to calculate above metrics is shown below.

$$\mathrm{RMSE} = \sqrt{\frac{1}{N}\sum_{i=1}^{N}(y_i - \tilde{y}_i)^2} \tag{19}$$

$$\mathrm{NSE} = 1 - \frac{\sum_{i=1}^{N}(y_i - \tilde{y}_i)^2}{\sum_{i=1}^{N}(y_i - \bar{y})^2} \tag{20}$$

$$\mathrm{MAE} = \frac{1}{N}\sum_{i=1}^{N}|y_i - \tilde{y}_i| \tag{21}$$

where y_i indicates the true value; $\tilde{y}$ indicates the forecasted value; $\bar{y}$ represents the average value of y_i; and N indicates an amount of the total data for y_i.

3. Results

3.1. SPEI Values on Different Timescales

The 1−, 3−, 6−, 12−, and 24−month timescale SPEI values of Hangjinhouqi, Huanxian, Taian, and Maqin were calculated using monthly meteorological data. The results are shown in Figure 5. Combined with the Mann−Kendall trend test (Table 3), it can be observed that the SPEI1, SPEI3, SPEI6, SPEI12, and SPEI24 of the four stations show a decreasing trend. In particular, the following show a significant decreasing trend: SPEI9, SPEI12, and SPEI24 of the Hangjinhouqi site; SPEI3, SPEI6, SPEI12, and SPEI24 of the Huanxian site; SPEI9, SPEI12, and SPEI24 of the Taian site; and SPEI24 of the Maqin site. The four stations show a high frequency of extreme droughts. In the past decade, the temperature of the Yellow River Basin has been increasing, and the runoff of the main and tributary streams has been decreasing since 1960 [8], which has caused the SPEI values to decrease.

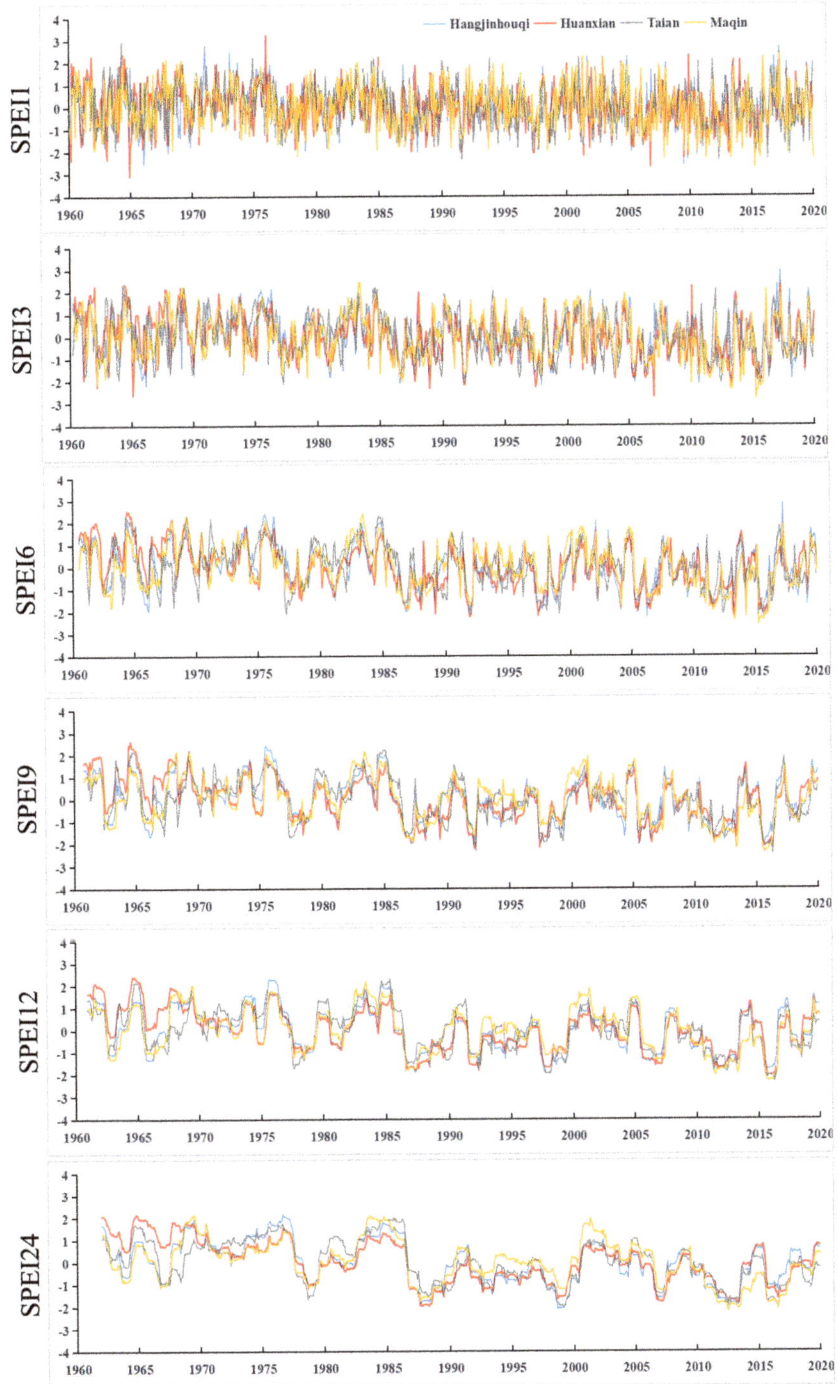

Figure 5. Observed SPEI values on different timescales of the example stations.

Table 3. Mann–Kendall trend test for SPEI.

Example Stations	SPEI Series	*p* Value	Trend
Hangjinhouqi	SPEI1	0.00055	decreasing
	SPEI3	1.31×10^{-6}	decreasing
	SPEI6	5.973×10^{-14}	decreasing
	SPEI9	0	decreasing
	SPEI12	0	decreasing
	SPEI24	0	decreasing
Huanxian	SPEI1	1.349×10^{-7}	decreasing
	SPEI3	0	decreasing
	SPEI6	0	decreasing
	SPEI9	0	decreasing
	SPEI12	0	decreasing
	SPEI24	0	decreasing
Taian	SPEI1	5.975×10^{-5}	decreasing
	SPEI3	1.372×10^{-10}	decreasing
	SPEI6	2.22×10^{-16}	decreasing
	SPEI9	0	decreasing
	SPEI12	0	decreasing
	SPEI24	0	decreasing
Maqin	SPEI1	3.162×10^{-5}	decreasing
	SPEI3	1.086×10^{-9}	decreasing
	SPEI6	6.443×10^{-13}	decreasing
	SPEI9	2.44×10^{-15}	decreasing
	SPEI12	2.22×10^{-16}	decreasing
	SPEI24	0	decreasing

3.2. Analysis of Model Prediction Results

Using multi-scale SPEI data from 1960–2007 as training data, the SPEI values of the four meteorological battle sites on multiple timescales were predicted using the LSTM, ARIMA, and Informer models for 2008–2020. A comparison of the prediction performance of the three models and the prediction evaluation indexes are shown in Figures 6–9 and Table 4. It is suggested that the Informer model accurately fit the predicted values to the true values compared to the ARIMA and LSTM models, and it effectively captured the variations in the SPEI values.

The predicted values of ARIMA, as well as those of LSTM, for the four meteorological stations were significantly different from the true SPEI values on the 1-month timescale. In particular, LSTM lost prediction ability in predicting SPEI1 for Hangjinhouqi. The differences between the predicted and actual values of ARIMA and LSTM decreased when predicting SPEI3, SPEI6, SPEI9, SPEI12, and SPEI24. In this study, the data of SPEI1 changed relatively fast and fluctuated more, which required more complex modeling methods to predict, and, therefore, the prediction was the worst on this timescale.

The Informer model predictions were more similar to the true SPEI values, and the predicted SPEI trends were consistent with the actual trends. In Figures 6–9, the Informer model shows better prediction results on SPEI3, SPEI6, SPEI9, SPEI12, and SPEI24. The Informer model is able to handle long sequences, and it performs better when dealing with long-term dependencies. It can model different timescales adaptively and, therefore, better capture relevance on different timescales. As a result, Informer has good performance in predicting SPEI for each meteorological station.

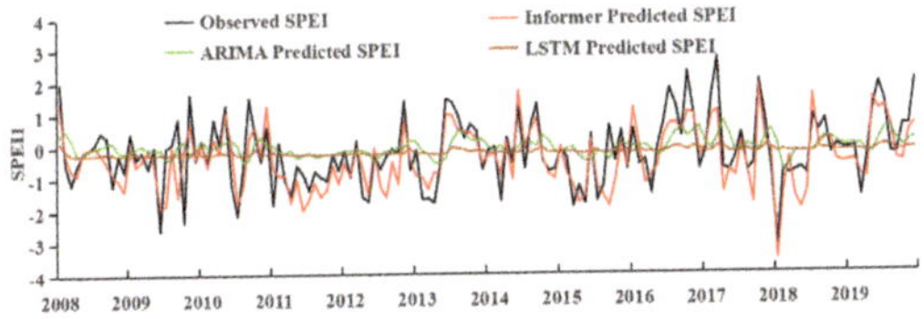

(**a**)

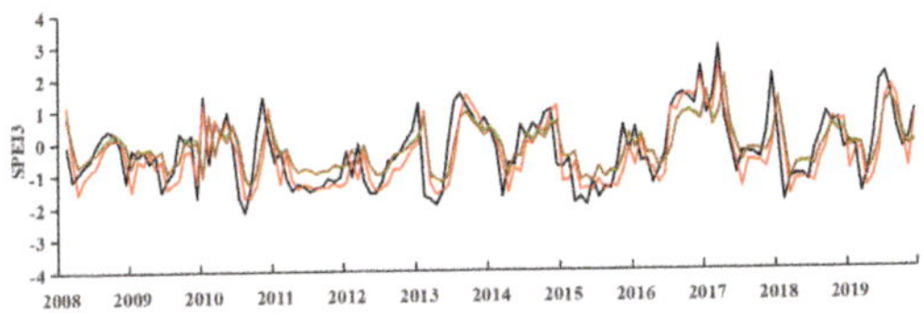

(**b**)

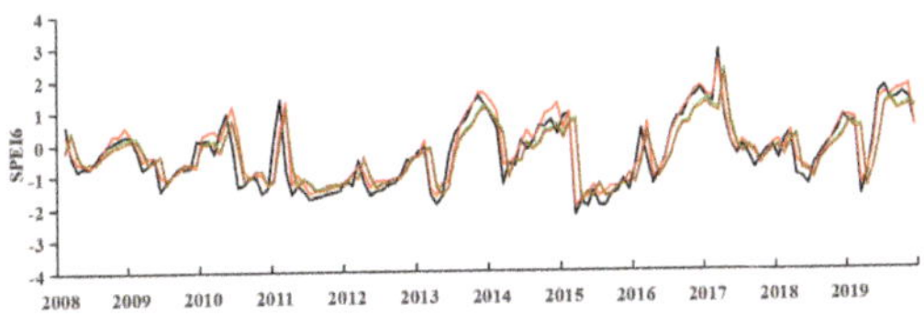

(**c**)

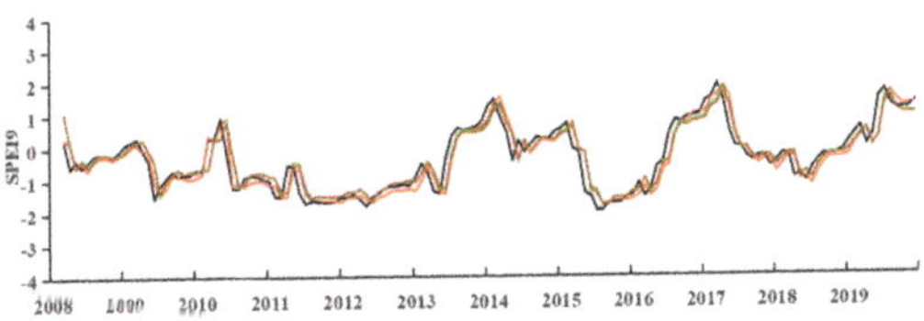

(**d**)

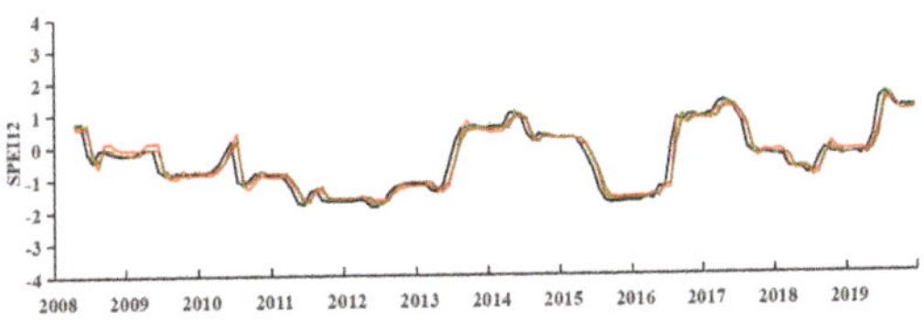

(**e**)

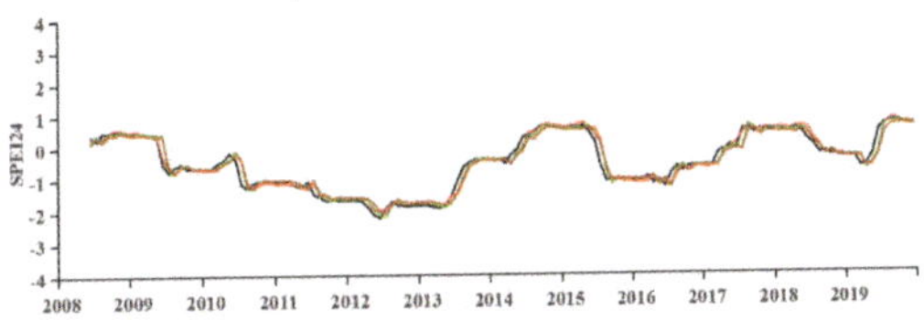

(**f**)

Figure 6. Prediction results of multi-timescale SPEI values of the ARIMA, LSTM, and Informer models at Hangjinhouqi: (**a**) 1−month timescale (**b**) 3−month timescale; (**c**) 6−month timescale; (**d**) 9−month timescale; (**e**) 12−month timescale; (**f**) 24−month timescale.

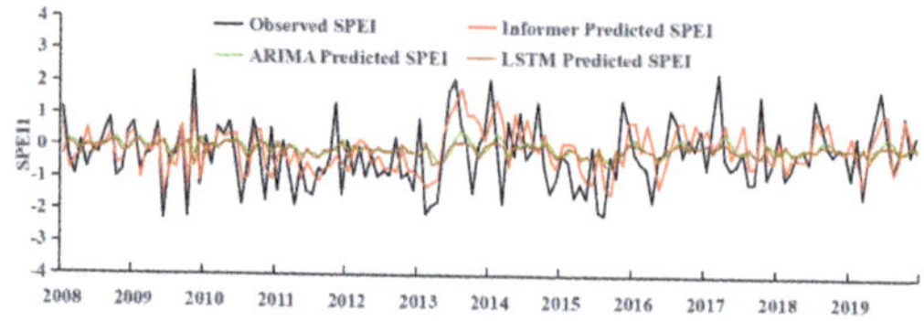

(**a**)

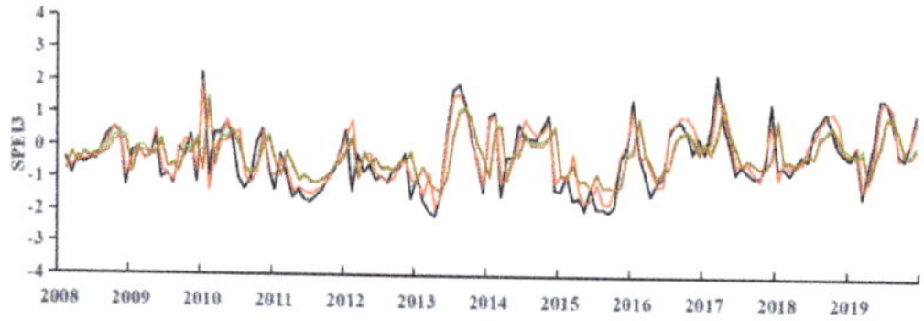

(**b**)

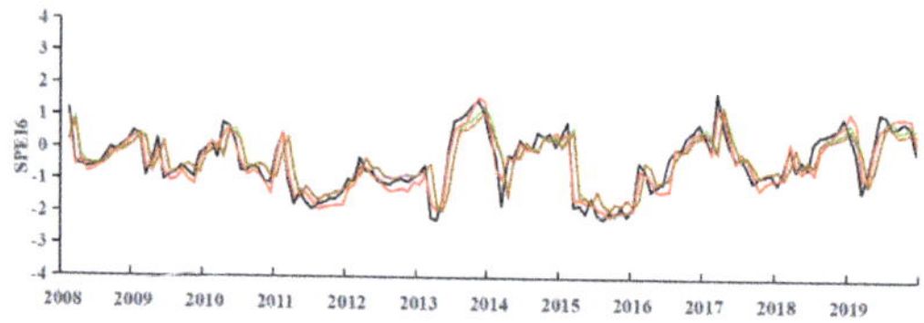

(**c**)

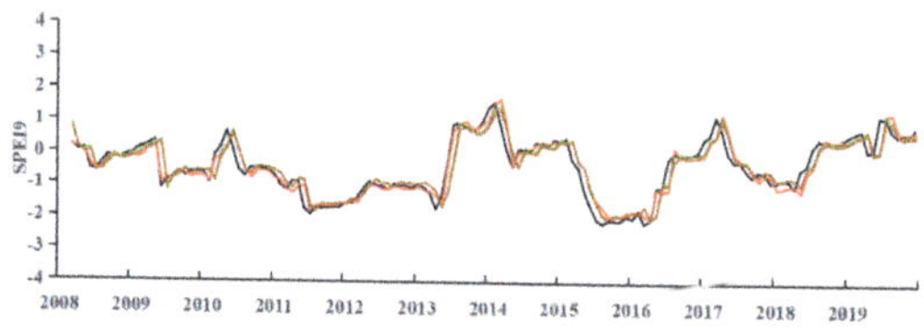

(**d**)

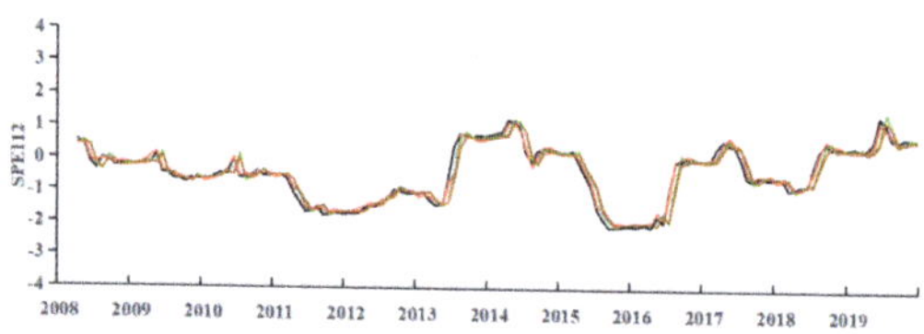

(**e**)

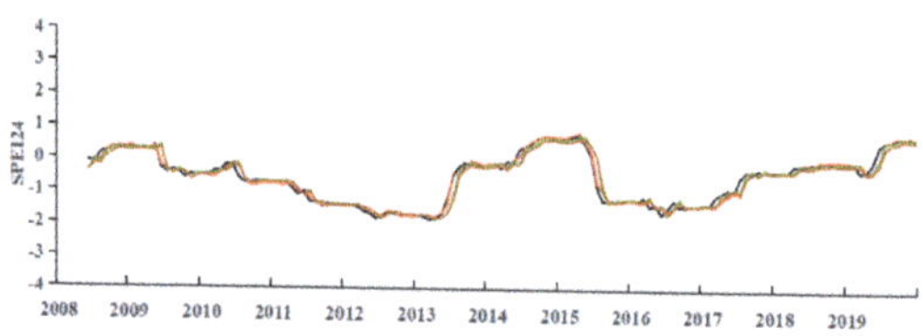

(**f**)

Figure 7. Prediction results of multi-timescale SPEI values of the ARIMA, LSTM, and Informer models at Huanxian: (**a**) 1−month timescale (**b**) 3−month timescale; (**c**) 6−month timescale; (**d**) 9−month timescale; (**e**) 12−month timescale; (**f**) 24−month timescale.

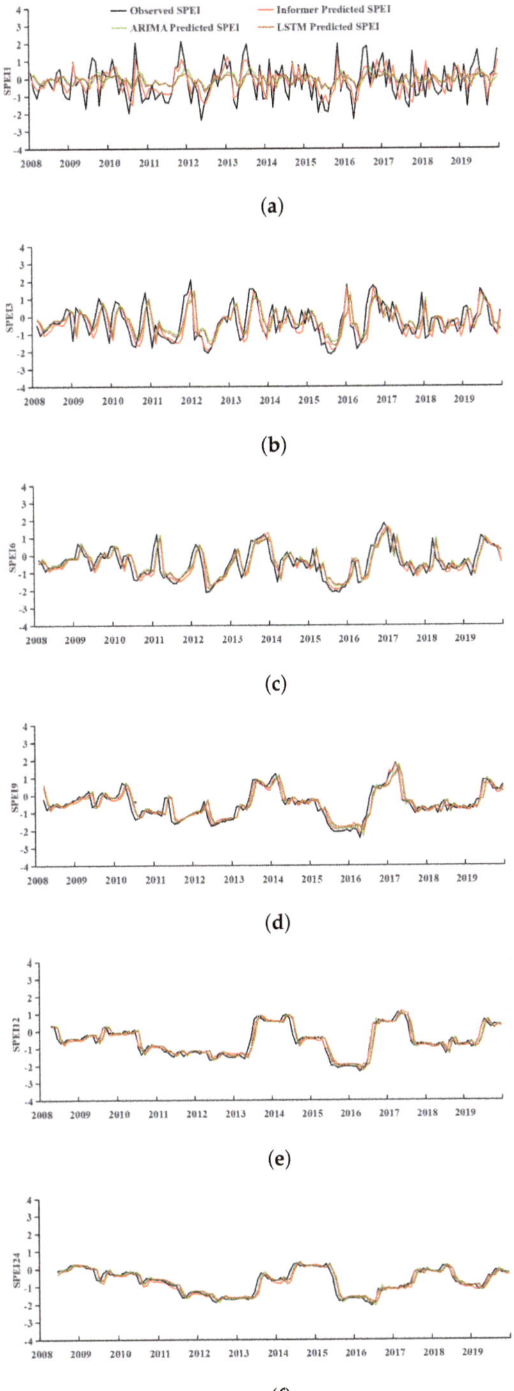

Figure 8. Prediction results of multi-timescale SPEI values of the ARIMA, LSTM, and Informer models at Taian: (**a**) 1−month timescale (**b**) 3−month timescale; (**c**) 6−month timescale; (**d**) 9−month timescale; (**e**) 12−month timescale; (**f**) 24−month timescale.

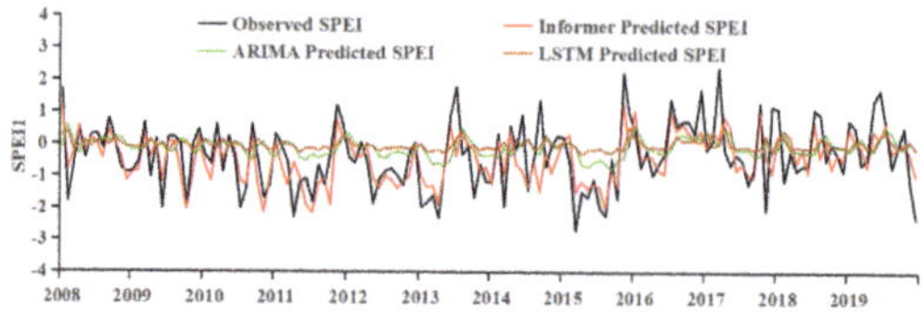

(**a**)

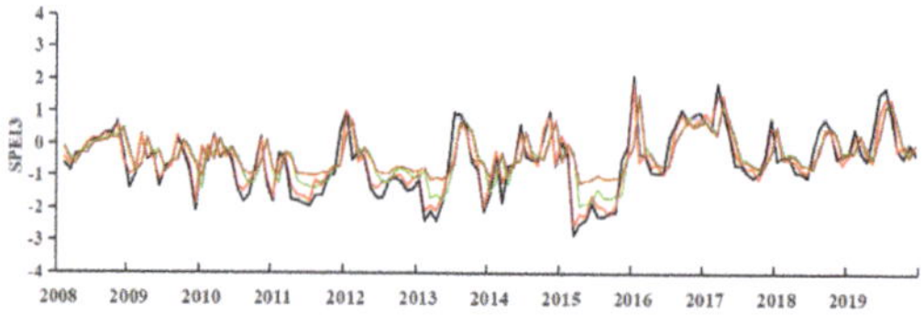

(**b**)

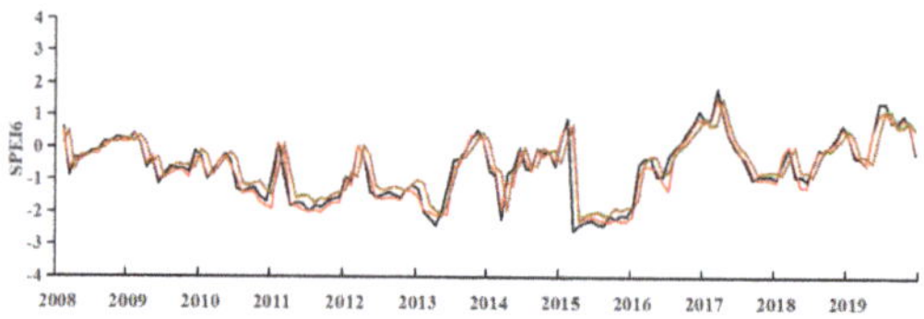

(**c**)

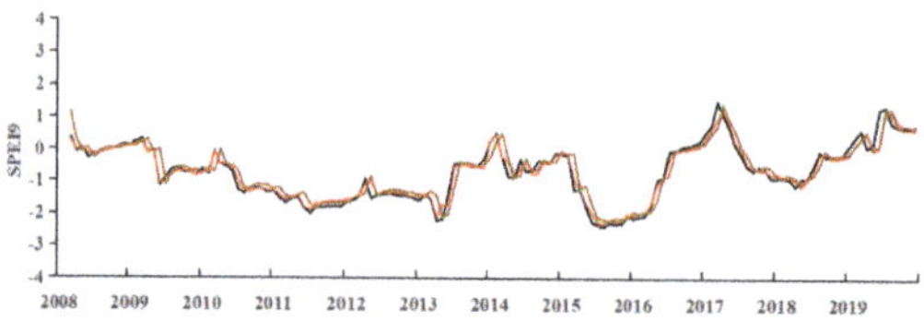

(**d**)

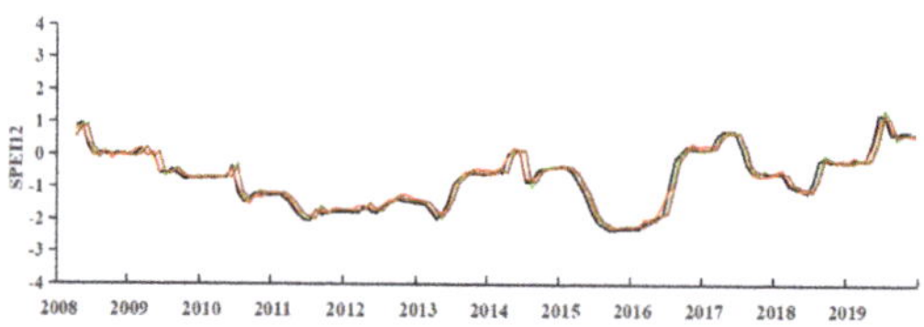

(**e**)

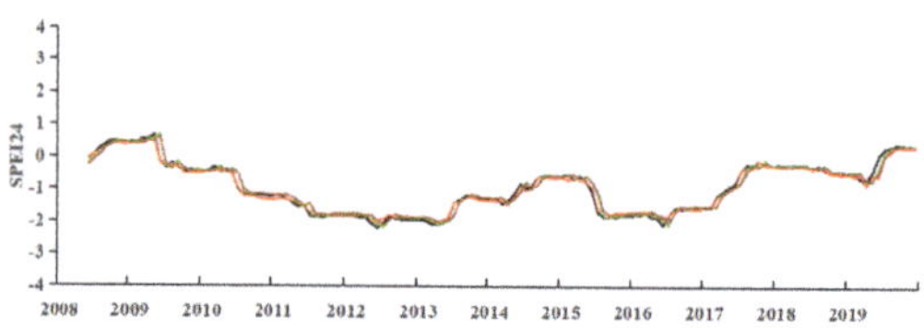

(**f**)

Figure 9. Prediction results of multi-timescale SPEI values of the ARIMA, LSTM, and Informer models at Maqin: (**a**) 1−month timescale (**b**) 3−month timescale; (**c**) 6−month timescale; (**d**) 9−month timescale; (**e**) 12−month timescale; (**f**) 24−month timescale.

Table 4. The statistical criteria of the ARIMA, LSTM, and Informer models.

Example Stations	SPEI Series	Model	MAE	RMSE	NSE
Hangjinhouqi	SPEI1	ARIMA	0.800	1.027	0.022
		LSTM	0.799	1.021	0.032
		Informer	0.531	0.688	0.561
	SPEI3	ARIMA	0.633	0.827	0.371
		LSTM	0.635	0.824	0.375
		Informer	0.388	0.521	0.434
	SPEI6	ARIMA	0.455	0.655	0.573
		LSTM	0.452	0.642	0.590
		Informer	0.277	0.416	0.828
	SPEI9	ARIMA	0.279	0.397	0.821
		LSTM	0.291	0.402	0.817
		Informer	0.271	0.382	0.835
	SPEI12	ARIMA	0.166	0.279	0.910
		LSTM	0.187	0.296	0.899
		Informer	0.182	0.287	0.905
	SPEI24	ARIMA	0.124	0.201	0.940
		LSTM	0.145	0.214	0.932
		Informer	0.123	0.190	0.968
Huanxian	SPEI1	ARIMA	0.804	1.006	−0.049
		LSTM	0.804	1.003	−0.042
		Informer	0.666	0.842	0.264
	SPEI3	ARIMA	0.628	0.826	0.250
		LSTM	0.617	0.812	0.276
		Informer	0.271	0.402	0.822
	SPEI6	ARIMA	0.423	0.594	0.580
		LSTM	0.415	0.581	0.598
		Informer	0.211	0.271	0.912
	SPEI9	ARIMA	0.243	0.354	0.842
		LSTM	0.254	0.361	0.836
		Informer	0.191	0.286	0.896
	SPEI12	ARIMA	0.166	0.255	0.915
		LSTM	0.176	0.272	0.904
		Informer	0.096	0.133	0.977
	SPEI24	ARIMA	0.109	0.177	0.945
		LSTM	0.127	0.193	0.936
		Informer	0.086	0.123	0.974
Taian	SPEI1	ARIMA	0.844	1.007	−0.013
		LSTM	0.835	0.994	0.014
		Informer	0.507	0.672	0.548
	SPEI3	ARIMA	0.0.619	0.791	0.289
		LSTM	0.620	0.792	0.288
		Informer	0.508	0.699	0.445
	SPEI6	ARIMA	0.401	0.552	0.575
		LSTM	0.413	0.554	0.573
		Informer	0.391	0.542	0.591
	SPEI9	ARIMA	0.270	0.387	0.789
		LSTM	0.277	0.397	0.777
		Informer	0.201	0.283	0.887
	SPEI12	ARIMA	0.193	0.295	0.876
		LSTM	0.202	0.316	0.858
		Informer	0.133	0.192	0.948
	SPEI24	ARIMA	0.137	0.202	0.909
		LSTM	0.148	0.216	0.897
		Informer	0.131	0.192	0.972

Table 4. *Cont.*

Example Stations	SPEI Series	Model	MAE	RMSE	NSE
Maqin	SPEI1	ARIMA	0.846	1.052	−0.047
		LSTM	0.857	1.059	−0.061
		Informer	0.543	0.738	0.484
	SPEI3	ARIMA	0.592	0.753	0.418
		LSTM	0.635	0.788	0.363
		Informer	0.245	0.335	0.884
	SPEI6	ARIMA	0.389	0.555	0.655
		LSTM	0.382	0.550	0.661
		Informer	0.162	0.329	0.879
	SPEI9	ARIMA	0.231	0.334	0.858
		LSTM	0.235	0.338	0.855
		Informer	0.124	0.193	0.952
	SPEI12	ARIMA	0.162	0.247	0.920
		LSTM	0.172	0.261	0.911
		Informer	0.101	0.145	0.972
	SPEI24	ARIMA	0.102	0.159	0.959
		LSTM	0.117	0.169	0.954
		Informer	0.064	0.092	0.986

As the timescale becomes smaller, the prediction abilities of the Informer, ARIMA, and LSTM models decrease, but Informer still performs better than ARIMA and LSTM, indicating that Informer can widely capture the information of the input series over time and is more effective in long-term prediction problems. In this paper, to assess the prediction performance of the ARIMA, LSTM, and Informer models, three evaluation metrics, MAE, RMSE, and NSE, are utilized (Table 4). The MAE values of ARIMA and LSTM are both above 0.7 at SPEI1 and below 0.2 at SPEI24. The MAE and RMSE values tend to decrease with an increasing timescale, while the values of NSE show the reverse trend. These trends suggest that the prediction accuracy of the ARIMA, LSTM, and Informer models improves with increasing timescales. The prediction performance of the Informer model is superior to that of the ARIMA and LSTM models on different timescales, indicating that the Informer model can significantly enhance the prediction accuracy of SPEI. The NSE values of the Informer model for the four meteorological stations on SPEI24 are 0.968, 0.974, 0.972, and 0.986. On all timescales, the Informer model is superior to the ARIMA and LSTM models in evaluating metric data for prediction results.

Informer solves the problem of the dependencies between the output and input being not well captured due to long distances when predicting long time series. Moreover, the Informer model optimizes the temporal and spatial sophistication of the attention mechanism in the Transformer model so that Informer can obtain higher prediction accuracy. From the analysis, it is obvious that the LSTM and ARIMA models have lower prediction accuracies due to their own structural limitations.

4. Discussion

Drought forecasting is crucial for mitigating risks and preparing measures to alleviate its impact [37]. In this paper, we used the newest time series prediction model, namely, Informer, to predict the drought in the Yellow River Basin, and we compared the prediction results with those of the ARIMA and LSTM models, which showed that the Informer model exhibits superior prediction accuracy compared to both the ARIMA and LSTM models on multiple timescales. Because the data of SPEI1 changed relatively fast and fluctuated more, the predicted values of ARIMA, as well as those of LSTM, for the four meteorological stations were significantly different from the true SPEI values on the 1-month timescale, which is consistent with the conclusion reached by Xu et al. [22]. In particular, LSTM lost prediction ability in predicting SPEI1 for Hangjinhouqi. As the timescale increased, the data series tended to be smooth, and the prediction accuracy of ARIMA and LSTM gradually improved. Xu et al. [2] found that the prediction accuracy was related to the timescale based on the ARIMA-SVR model for multi-scale SPI prediction, and the prediction precision gradually improved with an increasing timescale. Hinge et al. [37] found that the hybrid WPT-MLR model has the potential to be employed for drought warnings in the study region, but the prediction accuracy decreased as the timescale increased. The predicted values of the Informer model were closer to the measured SPEI values, and the predicted SPEI trends aligned with the actual trends. The Informer model can model different timescales adaptively and, therefore, better capture relevance on different timescales. The NSE values of the Informer model for the four meteorological stations on SPEI24 were 0.968, 0.974, 0.972, and 0.986.

The Informer model provides various advantages for capturing long-term dependencies in time series data using a self-attentive mechanism [38], which enables the prediction of droughts over a longer term. In addition, the Informer model adopts the adaptive length idea, which can automatically adapt to different timescales and data features with high flexibility and adaptability [39]. The Informer model is also able to process multiple time series in parallel using the multi-headed self-attentive mechanism, which improves the training and prediction efficiency of the model, and there is no need to manually perform feature engineering, which can automatically extract important features in time series with better generalizability and interpretability. Applying Informer to drought prediction in the Yellow River Basin can improve the accuracy and reliability of drought prediction [40], which, in turn, can improve the efficiency and quality of water resources management and agricultural production [24].

Although the Informer model in this study outperforms that in existing studies in the accuracy of small-scale SPEI prediction, the fit of small-scale prediction results is still not as good as that of a large timescale. In the future, the predictive capability of Informer for different timescales can be improved by combining it with the multi-scale method. In addition, multi-source data and deep learning techniques can be brought in to build deep drought prediction models to better predict the evolution and trends of drought [41]. These measures are expected to improve the timescale of Informer's performance in drought prediction and further refine its role in practical applications.

There are some aspects of the Informer model that can still be improved to further enhance prediction precision. Future improvements of the Informer model for drought prediction in the Yellow River Basin include adding multi-scale mechanisms to better capture multiple patterns and periodicity in the time series; integrating domain knowledge, such as meteorological and hydrological data, to improve prediction accuracy and interpretability; combining other traditional time series models, such as LSTM and GRU, to build a powerful integrated model; and integrating multiple target prediction problems to deal with multiple indicators and factors in drought prediction to improve prediction accuracy and comprehensiveness. The next steps in research on using the Informer model to predict small-scale SPEI drought could include exploring the use of additional data sources to improve prediction accuracy, such as combining meteorological or remote sensing data. In addition, further investigation into the model's limitations on larger timescales could be carried out to improve its performance. Other areas of future research could include expanding the model's application to other meteorological forecasting domains, studying prediction uncertainty, and improving the model's overall reliability and accuracy.

5. Conclusions

In this paper, multi-scale SPEI was calculated using meteorological station monitoring data in the Yellow River Basin; the SPEI values were predicted using the Informer, ARIMA, and LSTM models; and the following conclusions were obtained from a comparative analysis of the prediction results:

(1) Because the data of SPEI1 changed relatively fast and fluctuated, the predicted values of ARIMA, as well as those of LSTM, for the four meteorological stations were significantly different from the true SPEI values on the 1-month timescale. The differences between the

predicted and actual values of ARIMA and LSTM decreased when predicting SPEI3, SPEI6, SPEI9, SPEI12, and SPEI24. The Informer model showed better prediction results on SPEI3, SPEI6, SPEI9, SPEI12, and SPEI24. This indicates that the Informer model is able to handle long sequences and performs better when dealing with long-term dependencies.

(2) The predicted values of the Informer model were closer to the measured SPEI values, and the predicted SPEI trends were consistent with the actual trends. The Informer model can model different timescales adaptively and, therefore, better capture relevance on different timescales, and it can capture sudden changes in SPEI values in a timely and effective manner.

(3) As the timescale became smaller, the prediction ability of the Informer, ARIMA, and LSTM models decreased, but Informer still performed better than ARIMA and LSTM, indicating that Informer can widely capture the information of the input series over time, that it is more effective in long-term prediction problems, and that it can be efficient in improving the prediction precision of SPEI. As a result, Informer has good performance in predicting SPEI for each meteorological station.

Drought prediction not only enables the assessment of drought risks but also guides water resource management, agricultural planning, and ecosystem management and facilitates climate change research. The accuracy and timeliness of drought forecasts empower decision-makers to take appropriate measures, mitigating the adverse impacts of drought on society, economy, and the environment and ensuring sustainable development and resource utilization goals.

Author Contributions: All authors contributed equally to this work. All authors have read and agreed to the published version of the manuscript.

Funding: This research was funded by the Major Science and Technology Project of Henan Province, China, grant numbers "221100210600", "201400211000", "201400211300", and "201400210600".

Institutional Review Board Statement: Not applicable.

Informed Consent Statement: Not applicable.

Data Availability Statement: The datasets from 1960 to 2019 were obtained from the National Meteorological Data Center (http://data.cma.cn/ accessed on 2 April 2021).

Acknowledgments: All model code programs in this paper were run on the computational platform of the National Supercomputing Zhengzhou Center.

Conflicts of Interest: The authors declare no conflicts of interest.

Abbreviations

The following abbreviations are used in this manuscript:

MDPI	Multidisciplinary Digital Publishing Institute
DOAJ	Directory of open-access journals
TLA	Three-letter acronym
LD	Linear dichroism

References

1. Rusca, M.; Savelli, E.; Di Baldassarre, G. Unprecedented droughts are expected to exacerbate urban inequalities in Southern Africa. *Nat. Clim. Chang.* **2023**, *13*, 98–105. [CrossRef]
2. Xu, D.; Zhang, Q.; Ding, Y.; Zhang, D. Application of a hybrid ARIMA-LSTM model based on the SPEI for drought forecasting. *Environ. Sci. Pollut. Res.* **2022**, *29*, 4128–4144. [CrossRef]
3. Yang, R.; Geng, G.; Zhou, H.; Wang, T. Spatial-temporal Evolution of Meteorological Drought in the Wei River Basin Based on SPEI_PM. *Chin. J. Agrometeorol.* **2021**, *42*, 962–974.
4. Abbas, A.; Waseem, M.; Ahmad, R.; Khan, K.A; Zhao, C; Zhu, J. Sensitivity analysis of greenhouse gas emissions at farm level: case study of grain and cash crops. *Environ. Sci. Pollut. Res.* **2022**, *29*, 82559–82573. [CrossRef]
5. Abbas, A.; Zhao, C.; Waseem, M.; Ahmed, K.K.; Ahmad, R. Analysis of Energy Input–Output of Farms and Assessment of Greenhouse Gas Emissions: A Case Study of Cotton Growers. *Front. Environ. Sci.* **2022**, *9*, 826838. [CrossRef]

6. Elahi, E.; Khalid, Z.; Zhang, Z. Understanding farmers'intention and willingness to install renewable energy technology: A solution to reduce the environmental emissions of agriculture. *Appl. Energy* **2022**, *309*, 118459. [CrossRef]
7. Elahi, E.; Khalid, Z.; Tauni, M.Z.; Zhang, H.; Xing, L. Extreme weather events risk to crop-production and the adaptation of innovative management strategies to mitigate the risk: A retrospective survey of rural Punjab, Pakistan. *Technovation* **2021**, *117*, 102255. [CrossRef]
8. Ren, Y.; Liu, J.; Shalamzari, M.J.; Arshad, A.; Liu, S.; Liu, T.; Tao, H. Monitoring Recent Changes in Drought and Wetness in the Source Region of the Yellow River Basin, China. *Water* **2022**, *14*, 861. [CrossRef]
9. Alahacoon, N.; Edirisinghe, M. A comprehensive assessment of remote sensing and traditional based drought monitoring indices at global and regional scale. *Geomat. Nat. Hazards Risk* **2022**, *13*, 762–799. [CrossRef]
10. Saeed, S.; Mohammadi, G.M.; Saviz, S. Spatial and temporal analysis of drought in various climates across Iran using the Standardized Precipitation Index (SPI). *Arab. J. Geosci.* **2022**, *15*, 1279.
11. Sergio, M.V.; Santiago, B.; Ji, L.-M. A Multiscalar Drought Index Sensitive to Global Warming: The Standardized Precipitation Evapotranspiration Index. *J. Clim.* **2010**, *23*, 1696–1718.
12. Palmer, W.C. *Meteorological Drought*; U.S. Department of Commerce Weather Bureau Research Paper: San Diego, CA, USA, 1965.
13. Narasimhan, B.; Srinivasan, R. Development and evaluation of Soil Moisture Deficit Index (SMDI) and Evapotranspiration Deficit Index (ETDI) for agricultural drought monitoring. *Agric. For. Meteorol.* **2005**, *133*, 69–88. [CrossRef]
14. Chen, H.; Sun, J. Changes in Drought Characteristics over China Using the Standardized Precipitation Evapotranspiration Index. *J. Clim.* **2015**, *28*, 5430–5447. [CrossRef]
15. Wei, J.; Wang, Z.; Han, L.; Shang, J.; Zhao, B. Analysis of Spatio-Temporal Evolution Characteristics of Drought and Its Driving Factors in Yangtze River Basin Based on SPEI. *Atmosphere* **2022**, *13*, 1986. [CrossRef]
16. Ma, X.; Zhu, X.; Zhao, J.; Zhao, N.; Shi, Y. Analysis of Drought Characteristics and Driving Forces in the Urban Belt Along the Yellow River in Ningxia Based on SPEI. *Res. Soil Water Conserv.* **2022**, *29*, 1986.
17. Wu, Z.; Lu, G.; Guo, H.; Kuang, Y. Drought monitoring technology based on simulation of soil moisture. *J. Hohai Univ. (Nat. Sci.)* **2012**, *40*, 28–32.
18. Li, Y.; Chang, J.; Fan, J.; Yu, B. Agricultural drought evolution characteristics and driving mechanisms in the Yellow River Basin under climate and land use changes. *Trans. Chin. Soc. Agric. Eng.* **2021**, *37*, 84–93.
19. Liu, J.; Ren, Y.; Tao, H.; Shalamzari, M. Spatial and Temporal Variation Characteristics of Heatwaves in Recent Decades over China. *Remote. Sens.* **2021**, *13*, 3824. [CrossRef]
20. Hu, X.; Zhao, A.; Xiang, K.; Zhang, X. Evaluating the application of LSTM model for drought forecasting in Beijing-Tianjin-Hebei region. *J. Xi'An Univ. Technol.* **2022**, *38*, 356–365.
21. Zhang, X.; Sun, C.; Wang, H.; Li, M. Assessment of the effectiveness of ensemble-based drought forecasting models in the Yellow River Basin, China. *Nat. Hazards* **2019**, *95*, 347–363.
22. Xu, D.; Ding, Y.; Liu H.; Zhang Q.; Zhang D. Applicability of a CEEMD—ARIMA Combined Model for Drought Forecasting: A Case Study in the Ningxia Hui Autonomous Region. *Atmosphere* **2022**, *13*, 1109. [CrossRef]
23. Liu, J.; Zhang, W. Climate changes and associated multi-scale impacts on watershed discharge over the upper reach of Yarlung Zangbo River Basin, China. *Adv. Meteorol.* **2018**, *2018*, 4851645. [CrossRef]
24. Khan, M.M.H.; Muhammad, N.S.; El-Shafie, A. Wavelet Based Hybrid ANN-ARIMA Models for Meteorological Drought Forecasting. *J. Hydrol.* **2020**, *590*, 125380. [CrossRef]
25. Xiang, D.; Zhang, P.; Xiang, S.; Pan, C. Multi-modal Meteorological Forecasting Based on Transformer. *Comput. Eng. Appl.* **2023**, *59*, 94–103.
26. Zhou, H.; Zhang, S.; Peng, J.; Zhang, S.; Li, J.; Xiong, H.; Zhang, W. Informer: Beyond Efficient Transformer for Long Sequence Time-Series Forecasting. *Proc. AAAI* **2021**, *35*, 11106–11115. [CrossRef]
27. Vaswani, A.; Shazeer, N.; Parmar, N.; Uszkoreit, J.; Jones, L.; Gomez, A.N.; Kaiser, L.; Polosukhin, L. Attention Is All You Need. In Proceedings of the Annual Conference on Neural Information Processing Systems 2017, Long Beach, CA, USA, 4–9 December 2017; pp. 5998–6008.
28. Dong, H.; Sun, L.; Ouyang, F. Prediction of PM2.5 Concentration Based on Informer. *Environ. Eng.* **2022**, *40*, 48–54.
29. Yu, F.; Koltun, V.; Funkhouser, T. Dilated residual networks. In Proceedings of the 2017 IEEE Conference on Computer Vision and Pattern Recognition, CVPR 2017, Honolulu, HI, USA, 21–26 July 2017; pp. 636–644.
30. Clevert, D.; Unterthiner, T.; Hochreiter, S. Fast and Accurate Deep Network Learning by Exponential Linear Units (ELUs). In Proceedings of the 4th International Conference on Learning Representations, ICLR 2016, San Juan, Puerto Rico, 2–4 May 2016.
31. Hochreiter, S.; Schmidhuber, J. Long Short-Term Memory. *Neural Comput* **1997**, *9*, 1735–1780. [CrossRef]
32. Ko, M.S.; Lee, K.; Kim, J.K.; Hong, C.W.; Dong, Z.Y.; Hur, K. Deep Concatenated Residual Network with Bidirectional LSTM for One-Hour-Ahead Wind Power Forecasting. *IEEE Trans. Sustain. Energy* **2021**, *12*, 1321–1335. [CrossRef]
33. Ding, Y.; Yu, G.; Tian, R.; Sun, Y. Application of a Hybrid CEEMD-LSTM Model Based on the Standardized Precipitation Index for Drought Forecasting: The Case of the Xinjiang Uygur Autonomous Region, China. *Atmosphere* **2022**, *13*, 1504. [CrossRef]
34. Li, Z.; Wang, X.; Zhang, E.; Yu, J. Research on the Drought Prediction Model of Large Irrigation Areas Based on VMD-GRU. *China Rural. Water Hydropower* **2023**, *3*, 130–137.
35. Li, Z.; Zou, H.; Qi, B.; Guo, J. A fitting model of annual precipitation prediction based on eemd-arima. *Comput. Appl. Softw.* **2020**, *37*, 46–50.

36. Wu, X.; Chen, Y.; Guan, Y.; Tian, X.; Hua, Y. A hybrid CEEMDAN-SE-ARIMA model and its application to summer precipitation forecast over Northeast China. *Trans. Atmos. Sci.* **2023**, *46*, 205–216.
37. Hinge, G.; Piplodiya, J.; Sharma, A.;Hamouda, M.A.; Mohamed M.M. Evaluation of Hybrid Wavelet Models for Regional Drought Forecasting. *Remote. Sens.* **2022**, *14*, 6381. [CrossRef]
38. Liu, F.; Dong, T.; Liu, Y. An Improved Informer Model for Short-Term Load Forecasting by Considering Periodic Property of Load Profiles. *Front. Energy Res.* **2022**, 10, 950912. [CrossRef]
39. Pande, C.B.; AIAnsari, N.; Kushwaha, N.L.;Srivastava, A.; Noor, R.; Kumar, M.; Moharir, N.; Elbeltagi, A. Forecasting of SPI and Meteorological Drought Based on the Artificial Neural Network and M5P Model Tree. *Land* **2022**, *11*, 2040. [CrossRef]
40. Banadkooki, F.B.; Singh, V.P.; Ehteram, M. Multi-timescale drought prediction using new hybrid artificial neural network models. *Nat. Hazards* **2021**, *106*, 2461–2478. [CrossRef]
41. Zhang Y.; Yang, H.; Cui, H.; Chen, Q. Comparison of the Ability of ARIMA, WNN and SVM Models for Drought Forecasting in the Sanjiang Plain, China. *Nat. Resour. Res.* **2020**, *29*, 1469–1470. [CrossRef]

atmosphere

MDPI

Article

Drought in Shanxi Province Based on Remote Sensing Drought Index Analysis of Spatial and Temporal Variation Characteristics

Yuanyuan Xu [1,2], Yuxin Chen [1,2], Jiajia Yang [1,2], Weilai Zhang [1,2], Yongxiang Wang [1,2], Jiaxuan Wei [1,2] and Wuxue Cheng [1,2,*]

1 The Faculty Geography Resources Sciences, Sichuan Normal University, Chengdu 610101, China; 20211101059@stu.sicnu.edu.cn (Y.X.); 20211101051@stu.sicnu.edu.cn (Y.C.); 20211101056@stu.sicnu.edu.cn (J.Y.); 20221101012@stu.sicnu.edu.cn (W.Z.); 20201104001@stu.sicnu.edu.cn (Y.W.); wjx513380025@stu.sicnu.edu.cn (J.W.)

2 Key Laboratory of Land Resources Evaluation and Monitoring in Southwest China, Sichuan Normal University, Chengdu 610066, China

* Correspondence: cwx714826@sicnu.edu.cn

Abstract: Drought is a natural disaster with long duration and which causes great harm. Studying the characteristics of drought evolution in Shanxi Province can grasp the regularity of drought occurrence and provide a basis for drought prevention and resistance. This study utilizes MODIS products to analyze and quantify the extent of drought in a specific area. The study calculates several indices, including the Crop Water Stress Index (CWSI), Vegetation Supply Water Index (VSWI), and Temperature Vegetation Dryness Index (TVDI), using variables such as the Normalized Difference Vegetation Index (NDVI), Land Surface Temperature (LST), Evapotranspiration (ET), and Potential Evapotranspiration (PET). Additionally, three drought indices are analyzed for correlation with the self-calibrated Palmer Drought Severity Index (sc-PDSI), and the most suitable drought index is selected through validation with typical drought events. Finally, the selected indices are used to investigate the spatiotemporal characteristics of drought in the study area from 2001 to 2020. The results show: (1) CWSI and sc-PDSI have a strong correlation both in terms of time and spatial analysis. Furthermore, CWSI has been shown to be more effective in monitoring significant drought events. (2) The multi-year mean values of CWSI range from 0.71 to 0.85, with a significant degree of spatial heterogeneity. In the study area, the percentage of the area affected by different levels of drought is in the following order: moderate drought > severe drought > mild drought > no drought. (3) The trend of CWSI changes shows that the drought situation in Shanxi Province has been alleviated from 2001 to 2020, and the overall spatial distribution indicates that the degree of drought alleviation in the southern region is greater than that in the northern region. The turning point from drought to wetness in the study area was in 2011, showing the overall characteristic of "dry in the north and wet in the south".

Keywords: Shanxi Province; MODIS data; drought index; temporal and spatial characteristics

Citation: Xu, Y.; Chen, Y.; Yang, J.; Zhang, W.; Wang, Y.; Wei, J.; Cheng, W. Drought in Shanxi Province Based on Remote Sensing Drought Index Analysis of Spatial and Temporal Variation Characteristics. *Atmosphere* **2023**, *14*, 799. https://doi.org/10.3390/atmos14050799

Academic Editors: Jinping Liu, Quoc Bao Pham, Arfan Arshad and Masoud Jafari Shalamzari

Received: 1 April 2023
Revised: 24 April 2023
Accepted: 25 April 2023
Published: 27 April 2023

1. Introduction

Drought is one of the natural disasters that does great harm to human beings [1–3], and its recurring and long-lasting nature causes serious environmental, social, and economic disasters worldwide. With global warming [4], economic losses due to drought amount to billions of dollars and affect more than two billion people every year [5,6], which is far more than the losses caused by other natural disasters. Sixty percent of China's regions are prone to drought, especially in the last three decades when droughts have become more frequent. Shanxi Province is located in the upstream of the Yellow River in North China, a typical arid and semi-arid region where most areas are severely affected by drought, with only a few areas experiencing mild droughts. In recent years, due to the rapid development of the local economy and the fragile ecological environment, extreme weather has occurred frequently, leading to noticeable warming and drying of the climate. Shanxi Province has a typical Southeast Asian monsoon climate and is far from the ocean, with uneven precipitation mainly concentrated in summer and autumn, resulting in frequent extreme weather events. In recent years, the frequency of extreme weather events in Shanxi Province has been increasing, with drought being particularly prominent, causing serious impacts on agriculture, the economy, and even people's livelihoods. Therefore, studying drought in Shanxi Province can help us understand the patterns of drought occurrence, and provide guidance and reference for agricultural production in drought-prone areas. It can also provide valuable insights for the government to propose disaster prevention and mitigation measures, playing a crucial role in safeguarding food security.

Traditional drought monitoring relies heavily on meteorological data collected from monitoring stations, which provides high accuracy but has certain limitations. Firstly, site data is sparse and unevenly distributed, making it challenging to obtain continuous spatial coverage with a certain lag in data acquisition. Secondly, traditional monitoring requires a significant amount of human and material resources, and the scope of application is small. With the advancement of remote sensing technology [7,8], the challenges associated with drought research have been addressed to a large extent. It is recorded that drought research has been carried out since 1861 based on precipitation. However, due to the intricate nature of the causes of drought and its susceptibility to human activities, researchers have frequently used the drought index as a means of describing this phenomenon. Palmer [9,10] developed the Palmer Drought Index (PDSI), a widely used drought index that is based on the relationship between water supply and demand, but the selection of its parameters was somewhat territorial. To address this issue, Wells [11] proposed the self-calibrated Palmer Drought Severity Index (sc-PDSI), which is based on the same principles as the PDSI but uses a self-calibration technique to standardize the parameters across different regions. This makes the sc-PDSI a more reliable and consistent measure of drought severity that can be used worldwide. Mckee et al. [12] proposed the Standardized Precipitation Index (SPI), which calculates the cumulative probability density function of precipitation based on precipitation information to access drought conditions. Carlson et al. [13] proposed the Vegetation Supply Water Index (VSWI), a composite index of drought conditions with a good response to drought conditions throughout the growing season. The Crop Water Stress Index (CWSI) proposed by Jackson et al. [14] is based on the heat balance principle and can reflect certain vegetation soil moisture conditions. Sandholt et al. [15] proposed the Temperature Vegetation Dryness Index (TVDI) based on the relationship between surface temperature and vegetation index. Wang Pengxin et al. [16] proposed the Vegetation Temperature Condition Index(VTCI) based on the scatter plot of NDVI and LST with a triangular regional distribution.

Using meteorological station data to calculate drought index is convenient, and the data is easily accessible. However, the observation data is greatly influenced by the relocation and uneven distribution of meteorological observation stations, which limits the monitoring of drought. Therefore, drought indices based on remote sensing monitoring data are used to quantitatively characterize the drought situation in Shanxi Province. The MODIS is a remote sensing satellite sensor used to obtain surface information worldwide, including vegetation coverage, land surface temperature, and other parameters. Drought research based on MODIS products has the characteristics of global coverage, high resolution, comprehensive multiple parameters, and timely data updates. CWSI, VSWI, and TVDI are not easily affected by other non-drought factors, are easy to calculate, their data is easily accessible, and they have been widely applied. To reveal the drought characteristics of Shanxi Province, this research utilizes the evapotranspiration products, vegetation indices, and surface temperature data provided by MODIS sensors. These data are used to calculate CWSI, VSWI, and TVDI, respectively. The research assesses the variability of each drought index on the drought monitoring ability of Shanxi Province from different perspectives, and conducts correlation analysis with the Palmer drought index to select the most suitable index. Finally, based on the optimized drought index, the spatiotemporal variation characteristics of drought in Shanxi Province from 2001 to 2020 were analyzed, providing a model for agricultural production and drought prevention in the region.

2. Materials and Methods

2.1. Overview of the Study Area

Shanxi Province is located in the northwestern region of China, positioned between 34°34′–40°44′ N and 110°14′–114°33′ E, in the upper reaches of the Yellow River, with a total area of 156,700 km^2 (Figure 1). It borders Hebei to the east, Inner Mongolia to the north, Shanxi to the west, and Henan to the south. Shanxi Province is a typical mountainous plateau covered by loess. Its topography is characterized by high elevations in the northeast and lower elevations in the southwest, with undulating terrain, rivers, and valleys. The province's landforms are complex and diverse, comprising hills, mountains, plains, and other types of landscapes. Shanxi Province is located in the eastern mid-latitude inland region of Asia and Europe, and belongs to the semi-arid-semi-humid region. The winters are long and cold, with dry weather prevailing, while the summers are longer in the south and shorter in the north, with concentrated precipitation. The province enjoys abundant sunshine and heat resources, but its weather can be quite unpredictable and sometimes disastrous.

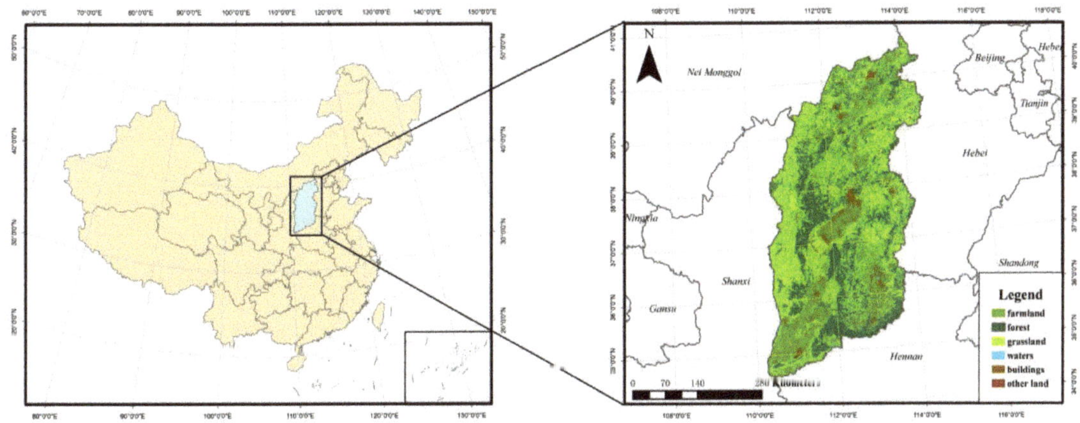

Figure 1. Study area land type zoning.

2.2. Data Sources and Research Methods

This article mainly applies RS and GIS technologies, based on MODIS digital data and sc-PDSI data, to analyze the spatiotemporal pattern of drought in the study area, under the premise of model verification. The technology roadmap illustrated in Figure 2 is shown below:

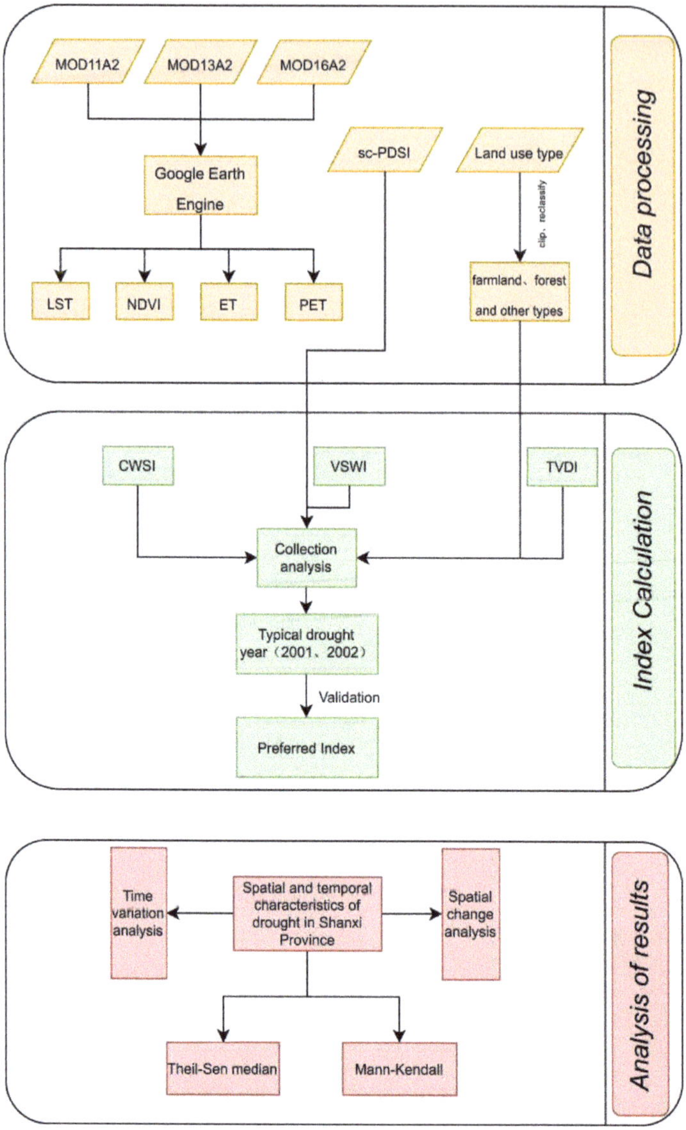

Figure 2. Technology roadmap.

2.2.1. Data Sources

The study uses evapotranspiration products, vegetation index, and surface temperature data from the MODIS sensor to calculate CWSI, VSWI, and TVDI, respectively. The images were obtained from NASA (https://ladsweb.modaps.eosdis.nasa.gov/ accessed on 1 October 2022).The 2001–2020 surface temperature (LST) was obtained from the NASA MOD11A2 sensor; the Normalized Difference Vegetation Index (NDVI) from the MOD13A1 sensor [17]; and the actual evapotranspiration (ET) and potential evapotranspiration (PET) from the MOD16A2 sensor. The sc-PDSI data from January to December of 2001–2020 are derived from climate research (https://crudata.uea.ac.uk/ accessed on 30 October 2022); land type data in Shanxi Province were obtained from the number of land types at 1 km resolution provided by the Resource and Environmental Science and Data Centre of the Chinese Academy of Sciences (http://www.resdc.cn/ accessed on 15 November 2022), which can be reclassified to obtain forest, grassland, and cropland types.A description of the relevant data is shown in Table 1:

Table 1. Data resolution and purpose used in this study.

Name of Data	Temporal Resolution	Usage
MOD11A2	8d	Calculation of VSWI and TVDI
MOD13A1	16d	Calculation of VSWI and TVDI
MOD16A2	8d	Calculation of CWSI

2.2.2. Research Methods

(1) CWSI

Remote sensing methods for soil moisture estimation typically provide information only on the soil surface, which may not accurately reflect the moisture levels at the root level of crops. To obtain more accurate measurements, it is necessary to measure the canopy temperature of vegetation and calculate the Crop Water Stress Index (CWSI). The CWSI is based on the principle of energy balance and monitors drought conditions in real time by considering soil moisture and evapotranspiration from farmland. In 1981, Idso et al. [18] proposed the CWSI based on the empirical relationship between canopy temperature and air vapor pressure deficit. Later, Jackson et al. [19] based their theoretical interpretation on the canopy energy balance and proposed the calculation of the CWSI, which they defined as [20,21]:

$$CWSI = 1 - \frac{ET}{PET} \tag{1}$$

where ET is the actual evapotranspiration and PET is the potential evapotranspiration. The value of $CWSI$ ranges from 0 to 1, with smaller values indicating wetter conditions and larger values indicating drier conditions.

(2) VSWI

The VSWI uses the ratio of vegetation index to surface temperature as an indicator of the extent of vegetation exposure to drought, and provides a better understanding of the drought condition in areas with high vegetation cover and strong vegetation transpiration [22,23]. Under normal conditions when crop water supply is adequate, the crop canopy temperature stays within a certain range. If there is a drought and the crop water supply is insufficient, the vegetation index from satellite remote sensing will decrease and at the same time the crop canopy temperature will increase. The VSWI drought monitoring model uses the Normalized Difference Vegetation Index (NDVI) and Channel 4 remote sensing bright temperature as factors, and is defined as [24]:

$$VSWI = \frac{NDVI}{T_C} \tag{2}$$

where $NDVI$ is the Normalized Difference Vegetation Index, and Tc is the canopy temperature of vegetation. Since it is difficult to obtain the canopy temperature, LST is used to replace it. The $VSWI$ takes on a value between 0 and 1, with smaller values indicating a drier region, and larger values indicating a more humid region.

(3) TVDI

In their study of soil moisture, Sandholt et al. [25] found a number of contours in the characteristic space of the normalized vegetation index and the surface temperature, based on which the TVDI was proposed. The defining equation is [26]:

$$TVDI = \frac{LST - LST_{NDVI,min}}{LST_{NDVI,max} - LST_{NDVI,min}} \tag{3}$$

$$LST_{NDVI,max} = a_1 + b_1 \times NDVI \quad (4)$$

$$LST_{NDVI,min} = a_2 + b_2 \times NDVI \quad (5)$$

where LST refers to Land Surface Temperature, $LST_{NDVI,max}$ and $LST_{NDVI,min}$ represent the minimum and maximum values of LST corresponding to a certain $NDVI$ value. a_1, a_2 and b_1, b_2 are the fitting coefficients for dry and wet edges. TVDI has a value between 0 and 1; a smaller TVDI value indicates a more humid region, while a larger value indicates a drier region.

(4) sc-PDSI

The Palmer Drought Severity Index (PDSI), developed by Palmer in 1956, is a widely used measure of accumulated deviation of surface moisture supply and demand on land. It incorporates the effects of temperature on precipitation and can accurately reflect the impact of climate on drought. However, its applications in analyzing drought in different spatial areas have limitations, and it may not be suitable for evaluating drought in diverse regions. To address these limitations, the self-calibrated PDSI (sc-PDSI) has been developed. The sc-PDSI dynamically calculates the monthly PDSI value and replaces the empirical constant of the original location. In this article, the sc-PDSI is used to analyze its correlation with different drought indices and to select the most suitable drought index for the study area. The findings of this study will help to enhance the accuracy of drought monitoring and prediction in the region.

(5) Other methods

The Pearson correlation coefficient is widely used to measure the degree of correlation between two variables. In order to test the drought monitoring ability of different remote sensing indices, the correlation index R between the three indices and sc-PDSI is analyzed, which represents the difference ratio between the different indices and the Pearson correlation coefficient, reflecting the dispersion degree of the drought index itself [27–29]. Overall, the correlation analysis between drought indices and sc-PDSI can help to identify which indices are most effective in monitoring drought conditions, and can provide valuable information for drought management and mitigation efforts.

The Theil-sen Median method and the Mann—Kendall (MK): The Theil-sen Median method is a robust non-parametric statistical trend calculation method [30]. This method has high computational efficiency and is insensitive to measurement errors and outliers. It is often used in trend analysis of long time series data. The MK trend test [31] is a non-parametric test for analyzing trends in time series [32,33]. It is essentially a non-parametric test that does not require the sample to follow a specific distribution and is not disturbed by a few outliers, but also has a wide detection range, a high degree of quantification, and a simple calculation process. Sen trend analysis and MK testing are often combined for analysis. First, the Sen trend value is calculated, and then the MK method is used to determine the significance of the trend. In this study, the Sen trend is used to analyze the trend of drought intensity in Shanxi Province, and the MK method is used to test the significance of the trend.

3. Results and Analysis

3.1. Validation of Integrated Drought Monitoring Models

3.1.1. Correlation Analysis

Using the above method, three indices are calculated. To verify the accuracy of the three remote sensing drought indices, Pearson correlation coefficient analysis is performed between the three indices and sc-PDSI data (Figure 3). Through statistical analysis, the correlation coefficients of CWSI, VSWI, and TVDI are −0.54, 0.35, and −0.16, respectively. Results show that overall CWSI and TVDI are negatively correlated with sc-PDSI, that is, the larger the CWSI and TVDI, the smaller the sc-PDSI and the drier the study area. VSWI is positively correlated with sc-PDSI, that is, the larger the VSWI, the larger the PDSI and the wetter the study area. The correlation between CWSI and sc-PDSI passed the significance test with a p value of less than 0.05 in most regions., indicating that CWSI is more sensitive to interannual changes in drought in the study area.

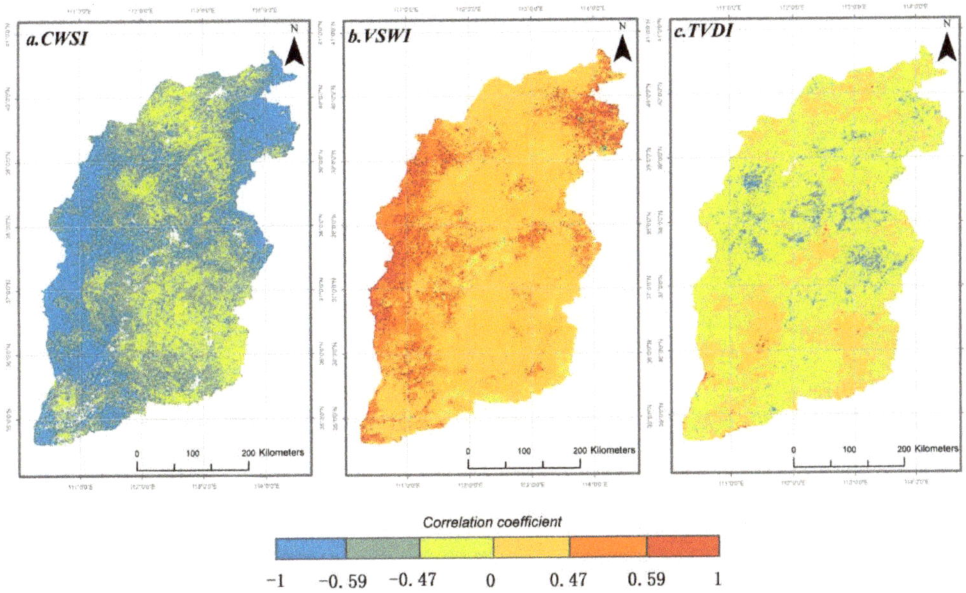

Figure 3. Correlation coefficient between remote sensing drought index and sc-PDSI in 2001–2020.

The monitoring results of droughts may differ in different land use types due to factors such as vegetation coverage, soil type, and terrain. Generally, urban areas are characterized by extensive impervious surfaces, which reduce the amount of water that can infiltrate into the soil. This can result in increased runoff and decreased soil moisture, making urban areas more susceptible to drought. Farmland typically has lower vegetation cover and higher evapotranspiration rates than natural land, which can result in lower soil moisture levels and increased susceptibility to drought. Grassland typically has lower evapotranspiration rates than forests, but higher rates than agricultural land. Overall, it is important to understand the characteristics of different land use types in order to assess their vulnerability to drought and develop effective drought mitigation strategies. For different land use types (as shown in Table 2), the higher the Pearson correlation coefficient, the better the fit and the more applicable the index. The correlation coefficients of CWSI are relatively high for different land use types, including farmland, forest, and grassland. Among them, the fitting degree of grassland is the highest, with a mean correlation coefficient of −0.55, which passed the significance test with $p < 0.05$. This is much higher than the other two index models. Through comprehensive analysis, it is concluded that CWSI has a greater advantage in drought monitoring and simulation in Shanxi Province.

Table 2. Average correlation coefficient between remote sensing drought index and sc-PDSI of each vegetation division from 2001 to 2020.

Index	Land Use Type			
	Farmland	Forest	Grassland	All
CWSI	−0.53	−0.54	−0.55	−0.54
VSWI	0.37	0.30	0.39	0.35
TVDI	−0.10	−0.24	−0.18	−0.16

3.1.2. Verification of Typical Drought Events

To further validate the accuracy of CWSI drought monitoring, it combines with typical drought events for verification. According to the Statistical Yearbook of Shanxi Province, between 1997 and 2002 Shanxi Province suffered from severe drought, which was caused by global warming, reduced precipitation, and a sharp decrease in water coming from rivers. From historical statistics, Shanxi Province suffers from a drought every 2.6 years, causing serious impacts on industries, agriculture, and other aspects. During the years 2001 and 2002, Shanxi Province experienced a relatively severe drought, where the maximum value of CWSI was 0.97 in 2001 (Figure 4a) and 0.96 in 2002 (Figure 4b); and its annual average values of CWSI were 0.85 and 0.82, respectively, which are the maximum values during the study period, further indicating the accuracy of CWSI in monitoring drought in Shanxi Province.

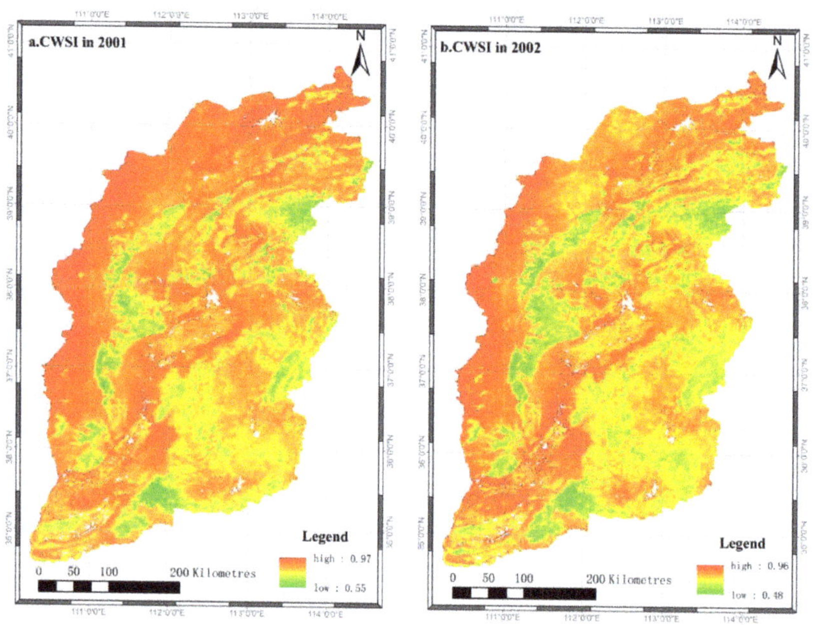

Figure 4. CWSI values in 2001 and 2002.

3.2. Drought Classification

The above correlation analysis shows that the CWSI index is more applicable in the study area than the other two indices. Therefore, the CWSI is selected to analyze the spatial and temporal characteristics of drought in the study area. However, before the analysis, the drought class criteria need to be classified. For this purpose, the research uses the sc-PDSI, and the CWSI value at the corresponding location for a one-dimensional linear regression analysis (Figure 5). During the study period, most of the sc-PDSI in Shanxi Province range from −3 to 1. Thus, the drought classes are classified into four levels according to the sc-PDSI criteria for classifying drought, while sample points are selected according to the area share of different land types in Shanxi Province; and one-dimensional linear regression is performed to obtain the classification thresholds of CWSI corresponding to different classes (Table 3).

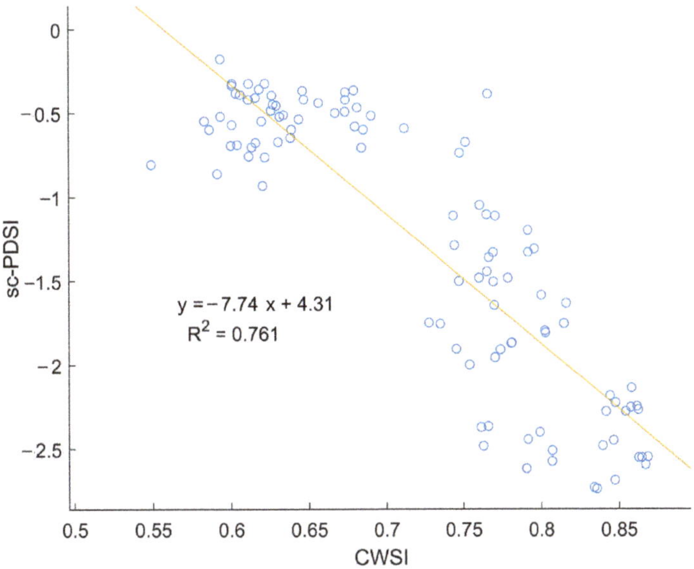

Figure 5. Univariate linear regression analysis of sc-PDSI and CWSI in Shanxi Province from 2001 to 2020.

Table 3. Drought grading.

Drought Rating	PDSI	CWSI
No drought	>0	0~0.56
Mild drought	0~−1	0.56~0.69
Moderate drought	−1~−2	0.69~0.81
Severe drought	−2~−3	0.81~0.92

3.3. *Temporal Variation Characteristics of Drought*

According to the interannual variation and cumulative anomaly of CWSI in Shanxi Province (Figure 6), the fluctuation range of CWSI has been small over the years, with a decreasing trend. The CWSI fluctuates between 0.71 and 0.85, with a multi-year average of 0.76, the maximum value in 2001 (0.85) and the minimum value in 2016 and 2018 (0.71). In 2001, the highest CWSI values are due to less rainfall, weaker actual evapotranspiration and stronger potential evapotranspiration, which led to higher CWSI values and more severe drought. In 2016 and 2018, the lowest CWSI values were due to abundant rainfall, lower temperatures, weaker actual evapotranspiration and stronger potential evapotranspiration, which led to lower CWSI values and less severe drought [34,35].

This study builds on previous research and identifies a turning point interval when the cumulative anomaly value is considered stable, i.e., when the trend change does not pass a significance test with $p < 0.05$. From the cumulative distance level values of CWSI during 2001–2020, it can be observed that the cumulative distance level values of CWSI show a significant increasing trend from 2001 to 2011, and start to decrease after the cumulative distance level reaches the highest value in 2011. This indicates that 2011 was a turning point from drought to wet conditions in the study area.

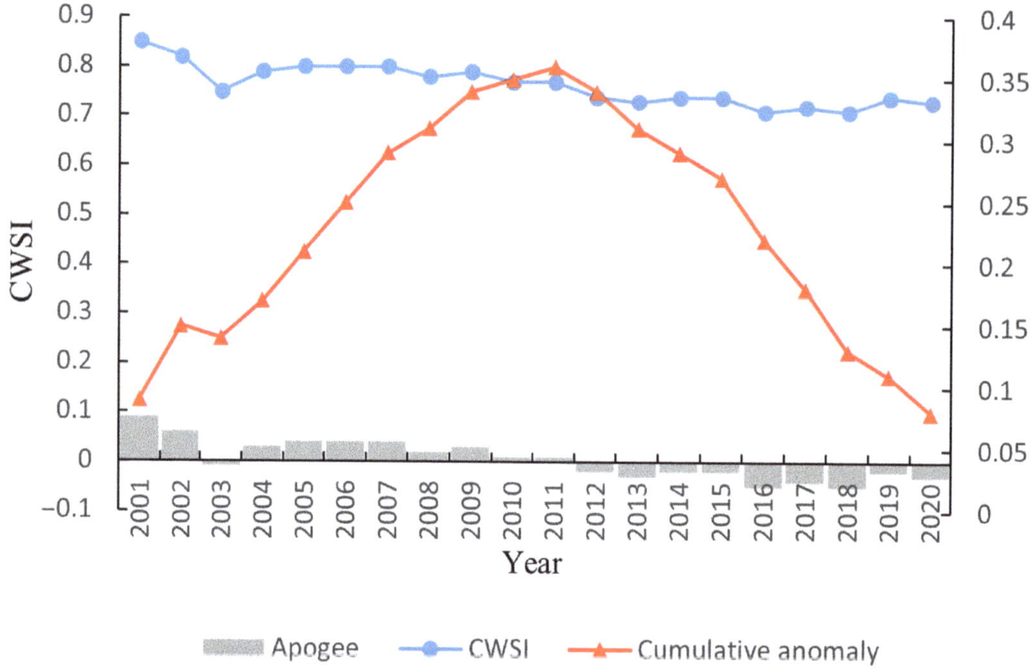

Figure 6. Interannual variation and cumulative distance level of CWSI.

3.4. Spatial Variation Characteristics of Drought

Figure 7 displays the spatial distribution of the annual average CWSI and its corresponding drought level in Shanxi Province. The average CWSI value for many years is between 0.71 and 0.85, and there is a noticeable spatial heterogeneity. The CWSI shows that the northwest is larger than the southeast, which means that the drought in the northwest is relatively severe, while the drought in the southeast is relatively mild. Multiple urban areas, including Datong City, Shuozhou City, Xinzhou City, and Taiyuan City have high CWSI values, indicating that drought is severe in these areas. In contrast, Jincheng City, Changzhi City, and other cities have lower CWSI values, indicating slight drought conditions. According to the drought grade map, Shanxi Province as a whole presents moderate drought, with severe drought regions mainly concentrated in the west of Shanxi Province, namely, Shuozhou City, Xinzhou City, and Luliang City. Moderate drought mainly concentrates in the southeast of Shanxi Province, namely, Jinzhong City and Changzhi City; it rarely distributes in drought-free areas. Mild drought accounts for 12% of the total area; moderate drought accounts for 62% of the total area; severe drought accounts for 25% of the whole area. In a comprehensive analysis, the percentages of the area occupied by drought levels in the study area in descending order: moderate drought > severe drought > mild drought > no drought.

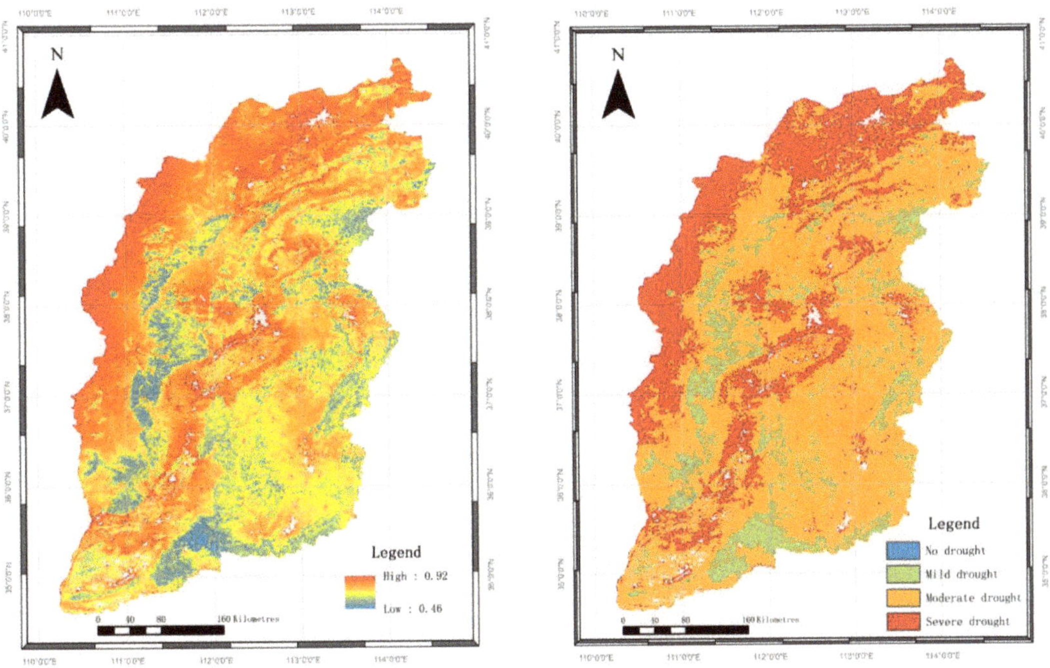

Figure 7. Drought changes in Shanxi Province from 2001 to 2020 based on CWSI.

According to the statistical interannual variation of the average annual Crop Water Stress Index in each city of Shanxi Province(Figure 8), it can be observed that fluctuations in 11 cities are not too significant, and the overall trend of each city is consistent with that of the province. The average value of CWSI in each city in the past 20 years is ranked from highest to lowest as follows: Shuozhou City (0.811), Datong City (0.793), Taiyuan City (0.776), Luliang City (0.774), Xinzhou City (0.772), Jinzhong City (0.755), Yangquan City (0.753), Linfen City (0.751), Yuncheng City (0.745), Changzhi City (0.742), and Jincheng City (0.722). Shuozhou City has the largest average CWSI of 0.811 over the past 20 years, indicating severe drought conditions. Jincheng City has the smallest average crop water deficit index at 0.722, indicating a relatively lower risk of drought.. The CWSI in Shuozhou City fluctuates between 0.72 and 0.82, with a multi-year average of 0.811, which is a severe drought; the rest of the urban areas, Datong City and Taiyuan City, have a multi-year average CWSI of less than 0.81, which is a moderate drought. The high CWSI in Shuozhou, Datong, and Taiyuan is mainly due to low precipitation and high evaporation, resulting in a relatively high risk of drought.

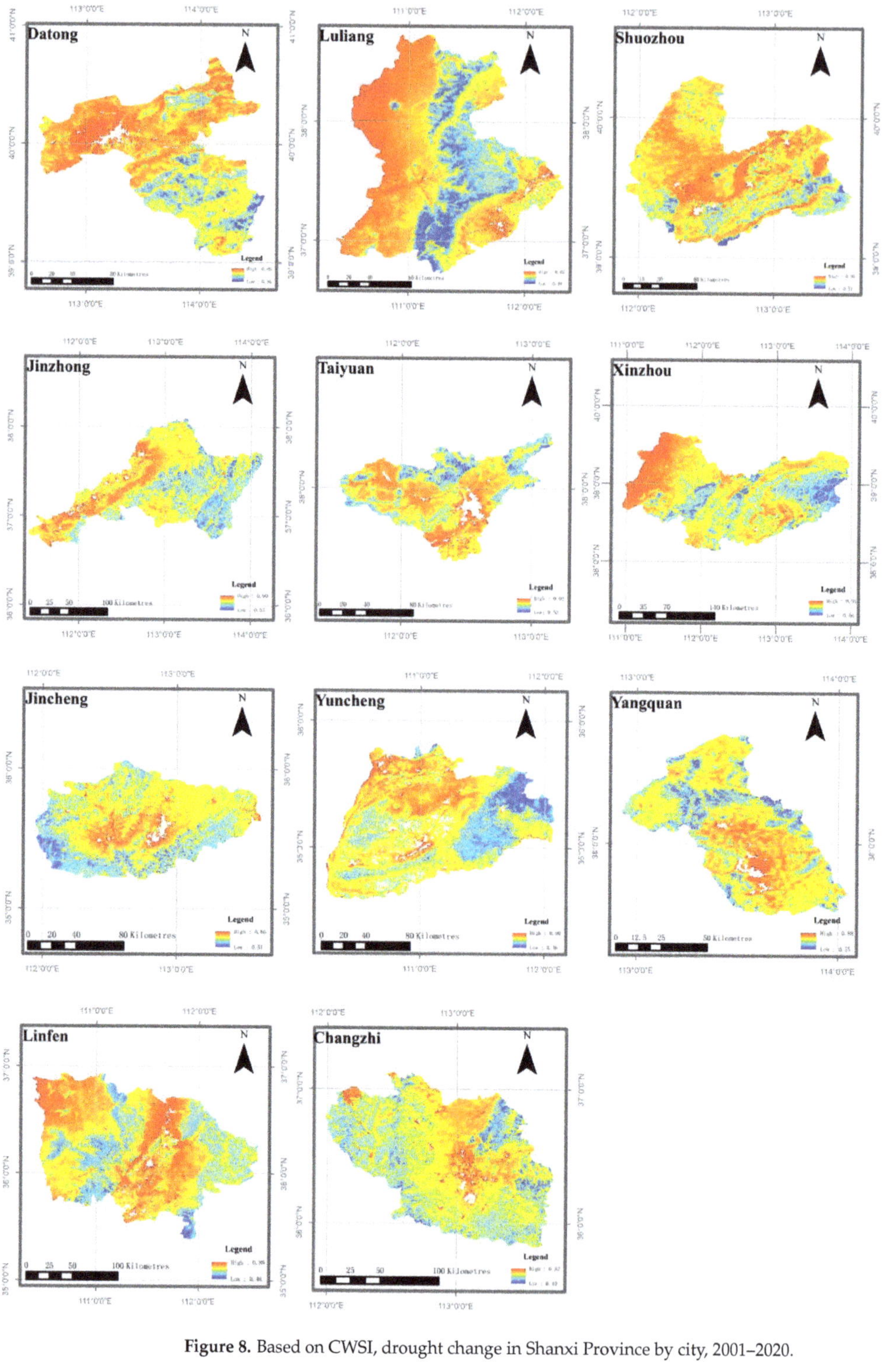

Figure 8. Based on CWSI, drought change in Shanxi Province by city, 2001–2020.

3.5. *Drought Change Characteristics of Different Land Use Types*

Different land use types will directly affect the growth of vegetation, as well as the change in evapotranspiration. Starting from different land use types, the characteristics of drought change are analyzed.

The interannual variation characteristics of drought conditions in different land use types in Shanxi Province from 2001 to 2020 are shown in Figure 9. In terms of interannual variation in annual average CWSI, the annual average CWSI for each land use type is, in descending order, buildings > unused land > farmland > grassland > forest. The CWSI of buildings fluctuates between 0.76 and 0.86, with severe drought in most years; the CWSI of other land fluctuates between 0.76 and 0.86, with moderate drought in most years; the CWSI of farmland fluctuates between 0.72 and 0.86, with moderate drought in most years; the CWSI of grassland fluctuates from 0.72 to 0.86, and most years it shows moderate drought; and the CWSI of the forest fluctuates between 0.72 and 0.86, with moderate drought in most years. Surface drought is mainly related to a number of factors, such as land cover type, geographical location, and climatic precipitation. The vegetation cover on building sites is generally low, and the risk of drought is high due to high temperatures and rapid water loss caused by the heat island effect. Conversely, forests are less at risk of drought, as they are generally located at higher altitudes, have abundant precipitation and a better ability to hold water, and their actual evapotranspiration is higher, making them relatively more resistant to drought. Most of the farmland is artificially vegetated and cultivated with crops such as rice, wheat, maize, and oilseed rape. The risk of drought on cultivated land is high because the harvesting of crops causes the annual mean ET to become smaller, resulting in larger CWSI values.

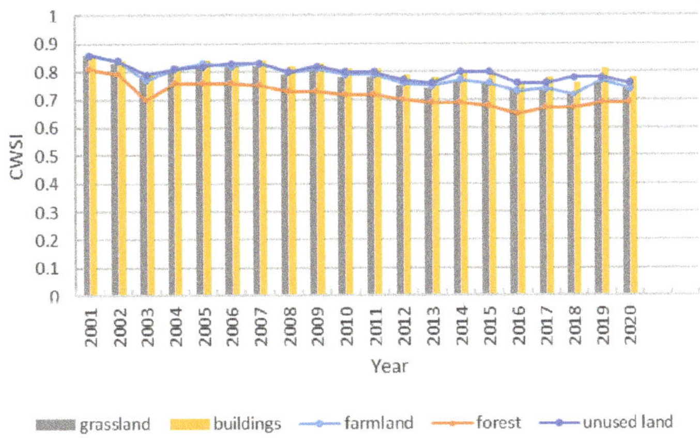

Figure 9. Drought conditions of various land use types in Shanxi Province from 2001 to 2020.

3.6. *Temporal and Spatial Evolution of Drought in Shanxi Province*

Sen slope estimates are used to calculate trend values and are usually used in conjunction with MK non-parametric test. In this study, the spatiotemporal change analysis method combining the Sen trend and the MK test is used to calculate the rate of change of Shanxi Province from 2001 to 2020 image by image; the slope of change image values greater than 0 indicated an increasing trend of the element, and less than 0 showed a decreasing trend. The spatial trends of CWSI in Shanxi Province and the spatial distribution of its significance are obtained as shown below (Figure 10a,b). Then, according to Table 4, the Sen trend analysis and MK test results are overlayed to obtain a detailed drought change map in Shanxi Province (Figure 10c).

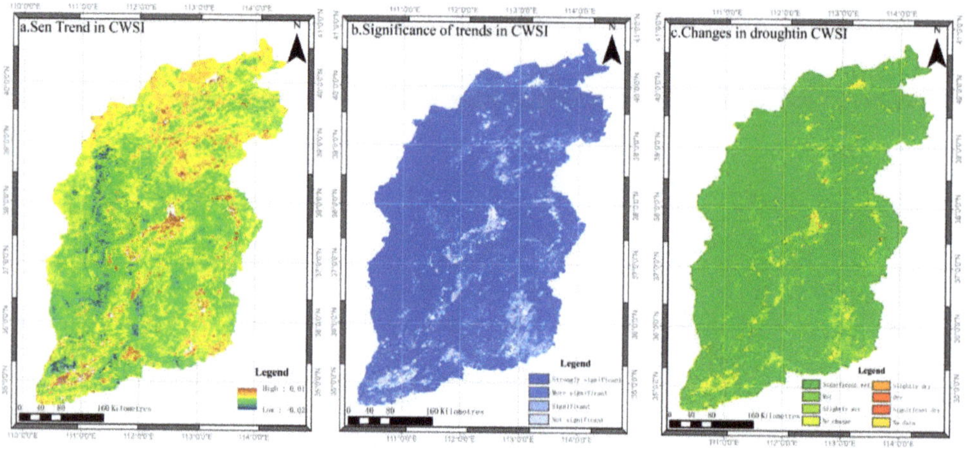

Figure 10. Spatial and temporal trends in drought evolution based on CWSI 2001–2020.

Table 4. Ranking of significance of drought trends.

$CWSI_{Slope}$	|Z|			
	|Z| ≤ 1.96	1.96 < |Z| ≤ 2.58	2.58 < |Z| ≤ 3.33	|Z| > 3.33
Slope ≤ −0.001	Stable and unchanged	Slightly wet	Wet	Significantly wet
−0.001 < Slope ≤ 0.001	Stable and unchanged	Stable and unchanged	Stable and unchanged	Stable and unchanged
Slope > 0.001	Stable and unchanged	Slightly dried	Dry	Significantly dried

From the figure, it can be seen that the change rate of CWSI is between −0.02 and 0.01. During the study period, the change rate of CWSI is mostly negative, indicating a decreasing trend in drought, and the overall spatial distribution shows that the drought alleviation degree in the south is greater than that in the north, while the drought in the north is severe. It is evident from the figure that about 87% of the total images are significantly wetted, while approximately 6% of the total images are wetted. Overall, the trend of wetter images accounts for around 94% of the total images, while the trend of drier images accounts for only about 0.1% of the total images. Therefore, a comprehensive analysis of the data suggests that the drought trend in the study area is generally becoming wetter, indicating that the overall drought condition in Shanxi Province has been continuously improving from 2001 to 2020, and the degree of drought in most areas has been alleviated.

4. Discussion

In recent years, several studies have investigated the spatiotemporal characteristics of drought in Shanxi Province using long-term meteorological observation data. However, the accuracy of these studies is affected by the limited number of observation stations, and their uneven distribution across the province. To overcome this limitation, the research uses MODIS data as the basis for drought research. MODIS data has a better temporal and spatial resolution [36] and is less affected by weather, making it widely used for calculating drought indices [37,38]. The research results indicate that CWSI is more suitable for drought monitoring in the study area, followed by VSWI. The MOD16 evapotranspiration data used by CWSI is derived from the Mu [39] Improved algorithm. The CWSI algorithm takes the transpiration of plants as the main pathway for water and energy exchange between plants and the environment. Then, the transpiration of plants is compared with the temperature and humidity of the surrounding environment to calculate the corresponding saturation vapor pressure of crop surface transpiration, which is used to obtain the CWSI index. CWSI can be used for real-time monitoring of crop water status, and to promptly detect and diagnose the degree of water stress that crops are subjected to, in order to take corresponding irrigation measures. CWSI can also combine historical data and meteorological forecast data for drought prediction to prevent and respond to drought risks in advance. CWSI can accurately detect the degree of crop water stress, provide highly accurate information, better describe soil moisture information, and its results are very easy to interpret and understand. Therefore, CWSI has a greater advantage in regional drought monitoring. TVDI evaluates the degree of soil drought using temperature and vegetation information. Although TVDI has some advantages in drought monitoring, its application is susceptible to weather conditions. It can only assess the degree of surface soil drought and cannot evaluate water evaporation. Therefore, it cannot reflect the water content inside the soil and crops very well.

According to the research results, the long-term average of CWSI is 0.76, and most areas in the region are considered arid. This is in good agreement with the research results of scholars such as Ma Zice and Li Lihong, once again verifying the applicability of CWSI in the study area. The drought is more severe in the northwestern part of the study area. Due to the combined effects of precipitation, temperature, and other factors [40], CWSI in Shuozhou City, located in the northwest direction, fluctuates between 0.72 and 0.82, with a long-term average of 0.811, which belongs to the severe drought category. The drought is relatively mild in the southeastern part of the study area. The southeastern part has generally lower elevations, and the drought intensity in these areas is relatively low. Overall, the monitoring results of CWSI indicate that the spatial and temporal evolution trend of drought in the study area is generally improving. This is mainly due to the fact that the Shanxi provincial government has taken a series of measures, such as building reservoirs, diverting water, and implementing soil and water conservation to increase the water resources in the region and improve the ecological environment.

In recent years, frequent droughts have occurred in Shanxi Province, causing serious economic and social impacts and affecting people's daily lives. Through scientific research, we can improve our understanding of drought phenomena and provide scientific basis for drought prevention and management in Shanxi Province. Research results can help relevant departments in Shanxi Province grasp the patterns of recent drought occurrences, develop more scientific and reasonable drought defense mechanisms, and adopt effective drought response measures. At the same time, research results can also provide important scientific support for agricultural production and water resources management in Shanxi Province. In summary, scientific research can provide important support for drought prevention and management, agricultural production, and water resources management in Shanxi Province, and contribute to its economic and social development.

This study constructed three drought indices and conducted a correlation analysis with the Palmer Drought Severity Index to select the most suitable drought index for analyzing the characteristics of drought changes in Shanxi Province. The study used multi-source remote sensing data for drought monitoring and simulation research, and although the overall monitoring effect was good, there were still some limitations in this study. Only considering ET and PET has certain limitations, and drought is an extremely complex natural phenomenon. Therefore, it is recommended to further consider the impact of factors such as vegetation phenology changes, temperature, precipitation, and human activities on drought.

5. Conclusions

This article calculates three drought indices based on ET, PET, NDVI, and LST data, respectively. The calculated results are correlated with the Palmer Drought Severity Index and validated with typical drought events in Shanxi Province to screen for a more suitable drought index for the study area, the CWSI. On this basis, the distribution of drought in Shanxi Province in the past 20 years is inverted, and the spatiotemporal variation characteristics of drought are analyzed. The following conclusions are drawn:

(1) The study has found that among the three drought indices (CWSI, VSWI, and TVDI) studied, CWSI is more effective in reflecting drought conditions in Shanxi Province. This

conclusion is based on the comparison of the relationship between these indices and the sc-PDSI. The study has shown that the correlation between CWSI and sc-PDSI is stronger than that of the other two indices, indicating that CWSI is more closely related to the actual drought conditions in the study area. Therefore, the CWSI is a more suitable index for monitoring and assessing drought in Shanxi Province.

(2) The temporal variation of drought in the study area: From 2001 to 2020, the average value of CWSI varied between 0.71 and 0.85, with an overall 20-year average of 0.76. The highest value was recorded in 2001 at 0.85, while the lowest values were observed in 2016 and 2018 at 0.71. The year 2011 was the turning point where the drought conditions started to shift towards wetter conditions.

(3) Spatial distribution pattern of drought in the study area: From 2001 to 2020, the overall drought in Shanxi province presented a "north dry and south wet" pattern, with significant spatial variability. The majority of the province was located in drought-prone areas, with the largest area experiencing moderate drought. In general, the northwest region was slightly more severe, specifically in the western areas of Shuozhou, Xinzhou, and Luliang. Moderate drought was mainly concentrated in the southeast of Shanxi province, specifically in Jinzhong and Changzhi. Areas without drought were rare. The areas of each drought level in descending order were moderate drought, severe drought, mild drought, and no drought.

(4) Land use types have a significant impact on the growth and distribution of vegetation, as well as on the changes in evapotranspiration in a region. In general, the severity of drought is closely related to land use type, with different land use types exhibiting different levels of vulnerability to drought. The research results indicate that in Shanxi Province, the drought severity of different land use types is in the following order: buildings, unused land, farmland, grassland, and forest.

(5) The overall trend of drought in the study area is improving, with most of the area experiencing relief from drought. Overall, the trend towards becoming wetter accounts for about 94% of the total area, while the trend towards becoming drier accounts for about 0.1% of the total area.

Author Contributions: Conceptualization, W.C.; methodology, Y.X.; software, Y.X. and Y.C.; validation, Y.X., J.Y. and W.Z.; formal analysis, Y.X., W.C. and J.W.; investigation, Y.X.; resources, W.C.; data curation, Y.W.; writing—original draft preparation, Y.X. and Y.C.; writing—review and editing, Y.X.; visualization, W.C.; supervision, W.C.; project administration, Y.X. and J.Y.; funding acquisition, W.C. All authors have read and agreed to the published version of the manuscript.

Funding: This research is funded by the Humanities and Social Sciences Foundation of the Ministry of Education of the People's Republic of China (approval number: 18YJC850004) and the National Natural Science Foundation of China (approval number: 32060370).

Institutional Review Board Statement: Not applicable.

Informed Consent Statement: Not applicable.

Data Availability Statement: The MODIS image data were obtained from the National Aeronautics and Space Administration (NASA) (https://www.nasa.gov/ accessed on 1 October 2022). Land use data of Shanxi Province were obtained from the Resource and Environmental Science and Data Center of the Chinese Academy of Sciences (https://www.resdc.cn/ accessed on 15 November 2022). The self-calibrated Palmer Drought Severity Index data was obtained from the Climatic Research Unit (http://www.cru.uea.ac.uk/data accessed on 30 October 2022).

Conflicts of Interest: The authors declare no conflict of interest.

References

1. Ji, L.; Peters, A.J. Assessing vegetation response to drought in the northern Great Plains using vegetation and drought indices. *Remote Sens. Environ.* **2003**, *87*, 85–98. [CrossRef]
2. He, B.; Lü, A.; Wu, J.; Zhao, L.; Liu, M. Drought hazard assessment and spatial characteristics analysis in China. *J. Geogr. Sci.* **2011**, *21*, 235–249. [CrossRef]

3. Mishra, A.K.; Singh, V.P. A review of drought concepts. *J. Hydrol.* **2010**, *391*, 202–216. [CrossRef]
4. Mohammad, A.H.; Jung, H.C.; Odeh, T.; Bhuiyan, C.; Hussein, H. Understanding the impact of droughts in the Yarmouk Basin, Jordan: Monitoring droughts through meteorological and hydrological drought indices. *Arab. J. Geosci.* **2018**, *11*, 103. [CrossRef]
5. Masih, I.; Maskey, S.; Mussá, F.E.F.; Trambauer, P. A review of droughts on the African continent: A geospatial and long-term perspective. *Hydrol. Earth Syst. Sci.* **2014**, *18*, 3635–3649. [CrossRef]
6. Ren, Y.; Liu, J.; Shalamzari, M.J.; Arshad, A.; Liu, S.; Liu, T.; Tao, H.J.W. Monitoring Recent Changes in Drought and Wetness in the Source Region of the Yellow River Basin, China. *Water* **2022**, *14*, 861. [CrossRef]
7. West, H.; Quinn, N.; Horswell, M. Remote sensing for drought monitoring & impact assessment: Progress, past challenges and future opportunities. *Remote Sens. Environ.* **2019**, *232*, 111291. [CrossRef]
8. Zhang, Y.; Ling, F.; Foody, G.M.; Ge, Y.; Boyd, D.S.; Li, X.; Du, Y.; Atkinson, P.M. Mapping annual forest cover by fusing PALSAR/PALSAR-2 and MODIS NDVI during 2007–2016. *Remote Sens. Environ.* **2019**, *224*, 74–91. [CrossRef]
9. Lampros Vasiliades, A.L. Hydrological response to meteorological drought using the Palmer drought indices in Thessaly, Greece. *ScienceDirect* **2009**, *237*, 3–21.
10. Sohrabi, M.M.; Ryu, J.H.; Abatzoglou, J.; Tracy, J. Development of Soil Moisture Drought Index to Characterize Droughts. *J. Hydrol. Eng.* **2015**, *20*, 15645. [CrossRef]
11. van der Schrier, G.; Barichivich, J.; Briffa, K.R.; Jones, P.D. A scPDSI-based global data set of dry and wet spells for 1901–2009. *Atmospheres* **2013**, *118*, 4025–4048. [CrossRef]
12. Naresh Kumar, M.; Murthy, C.S.; Sesha Sai, M.V.R.; Roy, P.S. On the use of Standardized Precipitation Index (SPI) for drought intensity assessment. *Meteorol. Appl.* **2009**, *16*, 381–389. [CrossRef]
13. Carlson, T.N.; Perry, E.M.; Schmugge, T.J. Remote estimation of soil moisture availability and fractional vegetation cover for agricultural fields. *Agric. For. Meteorol.* **1990**, *52*, 45–69. [CrossRef]
14. Jackson, R.D.; Idso, S.B.; Reginato, R. Canopy temperature as a crop water stress indicator. *Water Resour. Res.* **1981**, *17*, 1133–1138. [CrossRef]
15. Sandholt, I.R.K.; Andersen, J. A simple interpretation of the surface temperature/vegetation index space forassessment of surpace moisture status. *Remote Sens. Environ.* **2002**, *79*, 213–214. [CrossRef]
16. Patel, N.R.; Parida, B.R.; Venus, V.; Saha, S.K.; Dadhwal, V.K. Analysis of agricultural drought using vegetation temperature condition index (VTCI) from Terra/MODIS satellite data. *Env. Monit. Assess* **2012**, *184*, 7153–7163. [CrossRef] [PubMed]
17. Mu, Q.; Heinsch, F.A.; Zhao, M.; Running, S.W. Development of a global evapotranspiration algorithm based on MODIS and global meteorology data. *Remote Sens. Environ.* **2007**, *111*, 519–536. [CrossRef]
18. Berni, J.A.J.; Zarco-Tejada, P.J.; Sepulcre-Cantó, G.; Fereres, E.; Villalobos, F. Mapping canopy conductance and CWSI in olive orchards using high resolution thermal remote sensing imagery. *Remote Sens. Environ.* **2009**, *113*, 2380–2388. [CrossRef]
19. Jackson, R.D.; Kustas, W.P.; Choudhury, B.J. A reexamination of the crop water stress index. *Irrig. Sci.* **1988**, *9*, 309–317. [CrossRef]
20. Çolak, Y.B.; Yazar, A.; Çolak, İ.; Akça, H.; Duraktekin, G. Evaluation of Crop Water Stress Index (CWSI) for Eggplant under Varying Irrigation Regimes Using Surface and Subsurface Drip Systems. *Agric. Agric. Sci. Procedia* **2015**, *4*, 372–382. [CrossRef]
21. Chen, J.; Lin, L.; Lü, G. An index of soil drought intensity and degree: An application on corn and a comparison with CWSI. *Agric. Water Manag.* **2010**, *97*, 865–871. [CrossRef]
22. Chen, S.; Chen, Y.; Chen, J.; Zhang, Z.; Fu, Q.; Bian, J.; Cui, T.; Ma, Y. Retrieval of cotton plant water content by UAV-based vegetation supply water index (VSWI). *Int. J. Remote Sens.* **2020**, *41*, 4389–4407. [CrossRef]
23. Zhou, L.; Zhang, J.; Wu, J.; Zhao, L.; Liu, M.; Lu, A.; Wu, Z. Comparison of remotely sensed and meteorological data-derived drought indices in mid-eastern China. *Int. J. Remote Sens.* **2011**, *33*, 1755–1779. [CrossRef]
24. McVicar, T.R.; Jupp, D.L.B. The current and potential operational users of remote sensing to aid decisions on drought exceptional circumstances in Australia:a review. *Agric. Syst.* **1998**, *57*, 399–468. [CrossRef]
25. Patel, N.R.; Anapashsha, R.; Kumar, S.; Saha, S.K.; Dadhwal, V.K. Assessing potential of MODIS derived temperature/vegetation condition index (TVDI) to infer soil moisture status. *Int. J. Remote Sens.* **2008**, *30*, 23–39. [CrossRef]
26. Gao, Z.; Gao, W.; Chang, N.-B. Integrating temperature vegetation dryness index (TVDI) and regional water stress index (RWSI) for drought assessment with the aid of LANDSAT TM/ETM+images. *Int. J. Appl. Earth Obs. Geoinf.* **2011**, *13*, 495–503. [CrossRef]
27. Cohen, I.; Huang, Y.; Chen, J.; Benesty, J.; Benesty, J.; Chen, J.; Huang, Y.; Cohen, I.J.N. Pearson correlation coefficient. In *Noise Reduction in Speech Processing*; Springer Science & Business Media: Berlin/Heidelberg, Germany, 2009; pp. 1–4.
28. Wang, X.; Li, B.; Chen, Y.; Guo, H.; Wang, Y.; Lian, L. Applicability Evaluation of Multisource Satellite Precipitation Data for Hydrological Research in Arid Mountainous Areas. *Remote Sens.* **2020**, *12*, 2886. [CrossRef]
29. Sharma, T.C.; Panu, U.S. Predicting return periods of hydrological droughts using the Pearson 3 distribution: A case from rivers in the Canadian prairies. *Hydrol. Sci. J.* **2015**, *60*, 1783–1796. [CrossRef]
30. Zhu, X.; Zhang, S.; Liu, T.; Liu, Y. Impacts of Heat and Drought on Gross Primary Productivity in China. *Remote Sens.* **2021**, *13*, 378. [CrossRef]
31. McLeod, A.I. Kendall Rank Correlation and Mann-Kendall Trend Test. *R Package Kendall* **2005**, *602*, 1–10. Available online: http://www.stats.uwo.ca/faculty/aim (accessed on 23 April 2023).
32. Andreadis, K.M.; Lettenmaier, D.P. Trends in 20th century drought over the continental United States. *Geophys. Res. Lett.* **2006**, *33*, 10. [CrossRef]

33. Wu, Z.; Yu, L.; Du, Z.; Zhang, H.; Fan, X.; Lei, T. Recent changes in the drought of China from 1960 to 2014. *Int. J. Climatol.* **2019**, *40*, 3281–3296. [CrossRef]
34. Maes, W.H.; Steppe, K. Estimating evapotranspiration and drought stress with ground-based thermal remote sensing in agriculture: A review. *J. Exp. Bot.* **2012**, *63*, 4671–4712. [CrossRef] [PubMed]
35. Feldhake, C.; Glenn, D.; Edwards, W.; Peterson, D.J.N.Z.J.o.A.R. Quantifying drought for humid, temperate pastures using the Crop Water Stress Index (CWSI). *N. Z. J. Agric. Res.* **1997**, *40*, 17–23. [CrossRef]
36. Ren, Y.; Liu, J.; Liu, S.; Wang, Z.; Liu, T.; Shalamzari, M.J.J.R.S. Effects of Climate Change on Vegetation Growth in the Yellow River Basin from 2000 to 2019. *Remote Sens.* **2022**, *14*, 687. [CrossRef]
37. Orvos, P.I.; Homonnai, V.; Várai, A.; Bozóki, Z.; Jánosi, I.M. Global trend analysis of the MODIS drought severity index. *Geosci. Instrum. Methods Data Syst.* **2015**, *4*, 189–196. [CrossRef]
38. Du, L.; Tian, Q.; Yu, T.; Meng, Q.; Jancso, T.; Udvardy, P.; Huang, Y. A comprehensive drought monitoring method integrating MODIS and TRMM data. *Int. J. Appl. Earth Obs. Geoinf.* **2013**, *23*, 245–253. [CrossRef]
39. Mu, Q.; Zhao, M.; Running, S.W. Improvements to a MODIS global terrestrial evapotranspiration algorithm. *Remote Sens. Environ.* **2011**, *115*, 1781–1800. [CrossRef]
40. Li, Q.; Cao, Y.; Miao, S.; Huang, X.J.L. Spatiotemporal characteristics of drought and wet events and their impacts on agriculture in the Yellow River Basin. *Land* **2022**, *11*, 556. [CrossRef]

atmosphere

Article

Drought Vulnerability Assessment of Winter Wheat Using an Improved Entropy–Comprehensive Fuzzy Evaluation Method: A Case Study of Henan Province in China

Binbin Yuan, Shidong Wang * and Linghui Guo

School of Surveying and Engineering Information, Henan Polytechnic University (HPU), Jiaozuo 454003, China; ybb0403@163.com (B.Y.); guolinghui@hpu.edu.cn (L.G.)
* Correspondence: wsd0908@163.com

Abstract: The percentage precipitation anomaly was used to index the effect of drought on winter wheat grown in Henan Province for the years 2011–2020. Of interest was the effect of drought on winter wheat yield and the accurate assessment of the damage done to winter wheat by drought events in order to improve the risk management of winter wheat in the context of drought hazards. The spatial and temporal variability of winter wheat drought risk in Henan Province was determined by analysis of climate data, winter wheat yield, cultivated area, and socio-economic data across three dimensions: exposure or susceptibility to drought, economic–environmental sensitivity to drought, and capacity to resist drought. A drought vulnerability assessment model, based on the entropy value method and a comprehensive fuzzy evaluation, was developed to assess the drought vulnerability of winter wheat in Henan Province compared with the percentage precipitation anomaly model. (1) There were significant spatial differences in the frequency of the five drought categories devised. (2) Areas in which there was a high frequency of mild drought events were mainly in northern and western Henan and southwestern Henan, with the frequency ranging from 17% to 29%. (3) Areas in which there was a high frequency of moderate drought events were mainly in northwestern, central, and southeastern Henan. (4) Areas in which there was a high frequency of severe and extreme drought were mainly in Anyang in northern Henan, Zhengzhou in central Henan, and Xinyang and surrounding areas in southern Henan, with the frequency ranging from 7% to 9.70%. (5) Winter wheat drought vulnerability shows an overall annually increasing trend. The susceptibility dimension had the greatest influence of the three dimensions, followed by economic–environmental sensitivity and then drought resistance, which had the least impact. The model created in this study shows the influence of drought on winter wheat production more intuitively than a conventional fuzzy synthesis, and the results can inform decision-making in winter wheat drought risk assessment and management.

Keywords: Henan Province; winter wheat; drought vulnerability; fuzzy integrated evaluation; risk zoning

Citation: Yuan, B.; Wang, S.; Guo, L. Drought Vulnerability Assessment of Winter Wheat Using an Improved Entropy–Comprehensive Fuzzy Evaluation Method: A Case Study of Henan Province in China. *Atmosphere* **2023**, *14*, 779. https://doi.org/10.3390/atmos14050779

Academic Editors: Jinping Liu, Quoc Bao Pham, Arfan Arshad and Masoud Jafari Shalamzari

Received: 22 March 2023
Revised: 23 April 2023
Accepted: 24 April 2023
Published: 25 April 2023

1. Introduction

Droughts are characterized by their duration, frequency of occurrence, and the area affected. The Sixth Assessment Report of the IPCC states that the frequent occurrence of extreme heat and precipitation events, in the context of global warming, has led to more complexity in the factors that influence the occurrence of droughts, and that these events have increased the frequency and intensity of droughts. In China, drought events have jeopardized economic development and agricultural production. Agricultural drought is one of the main problems faced by agricultural production in China, and studying its vulnerability is the premise of scientific responses. As a major agricultural province, winter wheat production in Henan Province is of great importance to national agricultural production.

Winter wheat is grown in Henan Province in the north China plain region. Winter wheat production in the area accounts for about 25% of national wheat production, and the region is the main grain-producing area in China [1]. Spring droughts occur in the northern part of Henan Province 30–40% of the time, and early summer droughts occur 40–50% of the time; in the most severe drought years, 70% of the province is affected by drought [2]. The delineation of drought vulnerability zones, the scientific management of drought preparation and response, and the mitigation of drought risks, together with pressure on food producers to increase yields and improve quality, are important national issues.

Agricultural droughts result from the interaction of agricultural activities with natural events, with the outcome being insufficient water available for plants and animals [3,4]. Research into agricultural drought and those vulnerable to it began in the 1990s. Two major perspectives have emerged, assessment of the vulnerability of farmers to drought at a micro-scale and assessment of regional agricultural drought vulnerability from a macro perspective. For example, Li et al. [5] analyzed the relationship between drought vulnerability and farmer behavioral response at the micro-scale using questionnaire data from farmers in the North China Plain. Brant et al. [6] analyzed the relationship between household factors and drought sensitivity among farmers in Brazil. Savari et al. [7] investigated the drought vulnerability of farmers in southeastern Iran using a mathematical model developed by Me-Bar and Valdez [8] that identified five dimensions of vulnerability (economic, sociocultural, psychological, technological environment, and infrastructure). Cheng [9] assessed agricultural drought vulnerability in Xiaogan City, Hubei Province using a fuzzy analytical hierarchy process with empirical data from a sample of farmers and created an agricultural drought index insurance model. Xie et al. [10] used weighted composite scoring of several factors to identify relationships between different farm household livelihoods and quantify societal vulnerability to drought.

Examples of the macro perspective on regional agricultural drought vulnerability include the following. Pei et al. [11] used data envelopment analysis to examine changes in agricultural drought vulnerability in China over the past 40 years. Wang et al. [12] selected 32 indicators, including ecological recharge, grain yield in the summer harvest, water-saving irrigation machinery, and dry field area, to draw a graded vulnerability zone map using principal component analysis. Li et al. [13] used game theory combined with the weighting of 10 indicators to calculate their values for five administrative regions in Guanzhong, Shaanxi Province, and quantified the contribution of each indicator to agricultural drought vulnerability. Pei [14] determined the footprint of the water cycle in Heilongjiang Province and quantified agricultural drought risk zones using an improved standardized precipitation–evapotranspiration index (SPEI). Zarei et al. [15], based on the relationship between the percent annual yield loss (AYL) of winter wheat (Triticum sativum) and three commonly used drought indices, i.e., the standardized precipitation evapotranspiration index (SPEI), reconnaissance drought index (RDI), and standardized precipitation index (SPI), evaluated the accuracy of these indices at 1-, 3-, 6-, and 12-month time scales. Based on natural disaster risk theory, Jia et al. [16] established a drought disaster risk assessment model for winter wheat in Gansu Province and carried out risk zoning for winter wheat in Gansu Province. Yan et al. [17] discussed the spatial and temporal distribution characteristics of winter wheat drought using the Z index and analyzed the effects of meteorological drought on winter wheat yield in Henan Province.

At present, researchers have various understandings of the concepts of drought and drought vulnerability. There are no standard indicators that can be selected to match the actual climate conditions in a study area, nor are there standards for creating a set of indicators or interpreting drought indicators [18–25]. Most current research that includes the evaluation and interpretation of drought indicators uses conventional weighted synthesis methods [26,27]. However, drought is not constrained by the hard boundaries of administrative divisions, and classifications of drought are necessarily arbitrary or fuzzy, and a practical assessment of drought vulnerability must recognize this. The assessment of drought vulnerability has been developed and refined in ongoing research, but it is a work in progress.

We take an approach based on the concept that agricultural drought affects a human–land system and examine winter wheat farming at the scale of municipal administrative units in Henan Province. We selected fifteen indicators that cover three dimensions of agricultural drought: exposure, sensitivity, and resistance. The indicators were grouped into three classes of risk factors: susceptibility, or degree of exposure, to drought in the drought-affected area; economic sensitivity to drought in the area; and drought resistance in the area. An overall drought vulnerability model for winter wheat in Henan Province was created using comprehensive fuzzy evaluation, and categories of drought were established and quantified to describe the overall drought vulnerability of winter wheat in Henan Province.

2. Overview of the Study Area and Data Sources

2.1. Overview of the Study Area

Henan Province occupies the middle and lower reaches of the Yellow River in central–east China, between latitudes 31°23′–36°22′ N and longitudes 110°21′–116°39′ E [28], as shown in Figure 1. The province is bordered by Anhui and Shandong to the east, Hebei and Shanxi to the north, Shaanxi to the west, and Hubei to the south. The total area of the province is 165,700 km^2. Henan Province has a continental monsoon climate that transitions from the northern subtropical zone to the warm temperate zone. Henan Province experiences four distinct seasons, with simultaneous rain and heat in the summer. The complexity and diversity of the climate combine to create frequent meteorological disaster events. The terrain is high in the west and low in the east, with the Taihang Mountains, Funiu Mountains, Tongbai Mountains, and Dabie Mountains circling the province at its north, west, and south boundaries. The Yellow Huaihai Alluvial Plain is in the east–center of the province, and the Nanyang Basin is in the southwest. The total water resource of the province is 40.353 km^3, or about 368 m^3/person. This last figure is <20% of the national average, making Henan a province with a severe water shortage [28].

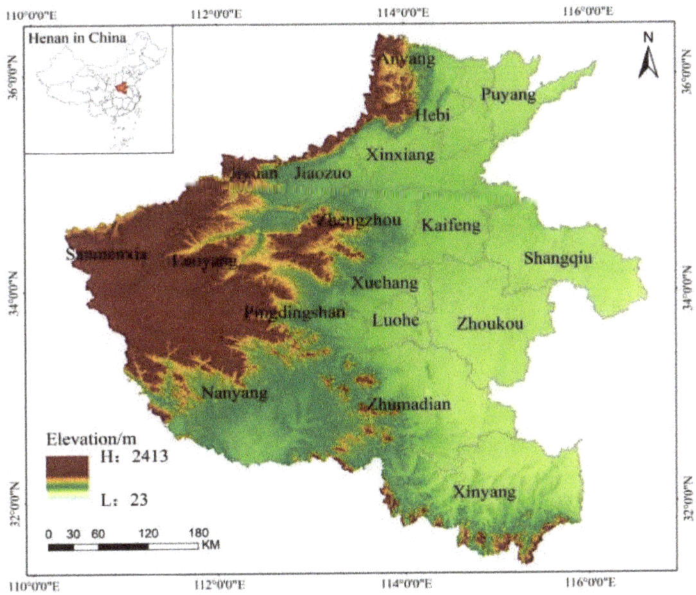

Figure 1. Map of the study area.

Henan Province is a major winter wheat-producing area in China. The national crop structure adjustment and increased market demand have expanded the winter wheat-planting area and production in the province. The published 2022 summer grain production data show that the total summer grain production in the province was 38.131 Gt, the planted area was 56,838 km^2, and the unit area yield was 6708.7 kg/ha. These figures show that winter wheat production in Henan Province is of key importance to total grain production in both the province and the country. It is therefore important to determine the spatial distribution and patterns of change in drought vulnerability in Henan to analyze the effects on winter wheat.

2.2. Data Sources and Processing

2.2.1. Data Sources

The datasets used in this study include the Raster data and Statistical data (Table 1). (1) Percentage precipitation anomaly and drought frequency for 1990–2021 were derived from monthly precipitation data from 10 meteorological observation stations in Henan Province from 1990–2021; precipitation data were obtained from the Agro-meteorological Big Data platform. (2) Topographic data at 30 m intervals in Henan Province were sourced from the Geospatial Data Cloud. (3) Socio-economic data were sourced from the Henan Provincial Statistical Yearbooks (2011–2021), including winter wheat production data, winter wheat planted area data, average annual temperature, unit area fertilizer application, winter wheat production value, per capita GDP, total primary industry share of GDP, effective irrigated area, pesticide use, rural electricity consumption, total agricultural machinery power, net farm income, and other data. Data for average annual precipitation, surface water resources, and underground water resources were obtained from the Henan Provincial Water Resources Bulletin (2011–2020).

Table 1. Data sources.

Data Type	Data Name	Years	Source
Raster data	Topographic data of Henan Province at 30 m intervals	2021	Geospatial Data Cloud
Statistical data	Socio-economic data	2011, 2014, 2017, 2020	Statistical Yearbook of Henan Province
	Average annual precipitation; surface water resources; groundwater resources	2011, 2014, 2017, 2020	Water Resources Bulletin of Henan Province
	Monthly precipitation data	1991–2021	Agro-meteorological big data platform

2.2.2. Data Normalization

Data normalization is used in cases where there are many factor indicators. Normalization makes the indicator data dimensionless, and the data are mapped onto the interval [0, 1] for subsequent calculations and analysis. We classified indicators as positive or negative, depending on their effect on drought vulnerability. Positive indicators were positively correlated with drought vulnerability and negative indicators were negatively correlated with drought vulnerability.

The equation for normalizing positive indicators is

$$D_{ij} = \frac{A_{ij} - min_i}{max_i - min_i} \tag{1}$$

The equation for normalizing negative indicators is

$$D_{ij} = \frac{max_i - A_{ij}}{max_i - min_i} \tag{2}$$

where j is the municipality; D_{ij} is the normalized value of indicator I for municipality j; A_{ij} is the value of indicator i for municipality j; and min_i and max_i are the minimum and maximum values of indicator i, respectively.

3. Research Methodology

3.1. Drought Vulnerability Assessment Indicator System for Winter Wheat

3.1.1. Selection of Assessment Indicators

Research into drought, drought vulnerability, and drought assessment is growing, and methods of quantifying drought factors are improving. Current methods include the standardized precipitation index (SPI) based on precipitation variability, the percentage precipitation anomaly, the Palmer drought severity index (PDSI) based on water demand status, the Z index, and various drought indexes that are combined with remote sensing and GIS data [29–32]. The percentage precipitation anomaly (Pa) is a traditional drought monitoring index, which shows the long-term average or normal precipitation percentage [33]. It is a representation of drought caused by precipitation anomalies when only precipitation is considered and is therefore widely used in drought monitoring and assessment [34]. Compared to other drought indices, such as SPI, SPEI (standard precipitation evapotranspiration index), and PDSI, the percentage of precipitation anomalies has the advantage of requiring simple information for calculation, being easily accessible and easy to calculate, as well as providing a better description of the degree of drought.

(1) Percentage precipitation anomaly

The difference between the precipitation in a particular year or month and the average for the year or month is known as the percentage precipitation anomaly (Pa). It is an important indicator of regional climate. A greater value of Pa indicates a greater deviation from the average for the year or month and thus a greater vulnerability to agricultural drought [28]. Pa is calculated by

$$PA = \frac{P - \overline{P}}{\overline{P}} \times 100\% \tag{3}$$

where PA is the percentage precipitation anomaly for a given period, P is precipitation for the period (mm), and $\overline{P}$ is average precipitation for the same period (mm), which is calculated by

$$\overline{P} = \frac{1}{n}\sum_{i=1}^{n} P_i \tag{4}$$

where the time range n is generally chosen to be 30 d (or some number of months or years), with $n = 31$ in this study; and P_i is precipitation (mm) for day, month, or year i. The P_a (annual scale) classifications used in this paper, using the national meteorological rating standards, are shown in Table 2.

Table 2. Drought classification based on percentage precipitation anomaly.

Class	Drought Type	Percentage Precipitation Anomaly Range (%) (Annual Scale)
1	No drought	>−15
2	Light drought	−15–−30
3	Moderate drought	−30–−40
4	Severe drought	−40–−45
5	Extreme drought	<−45

(2) Frequency of drought

The frequency of drought occurrence in a given time period is the number of months of drought at a site in the time period as a percentage of the total number of months in the time period that is used to calculate Pa. The frequency is often used to quantify the extent of the impact of drought in a given time period. The calculation is

$$F_i = \frac{N_i}{M_i} \times 100\% \tag{5}$$

where F_i is the total number of months or years in which a certain type of drought (light, moderate, severe, or extreme) occurred at weather station i, and M_i is the total number of months or years for which meteorological data was recorded at meteorological station i.

3.1.2. Indicator System

Winter wheat in Henan Province was the research object. We analyzed the spatial and temporal drought vulnerability of winter wheat in Henan Province in terms of severity using the percentage precipitation anomaly. Using drought vulnerability and drawing from related studies [13–19], we identified fifteen basic indicators of winter wheat vulnerability in Henan Province. The indicator selection process ensured the comprehensiveness, systematization, and operability of indicators, and selection was based on the analysis of indicator correlation with winter wheat growth characteristics along three dimensions (susceptibility to drought, environmental–economic sensitivity, and drought resistance). The fifteen indicators were as follows: area planted with winter wheat, yield of winter wheat, average annual temperature, average annual precipitation, fertilizer application per unit area, production value of winter wheat, per capita GDP, urbanization rate, total primary industry share of GDP, surface water resources, underground water resources, effective irrigation area, pesticide use, rural electricity consumption, total agricultural machinery power, and net farmer income. The indicator values were analyzed along the three dimensions, which were, in turn, each analyzed in terms of their spatial distribution and growth patterns of winter wheat in Henan Province. The fifteen indicators were weighted using the entropy weighting method in order to classify and analyze each dimension (Table 3).

Table 3. Drought vulnerability index system for winter wheat in Henan Province.

Target Level	Dimension	Indicator	Characteristic
Drought vulnerability	U1. Degree of susceptibility, or exposure, of the drought-affected area	Wheat planted area	+
		Wheat yield	+
		Average annual temperature	+
		Average annual precipitation	−
	U2. Environmental–economic sensitivity of the drought-affected area	Winter wheat production value	+
		Per capita GDP	−
		Urbanization rate	−
		Percentage of total primary industry (GDP)	+
	U3. Drought resistance of the affected area	Surface water resources	−
		Amount of underground water resources	−
		Effective irrigated area	−
		Amount of pesticide use	−
		Rural electricity consumption	−
		Total power of agricultural machinery	−
		Net farmer income	+

Note: + is a positive indicator, − is a negative indicator.

3.1.3. Determination of Indicator Weights

Entropy weighting uses normalized data to determine the degree of dispersion of an indicator. A greater degree of dispersion indicates less entropy in the indicated information and therefore a greater influence of the indicator on the outcome. We present an improved method for weighting each indicator using entropy weighting and comprehensive fuzzy evaluation. A comprehensive fuzzy evaluation model was created and used to predict winter wheat in Henan Province. The results show that this method had high accuracy and great reliability, and it provides an objective method of determining drought vulnerability. The calculation process is as follows.

(1) Calculate Q_{ij} using the normalized data:

$$Q_{ij} = \frac{Y_{ij}}{\sum_{i=1}^{x} Y_{ij}} \quad (6)$$

where x is the number of samples, and Q_{ij} is the weight of sample i of indicator j.

(2) Calculate the entropy value e_j for indicator j:

$$e_j = -k \sum_{i=1}^{x} Q_{ij} ln Q_{ij} \quad (7)$$

$$k = \frac{1}{lnx} > 0 \quad (8)$$

(3) Calculate the utility of indicator E_j:

$$E_j = 1 - e_j \quad (9)$$

(4) Calculate the weight Wj for indicator j:

$$W_j = \frac{E_j}{\sum_{j=1}^{z} E_j} \quad (10)$$

where z is the number of indicators.

3.2. *Comprehensive Fuzzy Evaluation-Based Drought Vulnerability Assessment Model for Winter Wheat*

A comprehensive fuzzy evaluation depends on the aggregation of indicators that contain uncertainty rather than meeting strict mathematical criteria. Fuzzy mathematics is widely used to evaluate systems, assess system effectiveness, and optimize systems. It combines qualitative and quantitative judgments [35–37]. The technical process of the model consisted of the following: firstly, determining the set of index factors and a total of 15 indicators according to the established drought vulnerability evaluation index system for winter wheat in Henan Province; secondly, ranking all indicators in order of importance by using the results of the weighting calculations; thirdly, using the fuzzy comprehensive evaluation method to determine the membership function and evaluation set; and, lastly, carrying out a normalization operation and comprehensive evaluation of the results.

The main steps of the comprehensive fuzzy evaluation process we used are as follows.

(1) Determine the set of evaluation factors that form the basis of the evaluation, which is the set consisting of u_1 (the *susceptibility* to drought of the drought-affected area), u_2 (the environmental–economic *sensitivity* of the drought-affected environment), and u_3 (the drought *resistance* of the affected area).

(2) Determine the set of evaluative criteria or grades, which are the various nonobjective evaluative judgments that an evaluator may make about the factor being evaluated. For example, the evaluative criteria we used to classify their influence on drought susceptibility, sensitivity, and resistance were v_1, *mild*; v_2, *average*; v_3, *moderate*; v_4, *severe*; and v_5, *extreme*.

(3) Create the fuzzy matrix $R_{m\times n}$ that consists of the fuzzy membership functions that map each evaluation factor (step 1) onto the set of evaluative grades (step 2):

$$R_{m\times n} = (r_{ij}) = \begin{bmatrix} r_{11} & \cdots & r_{1n} \\ \vdots & \ddots & \vdots \\ r_{m1} & \cdots & r_{mn} \end{bmatrix} \quad (11)$$

where r_{ij} is the membership degree of factor u_i ($i = 1, \ldots, m$) to the evaluative grade j (e.g., v_j; $j = 1, \ldots, n$). Each row of R therefore represents the (sub)set of degrees of membership of factor u_i in the set of evaluative grades.

(4) Determine the weight vector W, which consists of the set of evaluation factor weights and represents the weight that each factor accounts for. The weight assigned to a factor represents the importance of the factor in influencing the outcome and it therefore has a significant effect on the final assessment. The weights of the susceptibility, sensitivity, and resistance indicators are determined using the entropy method.

$$B = (a_1, a_1, \cdots a_n) \tag{12}$$

where a_i ($i = 1, \ldots, m$) is the weight of each evaluation factor value in the overall evaluation.

(5) A fuzzy operator is selected, and the evaluations are calculated. We used the weighted average operator, and the comprehensive fuzzy evaluation is given by

$$B = W \cdot R = (b_1, b_2, \cdots b_n) \tag{13}$$

where b_i is the degree of membership of the evaluated object in a fuzzy subset of the set of all v_j.

(6) Analyze the comprehensive evaluation vector B. Determine the rank of the evaluation object according to the principle of maximum subordination.

3.3. Drought Vulnerability Model Testing for Winter Wheat

A greater risk of drought indicates a greater susceptibility index, which, in turn, indicates a greater likelihood of a reduction in winter wheat yield. To validate the winter wheat drought vulnerability model, we compared the average yield reduction of winter wheat due to drought in Henan Province from 2010 to 2020 with the average drought vulnerability index calculated by the model for the same period. Crop yields can be interpreted as trends in fluctuating yields, with technological advances and improvements in agricultural production techniques being the main reasons for annual increases in winter wheat yields, and uncertain conditions (mainly meteorological hazards) causing fluctuations in yields, i.e., the climate yield of the crop. The variation in winter wheat yield was used to represent the climate yield of winter wheat in Henan Province [38], and the data were normalized. The relative climate yield of winter wheat is calculated by

$$Y_w = Y - Y_t \tag{14}$$

where Y_w is the relative climate yield of winter wheat, Y is the actual yield for the year, and Y_t is the linearly fitted yield calculated by linear fitting of the winter wheat yield for 2010–2020.

4. Results and Analysis

4.1. Assessment of the Characteristics of the Climate Drought Index for Winter Wheat in Henan Province

The 31 d Pa from 1991 to 2021 was calculated for each meteorological station in Henan Province using Equations (3) and (4). The frequency of drought occurrence for each meteorological station was calculated using the drought classification based on percentage precipitation anomaly (Table 2) using Equation (5). Pa was used to determine the type of drought for winter wheat in Henan Province in different years and to identify the characteristics of winter wheat drought. In order to ensure that the meteorological stations can scientifically reflect the meteorological situation in Henan Province, this paper selects an out-of-province meteorological station at the border of Henan Province in the eastern part of the province where meteorological stations are lacking (Figure 2). We used the drought frequency data from these ten meteorological stations as a basis and the inverse distance weighting method to spatially interpolate the missing cities to obtain a raster surface for Henan Province.

The frequency of the occurrence of light droughts varied significantly across space (Figure 3). Areas with high frequencies of light droughts (Figure 3a) were concentrated in northern and western Henan and southwestern Henan, where the frequency ranged from 17% to 29%. The maximum frequencies (29%) were in Anyang, Sanmenxia, and Zhumadian. In Shangqiu, parts of Nanyang, and Xinyang, the frequency ranged from 13–17%, and parts of Zhengzhou and Zhoukou also had a high frequency of light droughts. Figure 3b shows that areas with high values of moderate drought frequency were concentrated in northwest central and southeast Henan. Zhengzhou, Luoyang, Xinyang, and parts of Zhumadian had the highest frequencies (9.70%). In contrast, Anyang and Sanmenxia were areas of low incidence of moderate droughts with frequency 0–1.80%. Figure 3c shows that areas with high values for the frequency of severe and extreme droughts, with frequencies in the range 7–9.70%, were concentrated in and around Anyang, Zhengzhou, and Xinyang. The frequency of droughts in Sanmenxia, Luoyang, Nanyang, and Zhumadian was also high, showing an alternating distribution, but the frequency of severe and extreme droughts in other areas was low. In general, the frequency of light and severe droughts was greater in northern and southern Henan and less in central and eastern Henan; the frequency of medium droughts was greater in southwestern Henan and less in eastern and northern Henan.

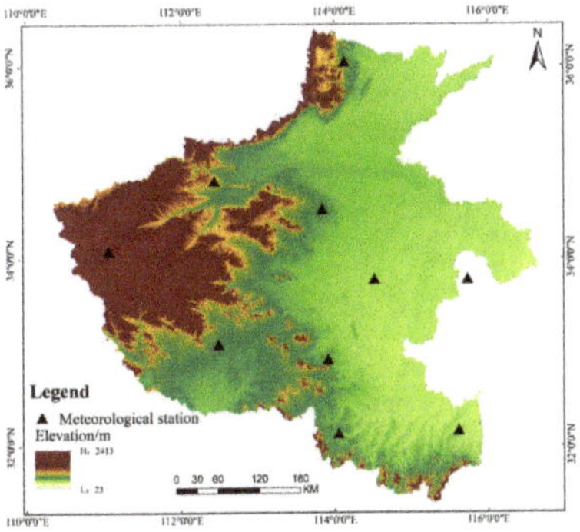

Figure 2. Map of meteorological stations.

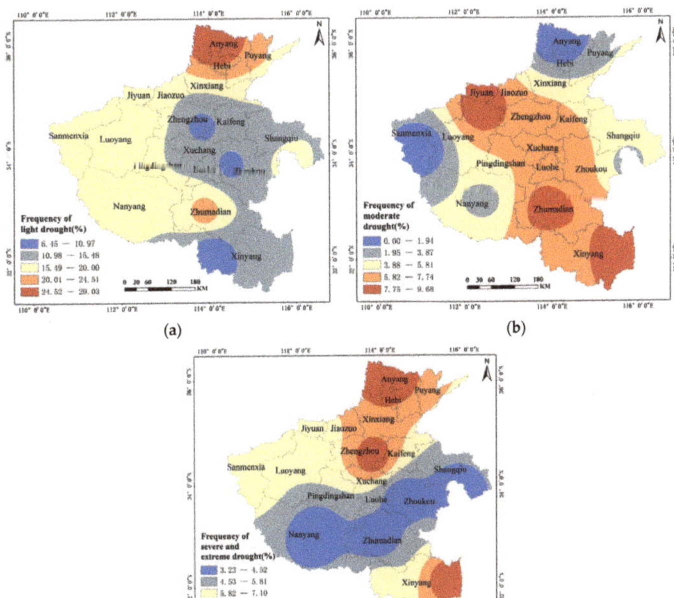

Figure 3. Frequency distribution of winter wheat drought in Henan Province. (**a**) Frequency of light drought. (**b**) Frequency of moderate drought. (**c**) Frequency of severe and extreme drought.

4.2. Drought Susceptibility Assessment and Zoning for Winter Wheat in Henan Province

4.2.1. Susceptibility Analysis of Winter Wheat in Henan Province

The susceptibility of winter wheat to drought can be used to determine winter wheat yield reduction due to drought damage. Winter wheat planted area, winter wheat yield, average annual precipitation, and average annual temperature were selected as indicators of susceptibility. The natural breaks method of ArcGIS software was used to map susceptibility zonally.

There was no significant change in the distribution of high or higher susceptibility during 2011–2020 (Figure 4). Higher susceptibility areas were concentrated mainly in the southern and eastern regions of Henan, such as Nanyang, Zhumadian, Zhoukou, and Shangqiu, and areas of medium susceptibility were Xinyang and Kaifeng, Xinxiang, Anyang, and Jiyuan. Susceptibility was generally low in northern Henan. When rainfall and acreage indicators were taken into account, in plain areas such as Zhumadian, a greater winter wheat planted area and greater winter wheat yield indicated greater susceptibility to extreme precipitation and warming events. Figure 4b shows a significant change in susceptibility due to the severe drought that occurred in 2014. Yields were generally low in northern Henan. The region has high precipitation, and when extreme precipitation or drought events are frequent, winter wheat yields in areas of high winter wheat planting, such as Anyang and Pingdingshan, may experience severe yield reductions.

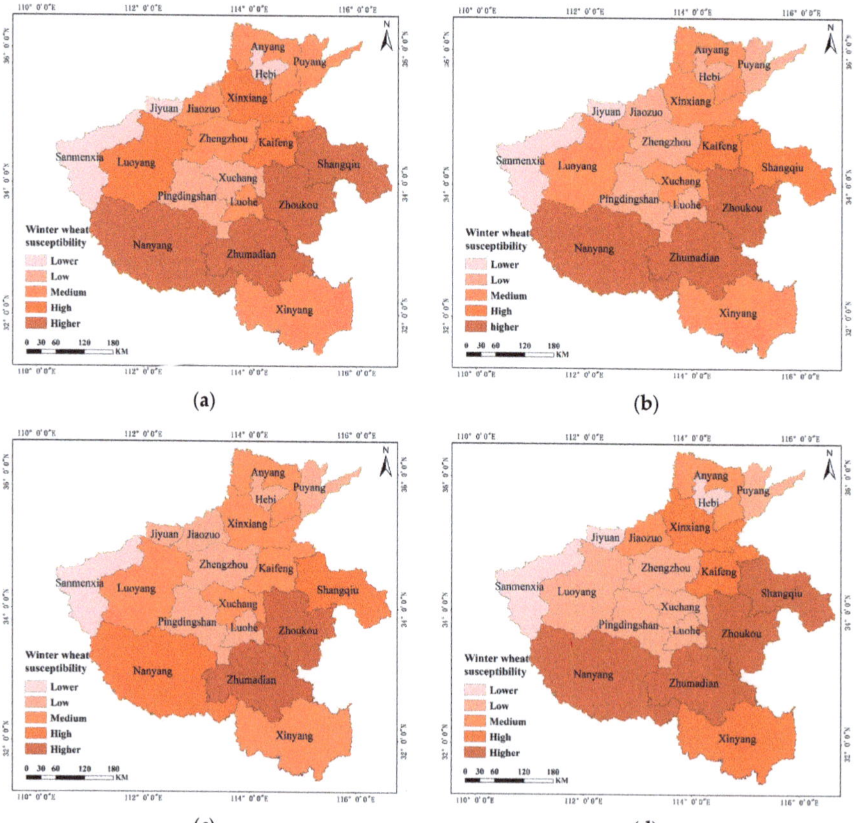

Figure 4. Distribution of winter wheat susceptibility in Henan Province. (**a**) Winter wheat susceptibility in 2011. (**b**) Winter wheat susceptibility in 2014. (**c**) Winter wheat susceptibility in 2017. (**d**) Winter wheat susceptibility in 2020.

4.2.2. Sensitivity of Winter Wheat in Henan Province

Sensitivity represents the extent of a regional response to drought and expresses the degree of dependence on agriculture in the region. When an agricultural drought occurs, a greater dependence on agriculture will result in greater sensitivity and greater vulnerability to drought.

Figure 5 shows that during the period 2011–2020, sensitivity was generally high and showed a decreasing trend year by year. Zhengzhou had low sensitivity, as did Jiaozuo, Luoyang, and Jiyuan, but other regions had medium or high sensitivity. The northern, western, and southern parts of Henan are surrounded by the Taihang, Funiu, Tongbai, and Dabie mountains around the provincial boundary and are prone to droughts, so areas such as Xinxiang, Anyang, Sanmenxia, Xinyang, and Nanyang, which are in mountainous and hilly areas, have a greater sensitivity than other areas. Luoyang and Jiyuan are in hilly areas but occupy the middle and lower reaches of the Yellow River and contain well-developed river networks, so they have the lowest sensitivity. Puyang is in the alluvial plain of the Yellow River and has a well-developed water net system, so it has low sensitivity. Zhengzhou is the capital city of the province and has a more developed economy than the rest of the province such as a higher urbanization rate, a higher per capita GDP and abundant surface and groundwater resources. Its low percentage of total primary industry (GDP) makes it less dependent on agriculture and less sensitive to drought. The Pingdingshan, Xuchang, Luohe, Kaifeng, Shangqiu, Zhoukou, and Zhumadian regions are located in the plains and have a high percentage of total primary industry (GDP), i.e., they are more dependent on agricultural development. Moreover, the value of winter wheat production is high in these areas, but yield losses due to droughts are also high, so their sensitivity is also high.

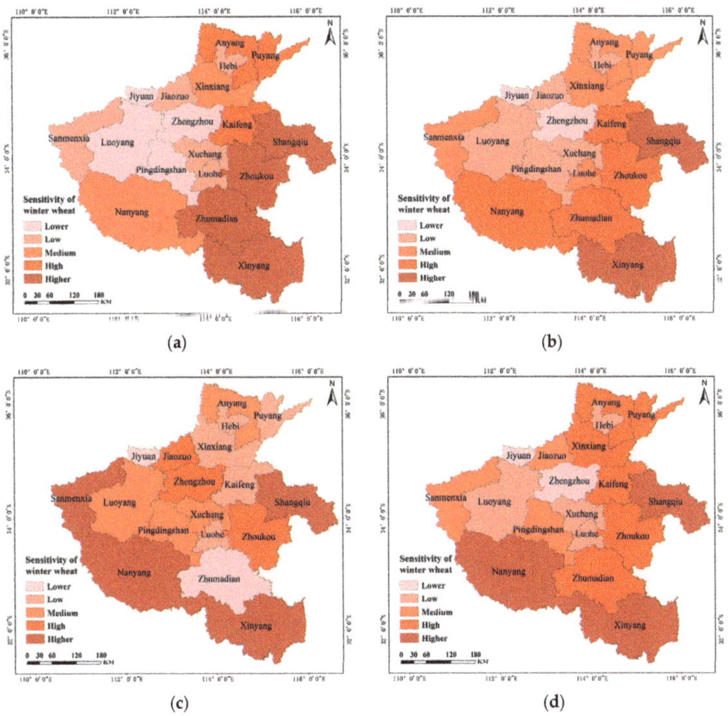

Figure 5. Distribution of winter wheat sensitivity in Henan Province. (**a**) Sensitivity of winter wheat in 2011. (**b**) Sensitivity of winter wheat in 2014. (**c**) Sensitivity of winter wheat in 2017. (**d**) Sensitivity of winter wheat in 2020.

4.2.3. Winter Wheat Drought Resistance

As research on drought has increased, resistance has become more closely linked to socio-economic concerns. Generally, greater resistance to drought in a region indicates less loss directly attributable to drought. Drought resistance is therefore inversely related to the risk of drought, and different equations are used to normalize resistance index data. Regions with lower resistance indexes are thus more likely to be resistant to disasters than regions with higher resistance indexes.

Figure 6 shows that drought resistance of winter wheat was generally high in Henan Province during 2011–2020. Drought resistance increased gradually over the period and showed an overall pattern of large areas of similar resistance and small areas of mixed resistance, but the resistance of plain areas was greater than that of hilly areas. Drought resistance was greater in Zhoukou, Shangqiu, Zhumadian, and Nanyang than in Hebi, Sanmenxia, Pingdingshan, Luohe, and Xinyang, where it was low. Zhengzhou is in a hilly area but has a high resistance, mainly because of its more developed economy and greater investment in agricultural irrigation facilities. In contrast, Hebi and Jiyuan are in the plain of the lower reaches of the Yellow River and have more developed water systems, so there is no need to greatly invest human and material resources into drought relief, which would lower the resistance to drought.

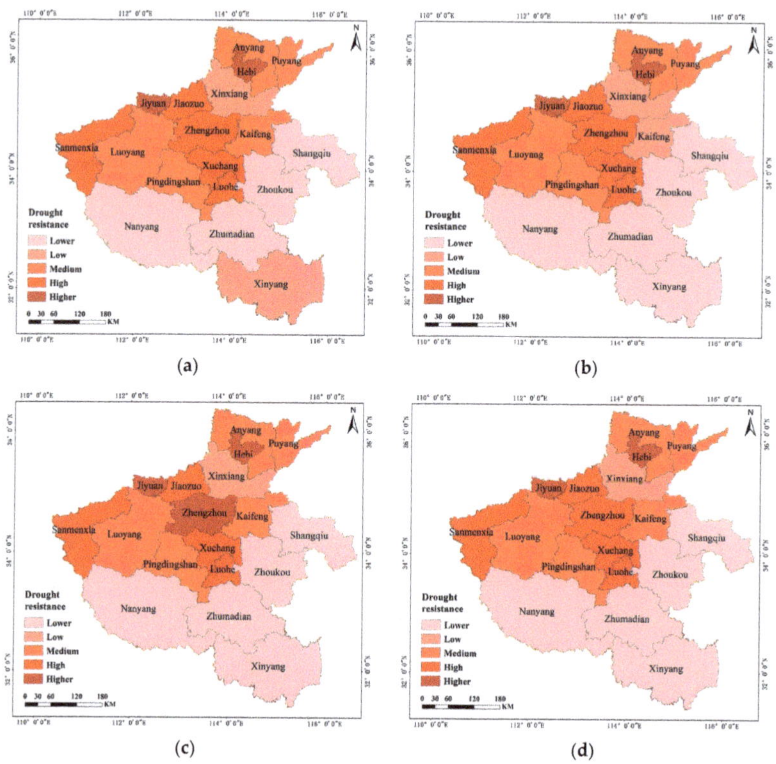

Figure 6. Distribution of drought resistance of winter wheat in Henan Province. (**a**) Drought resistance in winter wheat in 2011. (**b**) Drought resistance in winter wheat in 2014. (**c**) Drought resistance in winter wheat in 2017. (**d**) Drought resistance in winter wheat in 2020.

4.2.4. Classification of Drought Vulnerability of Winter Wheat in Henan Province

Data for the indicators shown in Table 3 were obtained, and the values of the vectors for indicator weights at the dimension level were calculated using the entropy weighting method. The fuzzy relationship matrix, weight vector, and fuzzy comprehensive evaluation were calculated using Equations (11)–(13), and the fuzzy transformation was used to predict the winter wheat drought vulnerability index in Henan Province from 2011 to 2020. The natural point interval method and reclassification functions of ArcGIS were used to weight the winter wheat drought vulnerability index of Henan Province.

The comprehensive fuzzy evaluation was conducted using SPSS software using the weighted mean type M(*,+) operator. The normalized weights of the degrees of membership of the three factor sets were obtained: 0.38, 0.36, and 0.26, as shown in Table 4. It can be seen from the table that the greatest weight was for susceptibility, indicating that susceptibility of an area to drought has the greatest influence on the winter wheat drought vulnerability index.

Table 4. Weights of drought vulnerability evaluation indicators for winter wheat in Henan Province.

Guideline Level	Weight
Susceptibility to drought	0.38
Environmental–economic sensitivity	0.36
Drought resistance	0.26

Table 5 shows the calculated area for each class of drought vulnerability of winter wheat in Henan Province between 2014 and 2017. In 2011, the cities with mild vulnerability to winter wheat drought in Henan Province were Pingdingshan, Sanmenxia, and Jiyan; the cities with average vulnerability were Zhengzhou, Hebi, Jiaozuo and Xuchang; the cities with moderate vulnerability were Luohe and Luoyang; the cities with severe vulnerability were Anyang, Xinxiang, Puyang, and Xinyang; and the cities with extreme vulnerability were Kaifeng, Shangqiu, Zhoukou, Zhumadian, and Nanyang. In 2014, the cities with mild vulnerability to winter wheat drought in Henan Province were Zhengzhou; cities with average vulnerability were Pingdingshan, Jiaozuo, Puyang, Sanmenxia, and Jiyuan; cities with moderate vulnerability were Luoyang, Hebi, Xuchang, and Luohe; cities with severe vulnerability were Anyang, Xinxiang, Shangqiu, and Xinyang; cities with extreme vulnerability were Kaifeng, Zhoukou, Zhumadian, and Nanyang. In 2017, the cities in Henan Province with mild vulnerability to winter wheat drought were Puyang; those with average vulnerability were Xinxiang, Zhumadian, and Jiyuan; those with moderate vulnerability were Kaifeng, Pingdingshan, and Luohe; those with severe vulnerability were Luoyang, Anyang, Hebi, Jiaozuo, Xuchang, Sanmenxia, and Xinyang; and those with extreme vulnerability were Zhengzhou, Shangqiu, Zhoukou, and Nanyang. In 2020, cities in Henan Province with mild vulnerability to winter wheat drought include Zhengzhou and Jiyuan; cities with average vulnerability include Luoyang, Hebi, Xuchang, Luohe, and Sanmenxia; cities with moderate vulnerability include Pingdingshan, Jiaozuo, and Puyang; cities with severe vulnerability include Kaifeng, Anyang, Xinxiang, Zhoukou, Zhumadian, and Xinyang; and cities with extreme vulnerability include Shangqiu and Nanyang.

Table 5. Winter wheat drought vulnerability by area in Henan Province (km^2).

Year	Areas of Mild Vulnerability	Areas of Average Vulnerability	Areas of Moderate Vulnerability	Areas of Severe Vulnerability	Areas of Extreme Vulnerability
2011	20,309	18,774	17,847	38,849	70,521
2014	7567	28,651	24,983	64,378	59,817
2017	4271	25,263	16,765	63,262	56,739
2020	9498	35,479	16,224	67,886	37,213

The table shows that the overall area of areas at all levels of drought risk for winter wheat in Henan Province decreased year by year during the period 2014–2017. This indicates that the government is beginning to pay attention to the issue of drought vulnerability of winter wheat. Cities in areas of extreme vulnerability, such as Kaifeng, Zhoukou, and

Zhumadian, gradually became cities in areas of severe vulnerability. This is shown by the area of extreme vulnerability decreasing to 37,213 km^2 in 2020 and the total area of severe vulnerability increasing to 67,886 km^2 in 2020. Cities in areas of moderate vulnerability gradually become cities in areas of average vulnerability, with the total area of moderate vulnerability decreasing to 16,224 km^2, and the area of average vulnerability increased to 35,479 km^2; cities in areas of average vulnerability gradually become cities in areas of mild vulnerability, with the area of mild vulnerability increasing from 4271 km^2 in 2017 to 9498 km^2 in 2020.

Figure 7 shows that the drought vulnerability index for winter wheat in Henan Province showed an annually increasing trend during the period 2011–2020. The category of vulnerability varied little, with most areas being average or moderate vulnerability. There was a real contiguity of vulnerability classes, as shown by Pingdingshan, Xuchang, and Luohe in central Henan and parts of Xinyang and Zhumadian in southern Henan. Severe and extremely vulnerable areas were scattered in hilly and plain areas, with the exception of Hebi and Puyang, and most such areas were in the hilly transition zone from the second to the third terraces. Mild vulnerability in southern Henan was due to the better-developed water system in the region and the warmer climate with abundant precipitation. The higher level of vulnerability in eastern Henan was due to the large area of winter wheat cultivation in the plains, in which vulnerability was mainly influenced by low drought resistance due to poor farmland infrastructure. The climate of northern Henan is drier and, although there are more natural water systems in that region, precipitation is lower, and there has been more investment in agricultural irrigation facilities to reduce the impact of drought on winter wheat yields.

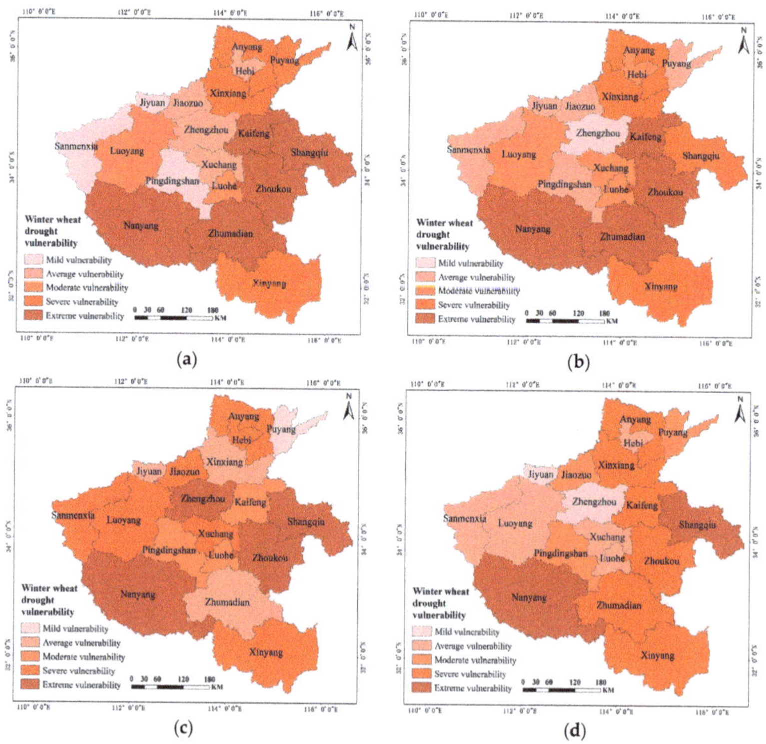

Figure 7. Distribution of levels of winter wheat drought vulnerability in Henan Province. (**a**) Winter wheat drought vulnerability in 2011. (**b**) Winter wheat drought vulnerability in 2014. (**c**) Winter wheat drought vulnerability in 2017. (**d**) Winter wheat drought vulnerability in 2020.

The distribution of winter wheat drought vulnerability classes was compared with the distribution of the winter wheat susceptibility index, economic–environmental sensitivity index, and drought resistance index. We found that the distribution of winter wheat drought vulnerability classes and the distribution of the winter wheat susceptibility index were essentially consistent. This is the same as the analysis presented in Table 4, where the winter wheat susceptibility index has a direct effect on winter wheat drought vulnerability. This result is almost the same as the results shown in the drought sensitivity graph for winter wheat yield in Henan Province produced by Wu et al. [2].

4.2.5. Validation of the Drought Vulnerability Model

The standardized values of the average winter wheat yield for 2010–2020 for each prefecture-level city in Henan Province were correlated with the average drought vulnerability index calculated by the drought vulnerability model and were then fitted using linear regression (Figure 8). The correlation coefficient R^2 obtained based on the improved entropy–fuzzy evaluation method is 0.44, specifically the drought vulnerability index explains 44% of the fluctuating yield of winter wheat. In contrast, a correlation analysis and fitting based on the traditional fuzzy integrated evaluation method yielded an R^2 of 0.26. The results show that the winter wheat drought vulnerability index in this study can effectively evaluate the drought vulnerability of winter wheat in Henan Province.

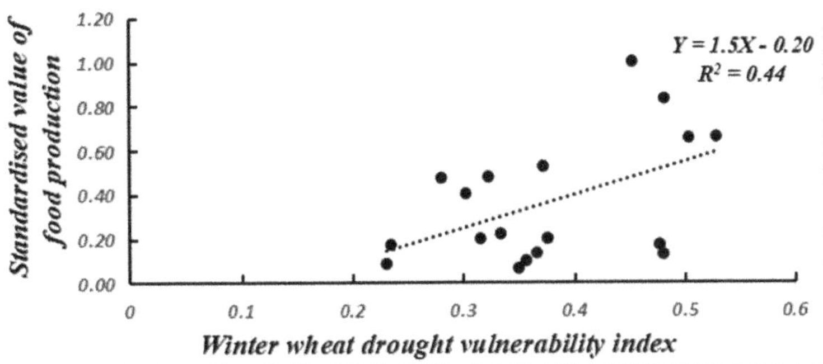

Figure 8. Relationship between average climate yield and average drought vulnerability index for winter wheat in Henan Province 2010–2020.

5. Conclusions

We used climate data, winter wheat production data, wheat cultivation area data, and socio-economic data of Henan Province to analyze the spatial and temporal variability of winter wheat drought vulnerability in Henan Province using the Pa index. We created a winter wheat drought vulnerability assessment model using the entropy method and comprehensive fuzzy evaluation for three dimensions of influence on winter wheat drought vulnerability: susceptibility to drought, economic–environmental sensitivity to drought, and drought resistance capacity. The principal conclusions are as follows.

(1) The determination of Pa and analysis of the different degrees or levels of drought effects on winter wheat in Henan Province showed that the frequency of occurrence of different levels of drought varied significantly spatially. The frequency of mild droughts ranged from 9.70% to 29%, the frequency of moderate droughts ranged from 0 to 9.70%, and the frequency of severe or extreme droughts ranged from 3.20% to 9.70%. Droughts have become more frequent in Henan Province over the study period; mild and extreme droughts occurred more often in the north, while moderate and severe droughts occurred more often in the south.

(2) The comprehensive fuzzy evaluation produced normalized weights of the susceptibility, sensitivity, and resistance dimensions that were, respectively, 0.38, 0.36, and

0.26. The susceptibility of winter wheat to drought has a large influence on the winter wheat drought vulnerability index. Areas of high susceptibility were concentrated in southern and eastern Henan, with susceptibility indices ranging from 0.59 to 0.80. Areas of high economic–environmental sensitivity were concentrated in parts of east, south, and west Henan, with sensitivity indices ranging from 0.52 to 0.70. High sensitivity was mainly due to the greater risk of drought in mountainous areas and the greater rate of winter wheat cultivation in the plain. Areas of higher drought resistance for winter wheat were concentrated around Zhengzhou and Hebi, with drought tolerance indices ranging from 0.44 to 0.62, due mainly to better conditions for agricultural production.

(3) There were also temporal changes between 2011 and 2020. Areas in Henan Province that were severely or extremely vulnerable showed a scattered distribution in 2011 and 2014 that became more blocked in 2017 and 2020. The distribution of winter wheat drought vulnerability classes' change indicated a trend towards a lower drought vulnerability index for winter wheat in Henan Province. The vulnerability to drought of winter wheat in Henan Province from 2011 to 2020 varied relatively little, with most areas showing average or moderate vulnerability, mainly due to local topography (higher elevations were more susceptible to drought), but well-developed farmland and water conservancy facilities with a well-developed social economy increased drought resistance, so the risk of drought remained low.

We have found that the actual drought losses are influenced by economic levels, irrigation facilities, and other conditions, and that there are significant regional or household differences. This provides basic theoretical support for the selection of wheat varieties, the development of disaster prevention and mitigation measures, and the risk zoning and control of winter wheat. The drought vulnerability model in this paper only selects some indicators to characterize drought exposure, sensitivity, and adaptation to winter wheat, which will result in a less precise zoning assessment. The next step is to further improve and optimize the assessment model by taking into account the physiological characteristics of winter wheat and specific on-farm production processes.

Author Contributions: Conceptualization, S.W.; data curation, L.G.; methodology, L.G. and S.W.; project administration, S.W.; data processing, B.Y.; writing—original draft, B.Y.; writing—review and editing, B.Y. All authors have read and agreed to the published version of the manuscript.

Funding: This research was jointly supported by the National Natural Science Foundation of China Key Projects (Grants No. U22A20620, Grants No. U21A20108) and the Natural Science Foundation of Henan Polytechnic University (Grant No. B2023-21).

Institutional Review Board Statement: Not applicable.

Informed Consent Statement: Not applicable.

Data Availability Statement: The data used in this study are available from the first author upon reasonable request.

Acknowledgments: The research was supported by the National Natural Science Foundation of China Key Projects (Grants: U22A20620, U21A20108), the Natural Science Foundation of Henan Polytechnic University (Grant: B2023-21). We would also like to express our respect and thanks to the anonymous reviewers and the editors for their helpful comments in improving the quality of this paper.

Conflicts of Interest: The authors declare no conflict of interest.

References

1. Chai, L.; Zhu, H. Evolutionary Trends of Regional Centralization of Grain Production in China. *J. Nat. Resour.* **2016**, *31*, 908–919.
2. Wu, H.; Guo, J.; Zheng, P. Drought Vulnerability of Winter Wheat in Henan Province. *Areal Res. Dev.* **2018**, *37*, 170–175.
3. Veisi, K.; Bijani, M.; Abbasi, E. A human ecological analysis of water conflict in rural areas: Evidence from Iran. *Glob. Ecol. Conserv.* **2020**, *23*, e01050. [CrossRef]
4. Sepulcre-Canto, G.; Horion SM, A.F.; Singleton, A.; Carrao, H.; Vogt, J. Development of a combined drought indicator to detect agricultural drought in Europe. *Nat. Hazards Earth Syst. Sci.* **2012**, *12*, 3519–3531. [CrossRef]

5. Li, X.; Yang, Y.; Liu, Y.; Liu, H. Impacts and effects of government regulation on farmers' responses to drought: A case study of North China Plain. *Geogr. Sci.* **2017**, *27*, 1481–1498. [CrossRef]
6. Brant, S. *Assessing Vulnerability to Drought in Ceara North-East Brazil*; University of Michigan: Ann Arbor, MI, USA, 2007.
7. Savari, M.; Damaneh, H.E.; Damaneh, H.E. Drought vulnerability assessment: Solution for risk alleviation and drought management among Iranian farmers. *Int. J. Disaster Risk Reduct.* **2022**, *67*, 102654. [CrossRef]
8. Me-Bar, Y.; Valdez, F. On the vulnerability of the ancient Maya society to natural threats. *J. Archaeol. Sci.* **2005**, *32*, 813–825. [CrossRef]
9. Cheng, J. A Study on Agricultural Drought Hazard Vulnerability and Risk Management: A Case of Xiaogan City in Hubei Province. Ph.D. Thesis, Huazhong Agricultural University, Wuhan, China, 2011.
10. Xie, Q.; Zhao, C.; Wang, Z. Evaluation of social vulnerability to drought disaster of farmers with different livelihoods: Take Bijie City in Wumeng Mountains area as an example. *J. Nat. Disasters* **2022**, *31*, 208–218.
11. Pei, H.; Wang, X.; Fang, S. Study on temporal-spatial evolution of agricultural drought vulnerability of China based on DEA model. *J. Catastrophol.* **2015**, *30*, 64–69.
12. Wang, Y.; Wang, J.; Yao, Y.B.; Wang, J.S. Evaluation of Drought Vulnerability in Southern China Based on Principal Component Analysis. *Ecol. Environ. Sci.* **2014**, *23*, 1897–1904.
13. Li, M.; Hui, Q.; Qiao, L. Evaluation of agricultural vulnerability to drought in Guanzhong Area. *Resour. Sci.* **2016**, *38*, 0166–0174.
14. Pei, W. Research on Regional Agricultural Drought Assessment and Temporal and SPATIAL Variation: A Case of Heilongjiang Province. Ph.D. Thesis, Northeast Agricultural University, Harbin, China, 2017.
15. Zarei, A.R.; Shabani, A.; Moghimi, M.M. Accuracy Assessment of the SPEI, RDI and SPI Drought Indices in Regions of Iran with Different Climate Conditions. *Pure Appl. Geophys.* **2021**, *178*, 1387–1403. [CrossRef]
16. Jia, J.; Han, L.; Wang, X.; Liu, W. Risk and Regionalization of Drought for Winter Wheat in Gansu Province. *Arid Zone Res.* **2019**, *36*, 1478–1486.
17. Yan, Y.; He, Z.; Zhu, K. Winter Wheat in Henan: Drought Index Characteristics and Risk Assessment. *J. Agric.* **2021**, *11*, 12–19.
18. Schwartz, C.; Ellenburg, W.L.; Mishra, V.; Mayer, T.; Griffin, R.; Qamer, F.; Matin, M.; Tadesse, T. A statistical evaluation of Earth-observation-based composite drought indices for a localized assessment of agricultural drought in Pakistan. *Int. J. Appl. Earth Obs. Geoinf.* **2022**, *106*, 102646. [CrossRef]
19. Mishra, A.K.; Desai, V.R. Drought forecasting using stochastic models. *Stoch. Env. Res Risk Assess.* **2005**, *19*, 326–339. [CrossRef]
20. Hughes, N.; Soh, W.Y.; Boult, C.; Lawson, K. Defining drought from the perspective of Australian farmers. *Clim. Risk Manag.* **2022**, *35*, 100420. [CrossRef]
21. Oyea, C.; Bja, B.; Amka, C. Global quantitative and qualitative assessment of drought research from 1861 to 2019. *Int. J. Disaster Risk Reduct.* **2022**, *70*, 102770.
22. Arun Kumar, K.C.; Reddy GP, O.; Masilamani, P.; Sandeep, P. Indices-based assessment of vulnerability to agricultural drought in the tropical semi-arid ecosystem using time-series satellite and meteorological datasets. *Arab. J. Geosci.* **2022**, *15*, 1022. [CrossRef]
23. Hoque MA, A.; Pradhan, B.; Ahmed, N.; Sohel, M.S.I. Agricultural drought risk assessment of Northern New South Wales, Australia using geospatial techniques. *Sci. Total Environ.* **2021**, *756*, 143600. [CrossRef]
24. Kar, S.K.; Thomas, T.; Singh, R.M.; Patel, L. Integrated assessment of drought vulnerability using indicators for Dhasan basin in Bundelkhand region, Madhya Pradesh, India. *Curr. Sci. A Fortn. J. Res.* **2018**, *115*, 338–346. [CrossRef]
25. Sahana, V.; Mondal, A.; Sreekumar, P. Drought vulnerability and risk assessment in India: Sensitivity analysis and comparison of aggregation techniques. *J. Environ. Manag.* **2021**, *299*, 113689. [CrossRef] [PubMed]
26. Li, Z.; Wang, G.; Yang, D.; Lu, Y. Vulnerability Analysis of Agricultural Drought in Henan Province. *Chin. Agric. Sci. Bull.* **2021**, *37*, 101–106.
27. Zhang, W.; Xu, L. Study on Drought Vulnerability of Summer Maize in Henan Province Based on ArcGIS. *Res. Soil Water Conserv.* **2018**, *25*, 228–234.
28. The People's Government of Henan Province. Available online: https://m.henan.gov.cn/2018/05-31/2408.html (accessed on 20 September 2022).
29. Domínguez-Castro, F.; Vicente-Serrano, S.M.; Tomás-Burguera, M.; Peña-Gallardo, M.; Beguería, S.; El Kenawy, A.; Luna, Y.; Morata, A. High spatial resolution climatology of drought events for Spain: 1961–2014. *Int. J. Climatol.* **2019**, *39*, 5046–5062. [CrossRef]
30. Dehghan, S.; Salehnia, N.; Sayari, N.; Bakhtiari, B. Prediction of meteorological drought in arid and semi-arid regions using PDSI and SDSM: A case study in Fars Province, Iran. *J. Arid Land* **2020**, *12*, 318–330. [CrossRef]
31. Ren, Y.; Liu, J.; Shalamzari, M.J.; Arshad, A.; Liu, S.; Liu, T.; Tao, H. Monitoring Recent Changes in Drought and Wetness in the Source Region of the Yellow River Basin, China. *Water* **2022**, *14*, 861. [CrossRef]
32. Kumar, S.A.; Meenu, R.; Kumar, P.R. Changes in drought characteristics based on rainfall pattern drought index and the CMIP6 multi-model ensemble. *Agric. Water Manag.* **2022**, *266*, 107568.
33. Yang, S.E.; Wu, B.F. Calculation of monthly precipitation anomaly percentage using web-serviced remote sensing data. In Proceedings of the 2010 2nd International Conference on Advanced Computer Control, Shenyang, China, 27–29 March 2010; IEEE: Washington, DC, USA, 2010; pp. 621–625.
34. Li, X.; Zhao, J.; Xu, J. An Analysis of Drought Characteristics in Anhui Province during the Past 50 Years Based on Precipitation Anomaly Percentage. *China Rural Water Hydropower* **2018**, *431*, 133–136+143.

35. Wang, Q. Construction and application of evaluation index system of social and economic carrying capacity in Xiong'an New Area. *J. Environ. Eng. Technol.* **2022**, *12*, 1272–1279.
36. Li, S.; Zhang, H.; Wang, L.; Zheng, G.; Guo, Y.; Gao, J. Winter Wheat Agrometeorological Drought Zoning and Analysis in Henan Province Based on Grid and Fuzzy Clustering Algorithm. *J. Henan Agric. Sci.* **2020**, *49*, 172–180.
37. Saha, S.; Kundu, B.; Paul, G.C.; Mukherjee, K.; Pradhan, B.; Dikshit, A.; Maulud, K.N.A. Spatial assessment of drought vulnerability using fuzzy-analytical hierarchical process: A case study at the Indian state of Odisha. *Geomat. Nat. Hazards Risk* **2021**, *12*, 123–153. [CrossRef]
38. Wang, C. *Impact Assessment and Risk Evaluation of Agricultural Meteorological Hazards*; Meteorological Publishing House: Beijing, China, 2010; pp. 82–86.

atmosphere

MDPI

Article

Spatiotemporal Characteristics of Watershed Warming and Wetting: The Response to Atmospheric Circulation in Arid Areas of Northwest China

Taohui Li [1], Aifeng Lv [2,*], Wenxiang Zhang [1] and Yonghao Liu [1]

1 Key Laboratory of Plateau Geographic Processes and Environment Change of Yunnan Province, Faculty of Geography, Yunnan Normal University, Kunming 650500, China
2 Key Laboratory of Water Cycle and Related Land Surface Processes, Institute of Geographic Sciences and Natural Resources Research, CAS, Beijing 100101, China
* Correspondence: lvaf@igsnrr.ac.cn

Abstract: The Tarim Basin is a large inland arid basin in the arid region of northwest China and has been experiencing significant "warming and wetting" since 1987. As a result, the purpose of this paper is to determine whether the climate transition phenomenon occurred in the Tarim Basin as well as the role of atmospheric circulation in this process. We use meteorological data and atmospheric circulation indexes to study the seasonal trends of climate change in this region from 1987 to 2020 to understand how they are affected by atmospheric circulation. The findings show that, from 1987 to 2020, the Tarim Basin experienced significant warming and wetting; with the exception of the winter scale, all other seasonal scales exhibited a clear warming and wetting trend. From the perspective of spatial distribution, most of the areas showed a significant warming trend, and the warming amplitude around the basin is greater than that in the central area of the basin. However, there are significant regional differences in precipitation change rates. Meanwhile, wavelet analysis shows that there is a significant oscillation period of 17–20 years between climate change and the atmospheric circulation index during 1987–2020. The correlation analysis shows that the Pacific decadal oscillation (PDO) and El Niño-Southern Oscillation (ENSO) are the main influencing factors of climate change in the Tarim Basin at different seasonal scales, while the teleconnection of the Arctic Oscillation (AO) and North Atlantic Oscillation (NAO) is low and the PDO dominates the summer and autumn temperature changes in the Tarim Basin. The research results of this paper show that, despite the warming and wetting trends since 1987 in the Tarim Basin, the climate type did not change. From 1987 to 2020, the main teleconnection factors of climate change in the Tarim Basin were PDO and ENSO.

Keywords: the Tarim Basin; climate change; spatiotemporal climate characteristics; atmospheric circulation; climate response

check for updates

Citation: Li, T.; Lv, A.; Zhang, W.; Liu, Y. Spatiotemporal Characteristics of Watershed Warming and Wetting: The Response to Atmospheric Circulation in Arid Areas of Northwest China. *Atmosphere* **2023**, *14*, 151. https://doi.org/10.3390/atmos14010151

Academic Editors: Jinping Liu, Quoc Bao Pham, Arfan Arshad and Masoud Jafari Shalamzari

Received: 23 December 2022
Revised: 31 December 2022
Accepted: 5 January 2023
Published: 10 January 2023

1. Introduction

The Tarim Basin in China has a typical continental desert climate and is thus very sensitive to the climate of Central Asia and to global climate change. According to current research, the arid region of northwest China has experienced a significant increase in temperature and humidity since 1987, mainly characterized by a significant increase in the frequency of extreme precipitation events [1–5]. The Tarim Basin is the most important inland basin in the arid region of northwest China and even in the whole arid region of Central Asia. A significant increase in extreme precipitation events in this region could destabilize the annual runoff in the Tarim River Basin. Seasonally specific climate change, in particular, will have a significant impact on the Tarim Basin's ecological environment, agricultural environment, and social economy [6–8]. Therefore, given the intensification of the "warming and wetting" phenomenon in the arid region of northwest China, the question arises of how climate change is manifested on the seasonal timescale and whether it significantly affects the Tarim Basin. In addition, the Tarim River is the major river in the arid continental region of northwest China and is important for the development of these areas, so the "wetting" phenomenon, especially the seasonally specific type, affects not only the water resources of the Tarim Basin but also the development of ecological resources. Moreover, it strongly affects the economic and agricultural development of this region [9–13]. Therefore, it is vital to understand how seasonally specific climate change affects the inland basins in these arid regions.

Numerous investigations, both domestic and international, have focused on the warming and wetting phenomena. For example, Shi et al. [1] reported a climate transition occurring in northwest China in the early 21st century, and, based on meteorological data, Chen et al. and Li et al. [14,15] confirmed that northwest China was experiencing a significant warming and wetting phenomenon through trend analysis. In addition, Wu et al. and Wang et al. [2,3] showed that the significant warming and wetting phenomenon in the arid region of northwest China was caused primarily by a significant increase in extreme precipitation events in the arid region. Similarly, by studying the intensity of the humidification index in arid areas, Yang et al. and Zhang et al. [16,17] showed that precipitation varied strongly in the different areas and seasons in northwest China, and Gessner et al. [18] confirmed the result. Numerous studies of these regions thus report that the main contributor to the warming and wetting phenomena is the significant increase in the frequency of extreme precipitation, which strongly affects local ecosystems [19–21]. However, to date, more studies have focused on climate change on the interannual scale in the arid region of northwest China, whereas little research has focused on seasonally specific climate change in this region.

Previous studies have shown that atmospheric circulation plays a significant role in climate change over different time scales. For example, the El Niño Southern Oscillation (ENSO) can cause extreme hydrological events [4]. Numerous studies have investigated the relationship between climate change and atmospheric circulation in arid areas, and the results confirm that atmospheric circulation helps determine the climate in arid areas on an inter-annual timescale [22–24]. However, at present, more studies focus on the interannual timescale than on the seasonal timescale. And the seasonal timescale is important for agriculture in the Tarim Basin because agriculture in this region is mainly rainfed, making it extremely dependent on seasonal precipitation. Thus, the present study considers the seasonal timescale not only to better analyze how atmospheric circulation affects the seasonal climate in the Tarim Basin but also to determine what seasonal agricultural adjustments should be made in this region.

Thus, the purpose of this study is to ascertain if the climate type in the arid Tarim Basin of northwest China has changed since 1987 and to explain the seasonal relationship between atmospheric circulation and climate change in this area. In addition, we investigate the spatial characteristics of the seasonally specific warming and wetting phenomena in the Tarim Basin by analyzing meteorological data and atmospheric circulation indexes from 1987 to 2020, and we discuss the relationship between seasonally specific climate change and atmospheric circulation. The results reveal seasonally changing climate patterns in this area and provide a basis for understanding how atmospheric circulation affects climate change on a seasonal timescale. Finally, the research results of this paper show that, despite the warming and wetting trends since 1987 in the Tarim Basin, the climate type did not change. On the seasonal timescale, the Pacific decadal oscillation (PDO) and the El Niño-Southern Oscillation (ENSO) are the main teleconnection factors of climate change in the Tarim Basin during 1987–2020.

2. Materials and Methods

2.1. Study Area and Data Sources

The Tarim Basin is an important inland arid basin in the arid region of northwest China (its geographical location is shown in Figure 1). It has a typical continental desert climate and is located between the Tianshan and Kunlun Mountains. It is 1100 km long from east to west and is one of the largest inland basins in the world. The Tarim River system is composed of four sources and one trunk. It is one of the more complicated river systems in the inland river basin [25,26]. The Tarim River is not only the only flowing water source for animals and plants in southern Xinjiang but also the water system on which the local economy and industry depend for development [18]. In this work, meteorological data is sourced from the Climatic Research Unit TS v.4.03 (CRU) database (http://www.cru.uea.ac.uk/data/, accessed on 22 December 2022) and the data cover the period 1987–2020. The atmospheric circulation index data are from the US National Oceanic and Atmospheric Administration (http://www.esrl.noaa.gov/psd/enso/, accessed on 22 December 2022). The atmospheric circulation indexes considered herein are the Arctic Oscillation (AO) index, the El Niño-Southern Oscillation (ENSO) index, the North Atlantic Oscillation (NAO) index, and the Pacific Decadal Oscillation (PDO) index, and the data cover the period 1987–2020.

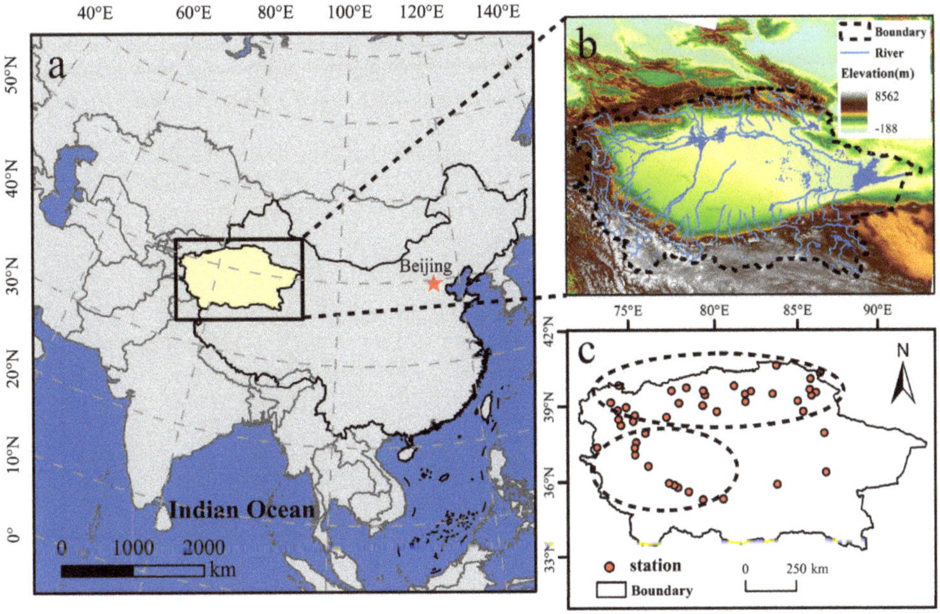

Figure 1. Location of the Tarim Basin.((**a**)—the location of the study area; (**b**)—the drainage systerm and elevation of Tarim Basin; (**c**)—distribution of meteorological stations in The Tarim Basin).

The results of Figure 1c show that meteorological stations are mainly distributed in the north and southwest, with an uneven spatial distribution. The CRU data sets are widely used in meteorological and hydrological research. Although the spatial resolution of the data set is relatively low, the spatial distribution in the area without monitoring sites is relatively excellent. For the above reasons, this data set is finally adopted in this paper. In this work, the meteorological data are rasterized using ArcGIS 10.8 software, following which the monthly data are counted as seasonal data by pixel. Finally, the rasterized seasonal data are extracted by region. The monthly atmospheric circulation indexes were calculated and integrated into the seasonal data by using the R programming language. The four seasons were divided according to the meteorological standard: spring ran from March to May; summer from June to September; autumn from September to November; and winter from December to February of the following year.

2.2. *Research Methods*

2.2.1. Analysis of Change Trend

Unitary linear regression is used in this paper to examine the temporal variation trend of meteorological element value in the study area [27,28]. It is expressed as follows:

$$S = \frac{n \times \sum_{i=1}^{n}(i \times x_i) \cdot (\sum_{i=1}^{n} i) \times \sum_{i=1}^{n}(x_i)}{n \times (\sum_{i=1}^{n} i^2) \cdot (\sum_{i=1}^{n} i)^2} \tag{1}$$

where x is the meteorological element value of each grid; n is the number of years in the research period, and n in this study is 10 a; S reflects the change rate of meteorological factor values over time. When $S > 0$ ($S < 0$), the meteorological factor values showed an increase-decrease trend, and the larger/smaller the value, the more significant the growth (reduction) rate.

2.2.2. Spatial Interpolation Method

Inverse distance weighted interpolation (IDW) is a relatively mature spatial analysis method used in the discipline of meteorology [29,30]. In this method, the distance between the interpolation point and the sample point is used as the weighted average weight. The closer the sample point is to the interpolation point, the greater the weight assigned to the sample point.

Suppose a series of discrete points are distributed on the plane. Because we know that all of the coordinates are X_i, Y_i, Z_i (i = 1,2,3, … ,n), the distance (D_i) between the discrete point (X_i, Y_i) and the grid point (X, Y) is:

$$D_i = \sqrt{(X_i - X)^2 + (Y_i - Y)^2} \tag{2}$$

The estimated value of grid points (X, Y) is:

$$Z = \frac{\sum_{i=1}^{n}\left(\frac{Z_i}{D_i^2}\right)}{\sum_{i=1}^{n}\left(\frac{1}{D_i^2}\right)} \tag{3}$$

2.2.3. Wavelet Analysis

To calculate the real part of the wavelet, this study uses the Morlet continuous complex wavelet as the basis function (i.e., the *comr* function) [31,32]. It is expressed as follows:

$$comr(x) = \frac{\sigma^{2i\pi \cdot F_e} \times \frac{x^2}{F_b}}{\sqrt{\pi \cdot F_b}} \tag{4}$$

where F_e is the center frequency and F_b is the frequency bandwidth. The wavelet square difference is denoted *Var* and can be obtained by integrating the square of the wavelet coefficient over the time translation domain b:

$$Var(\alpha) = \int_{-\infty}^{\infty} \omega_f|(a,b)|^2 db \tag{5}$$

2.2.4. Correlational Analysis

Correlation analysis is a statistical method to analyze the correlation between variables and is widely used in hydrometeorology. Consider two time series, x and y; we use statistical methods to calculate the correlation coefficient between the two, as follows:

$$r = \frac{\sum_{i=1}^{n}(x_i - \overline{x})(y_i - \overline{y})}{\sqrt{\sum_{i=1}^{n}(x_i - \overline{x})^2 \sum_{i=1}^{n}(y_i - \overline{y})^2}} \tag{6}$$

where r is the correlation coefficient, which ranges from −1.0 to 1.0. When $r > 0$ ($r < 0$), the two time series are positively (or negatively) correlated. The correlation is stronger for larger $|r|$ [33,34].

3. Results

3.1. Temporal Characteristics of Climate Change in the Tarim Basin

This study investigates the interannual variation of temperature and rainfall in the Tarim Basin (Figure 2). For all seasons, temperatures and precipitation rise during the study period. The warming trend in spring is the most significant, with a warming rate of 0.443 °C/decade. The warming trend in winter is not significant, with a warming rate of 0.008 °C/decade. The autumn precipitation increases at the greatest rate (1.34 mm/decade). The increasing rate of precipitation in winter is 0.05 mm/decade. Overall, the warming and wetting phenomenon has been significant in the Tarim Basin over the last 30 years in the spring, summer, and autumn but less so in the winter. This variation is consistent with the typical continental desert climate characteristics of the Tarim Basin, which involve significant changes in climate between the four seasons, with winters being cold and dry [35].

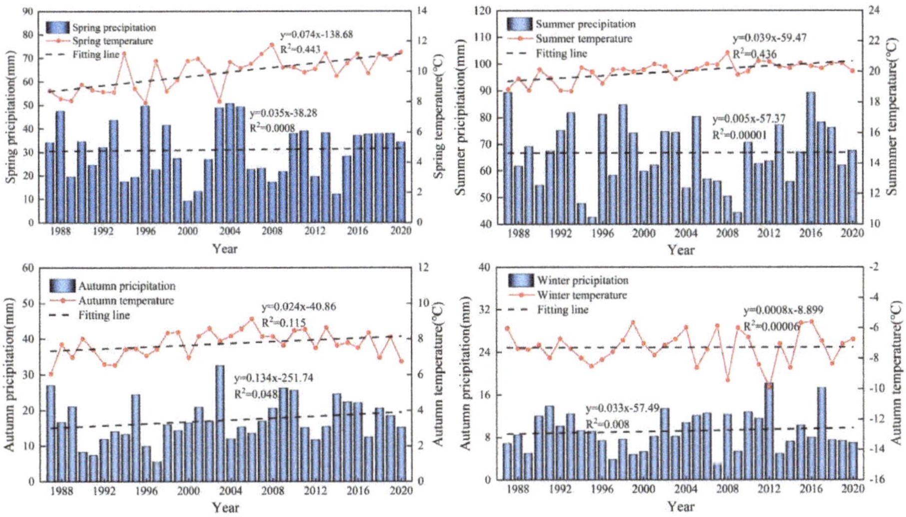

Figure 2. Temporal characteristics of seasonal climate change in the Tarim Basin. Panels show the climate change in spring, summer, autumn, and winter, respectively.

A Morlet wavelet analysis is used to study the seasonal climate cycles in the Tarim Basin from 1987 to 2020 (Figure 3). The results show that the temperatures and precipitation from 1987 to 2020 in the Tarim Basin alternate between a positive phase and a negative phase in the 17–20 year weekly period, which indicates that, from 1987 to 2020, the temperature and precipitation in the Tarim Basin undergo seasonally periodic variations. Calculating the wavelet square difference of temperature and precipitation for each season (results not shown) indicates that, for each season, the temperature and precipitation oscillate with a period of 17–20 years. This phenomenon indicates that the warming and wetting phenomena in the Tarim Basin from 1987 to 2020 followed the same pattern, as shown by the wavelet analysis of the various atmospheric circulation indexes (AO index, ENSO index, NAO index, and PDO index) from 1987 to 2020. From 1987 to 2020, the atmospheric circulation indexes also oscillate significantly with a 17–20-year period, which indicates that the warming and wetting phenomena of the Tarim Basin are related to the atmospheric circulation because they oscillate with the same period. However, how do atmospheric circulation and climate change in space in the Tarim Basin? Further research is needed to address this question.

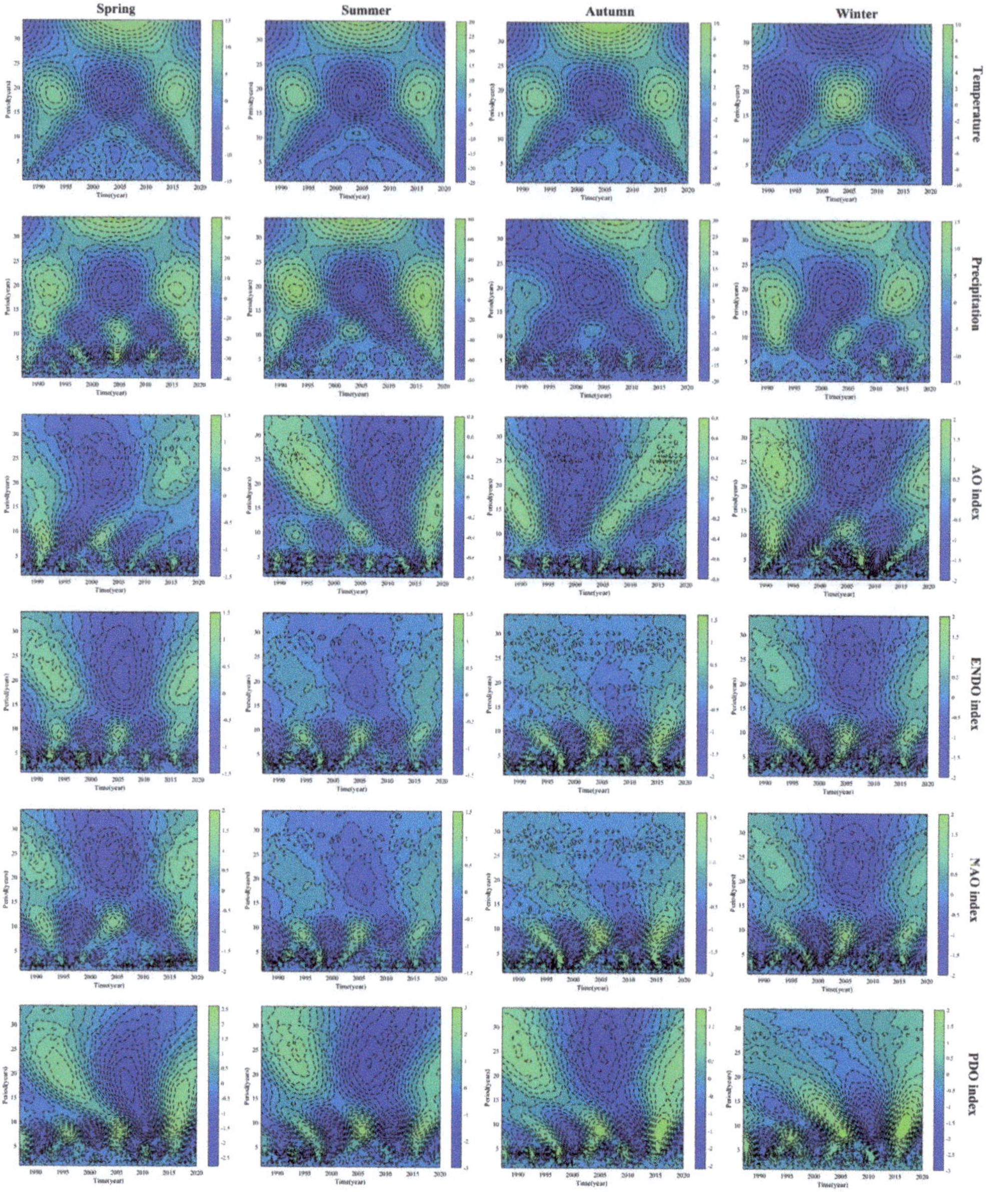

Figure 3. The wavelet transforms of climate change and atmospheric conditions in the Tarim Basin from 1987 to 2020 (Different factors of climate change and atmospheric circulation are listed on the horizontal and at different seasonal scales on the vertical).

3.2. Spatial Characteristics of Climate Change in the Tarim Basin

The warming and wetting rates of the Tarim Basin are calculated based on the rate of change of the climate, and spatial interpolation (inverse distance weighting) is used to analyze the spatial distribution of climate trends for each season in the Tarim Basin [27,28].

The results in Figure 4 show that the warming trend occurred in most regions of the Tarim Basin for all seasons; the warming amplitude around the basin is greater than that in the central area of the basin. In spring, the temperature increased in the Tarim Basin by 0.827 °C/decade, which was the fastest of all seasons. In winter, the temperature increased in the Tarim Basin by −0.088 °C/decade, which was the slowest of all seasons. The rate of change in precipitation in the Tarim Basin depends on the season and location, with significant increases in some areas and significant decreases in others. The regional rate of change in rainfall in spring is like that in autumn. Precipitation increased significantly in the southeastern parts of the basin, with the highest rate of change in precipitation being 5.062 mm/decade. The regional rate of change of precipitation in the summer varied the most, decreasing from the northwest part of the basin to the southeastern parts of the basin. The maximum rate of change was 6.965 mm/decade. The regional rate of change in precipitation varied the least in winter, reaching −2.319 mm/decade.

3.3. Climate Change and Atmospheric Circulation in the Tarim Basin

On the interannual scale, the wavelet analysis of climate change shows that the warming and wetting phenomena in the Tarim Basin are related to atmospheric circulation (Figure 3). The results of this paper are consistent with those of Wu et al. and Lv et al. at multiple scales. However, the spatial evolution of teleconnection between warm and humid phenomena is not clear from seasonal atmospheric circulation. For this reason, we use the monthly circulation indexes (AO index, ENSO index, NAO index, and PDO index) that describe the atmospheric circulation from 1987 to 2020 to calculate the seasonal data and determine if seasonal temperature and precipitation are correlated with these indexes in the Tarim Basin. There is a remote correlation between the seasonal temperature and the atmospheric circulation in Section 3.3.1 and a remote correlation between the seasonal precipitation and the atmospheric circulation in Section 3.3.2.

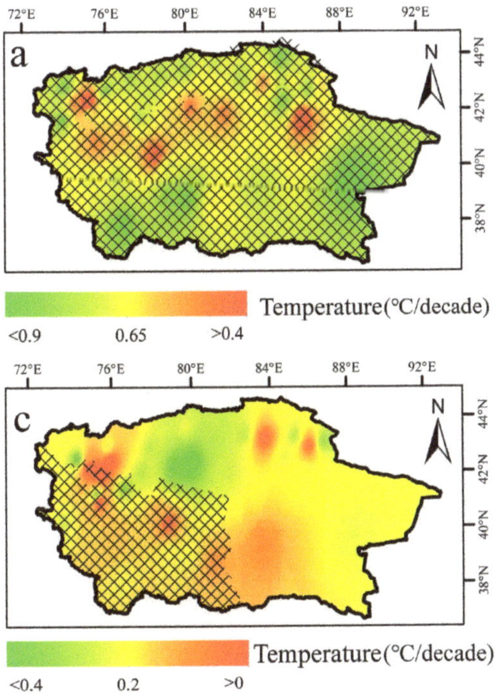

Figure 4. *Cont.*

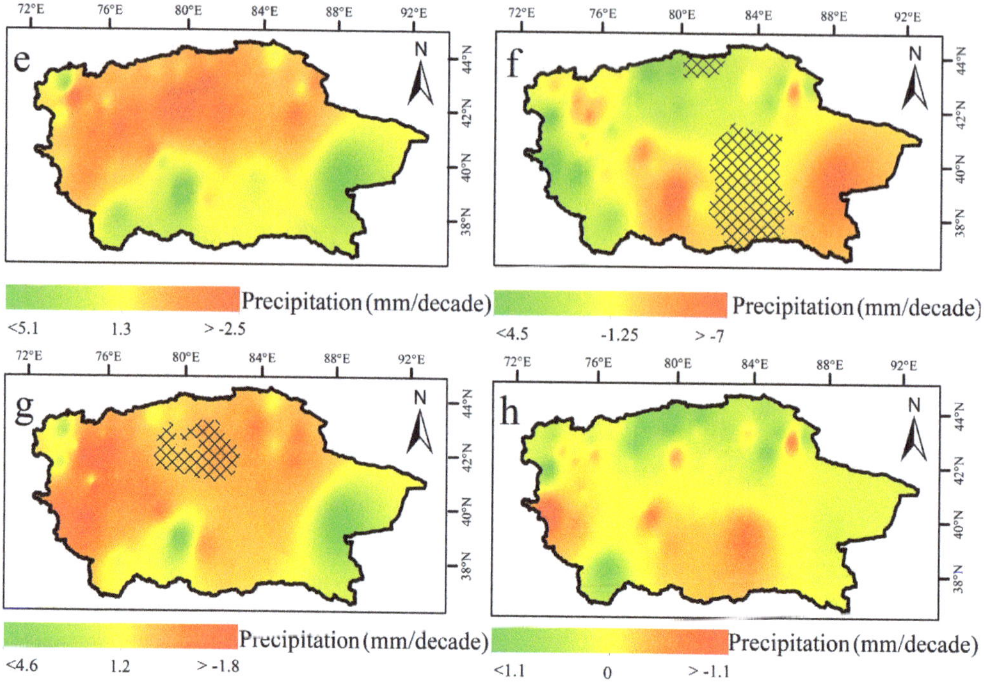

Figure 4. Spatial variations in seasonal climate change in the Tarim Basin. (**a**–**d**) Temperature variations in spring, summer, autumn, and winter, respectively. (**e**–**h**) Precipitation variations in spring, summer, autumn, and winter, respectively (The black grid indicates that the trends are statistically significant at the 0.05 level).

3.3.1. The Seasonal Temperature and Atmospheric Circulation in the Tarim Basin

In most Tarim Basin regions, seasonal temperature and atmospheric circulation indexes are relatively low ($\alpha = 0.05$) as shown in Figure 5. In spring, the PDO and ENSO indices showed a significant negative correlation with the temperature in some regions. In the summer (autumn), the PDO index had a significant negative correlation with Tarim Basin temperature, whereas the other indexes had no statistical significance. In winter, there is a significant positive correlation between the ENSO index and the southwest Tarim Basin. The results in Figure 5 show that the climate change in the Tarim Basin during 1987–2020 is related to the change of atmospheric circulation. PDO is the leading factor of summer and autumn climate change in the Tarim Basin, and ENSO also plays an important role in spring and winter climate change.

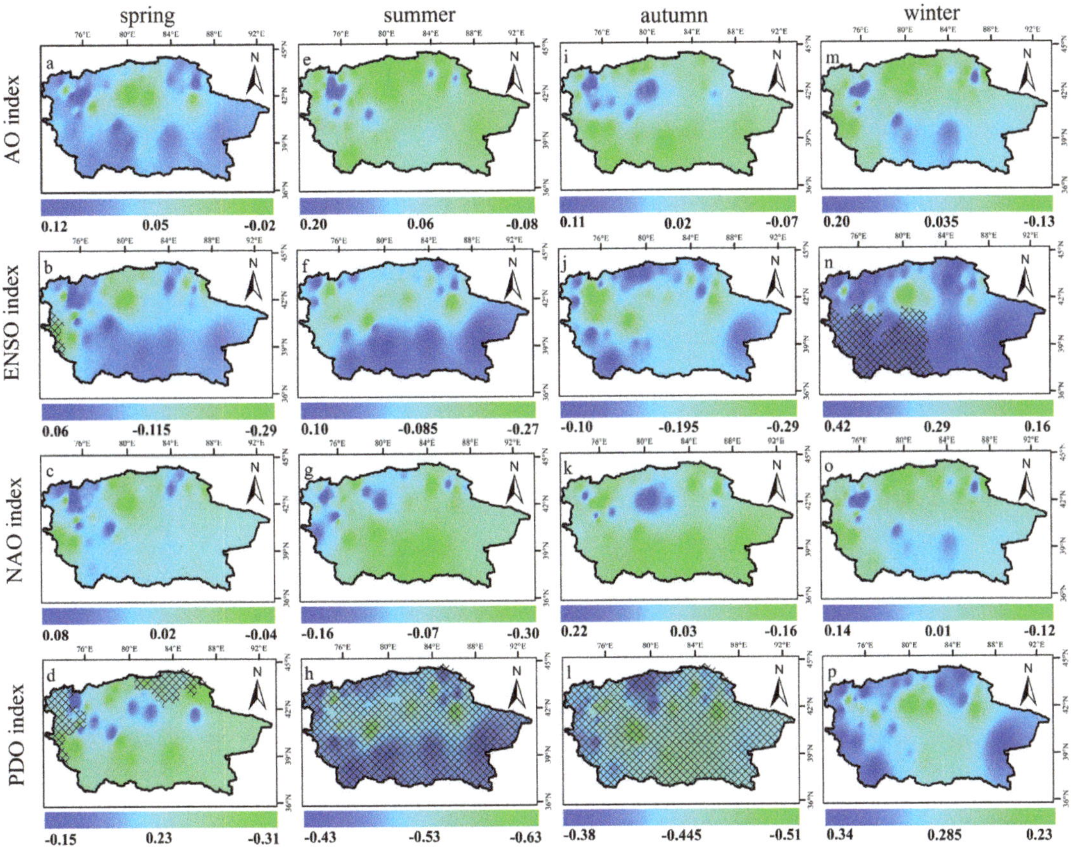

Figure 5. Spatial distribution of the correlation between average annual temperature and atmospheric circulation in the Tarim Basin. Each column corresponds to a season, and each row corresponds to an atmospheric circulation index (the black bars on the grid indicate that the trends are statistically significant at the 0.05 level).

3.3.2. The Seasonal Precipitation and Atmospheric Circulation in the Tarim Basin

Seasonal precipitation and the corresponding atmospheric circulation indexes are relatively low (α = 0.05) in most regions of the Tarim Basin, as shown in Figure 6. In spring, the AO index showed a significant negative correlation with precipitation in some regions, while the ENSO index (PDO index) showed a significant positive correlation with precipitation in some regions. In summer, the NAO index showed a significant positive correlation with precipitation in eastern parts of the basin, while other atmospheric circulation indexes showed no statistical significance. In autumn, there is a significant positive correlation between the ENSO index and precipitation in the southeast Tarim Basin. In winter, the correlation between the atmospheric circulation index and precipitation in the Tarim Basin is not statistically significant. The results of Figure 6 show that the influence of atmospheric circulation on the teleconnection of the Tarim Basin is relatively small. In this paper, it is concluded that atmospheric circulation is not the main factor affecting precipitation in the Tarim Basin at a seasonal scale and that the main controlling factors of precipitation seasonal variation need to be explained in combination with other factors.

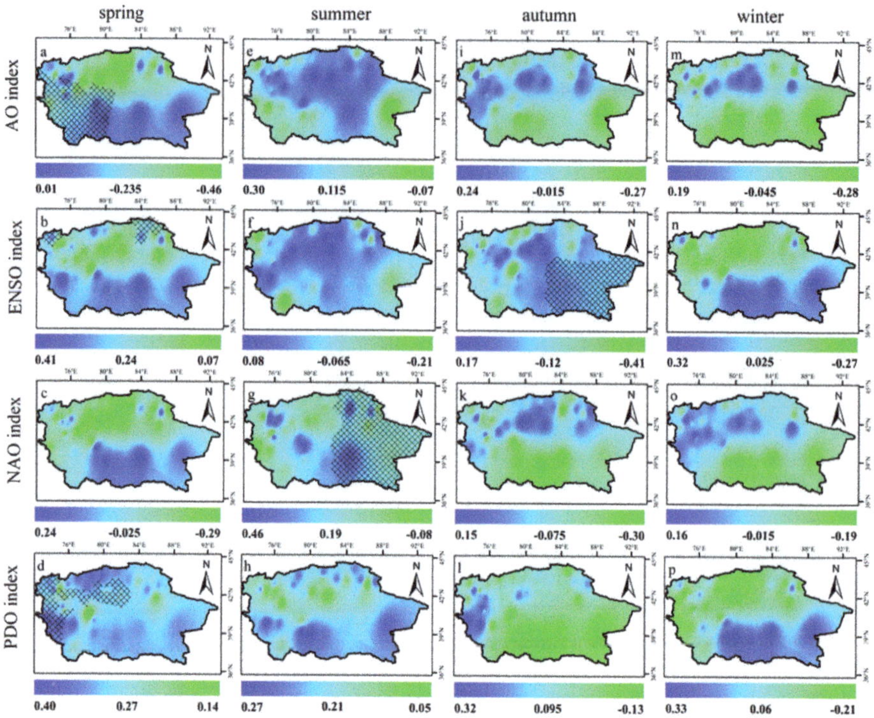

Figure 6. Spatial distribution of correlation between annual precipitation and atmospheric circulation in the Tarim Basin. Each column in the figure corresponds to a season, and each row corresponds to an atmospheric circulation index (The black grid indicate that the trends are statistically significant at the 0.05 level).

4. Discussion

The Tarim Basin has a typical continental desert climate and is in the extremely arid region of northwest China [20,21]. The Tarim Basin is the only inland basin in the arid region of northwest China and is of great significance for research into climate change, the ecological environment, and the social economy of the entire northwest region of China and even the entire arid region of Central Asia [6–8,36,37]. However, our current knowledge about the spatial distribution of the warming and wetting phenomena in the Tarim Basin on a seasonal timescale remains poor. Most studies of this phenomenon focus on the spatiotemporal evolution of water resources and drought in the Tarim Basin or analyze the spatial characteristics of climate change on a long-term scale in the arid region of northwest China [35,38,39].

4.1. Spatiotemporal Characteristics of Climate Change in the Tarim Basin

To elucidate the warming and wetting phenomenon that currently envelops the arid region of northwest China, we analyze here its seasonally specific spatial evolution in the Tarim Basin. The results show that the warming trend occurred in most regions of the Tarim Basin for all seasons; the warming amplitude around the basin is greater than that in the central area of the basin. At the same time, the wetting trend differs significantly between different zones of the region. These results are consistent with the climate of the Tarim Basin, namely, the significant seasonal differences in climate and the copious evaporation led to spatial variations in the wetting trend [11,12]. This indicates that the warming and wetting phenomenon in the arid Tarim Basin has not forced the spatiotemporal evolution of the climate on the seasonal scale to depart from the original climate type, which means that this phenomenon has not significantly changed the climate type of the Tarim Basin [13,40].

In this paper, the temporal and spatial characteristics of temperatures and precipitation on the seasonal scale are analyzed by statistical methods. The results show that the climate types in the Tarim Basin have not changed at the seasonal scale, and the climate changes in the Tarim Basin are still consistent with the original climate types. Based on previous research, we conclude that the warming and wetting phenomenon in the Tarim Basin is linked to global warming, extreme precipitation events, and seasonal agricultural activities. A key question now is: how long will the warming and wetting phenomena continue in the future? This is the reason why the climate type of the Tarim Basin will change in the future. Again, this is our next research topic.

4.2. Climate Evolution in the Tarim Basin: Effect of Atmospheric Circulation

A wavelet analysis shows that the periodic characteristics of climate in the Tarim Basin from 1987 to 2020 and atmospheric circulation are similar, which indicates that atmospheric circulation plays a role in generating the warming and wetting phenomena in the Tarim Basin. An analysis of the seasonal correlations between climate change and atmospheric circulation shows that PDO and ENSO are the main influencing factors of climate change in the Tarim Basin at different seasonal scales, while the teleconnection of AO and NAO is low. Among them, the PDO dominates the summer and autumn temperature changes in the Tarim Basin. However, the teleconnection effect of atmospheric circulation on precipitation in the Tarim Basin is relatively low. These results indicate that atmospheric circulation is not the only factor contributing to the warming and wetting phenomena in the Tarim Basin; internal variations in climate factors can also be important. The phenomenon may also be promoted by external stress factors other than atmospheric circulation [41,42].

Many studies have shown that atmospheric circulation is the dominant factor in climate change in northwest China [5,41,42]. However, due to the Tarim Basin's uniqueness in comparison to northwest China, for example, it has the Tarim River system and the Taklimakan Desert. Therefore, the actual physical mechanism in the Tarim Basin is complicated. In future studies, many factors should be considered to analyze the seasonally specific spatiotemporal characteristics of the warming and wetting phenomena in the Tarim Basin [43,44].

5. Conclusions

The seasonally specific spatiotemporal variations of the Tarim Basin's climate were studied using climate inclination rate, wavelet analysis, and correlation analysis, and the correlations ($\alpha = 0.05$) between these climate indexes and atmospheric circulation were examined. The results lead to the following conclusions:

(1) The Tarim Basin experienced a significant, seasonally specific warming and wetting phenomenon from 1987 to 2020. All areas of the Tarim Basin warmed significantly in all seasons over this period, whereas the precipitation differed significantly across the seasons. The climate of the basin and the atmospheric circulation both oscillated over a period of 17–20 years, which indicates that the atmospheric circulation is involved in the generation of the wetting phenomenon of the Tarim Basin. Moreover, the spatiotemporal evolution of climate change in the Tarim Basin still follows its original climate type, despite experiencing a warming and wetting phenomenon over the study period. Because the two indices of temperature and precipitation were studied by statistical methods in this paper, the research results are weak in explaining the mechanism. In future studies, we need to consider using different indicators (such as the Drought Index, Drought Frequency, and Normalized Difference Vegetation Index) to explore the characteristics of climate change in the Tarim Basin at the seasonal scale.

(2) Seasonal temperature (precipitation) and the corresponding atmospheric circulation indexes are relatively low ($\alpha = 0.05$) in most regions of the Tarim Basin. PDO is the leading factor of summer and autumn climate change in the Tarim Basin, and ENSO

also plays an important role in spring and winter climate change. However, the teleconnection effect of atmospheric circulation on precipitation in the Tarim Basin is relatively low. The results show that atmospheric circulation is only one of the dominant factors contributing to the warm and wet phenomenon in the Tarim Basin. For example, extreme precipitation events may be the main cause of the wetting phenomenon in the Tarim Basin. Therefore, because the actual physical mechanism in the Tarim Basin is complicated, many internal factors need to be considered in future research.

Author Contributions: Conceptualization, T.L.; methodology, A.L. and T.L.; software, Y.L. and T.L.; validation, T.L. and Y.L.; formal analysis, A.L. and T.L.; investigation, T.L. and Y.L.; resources, A.L. and W.Z.; data curation, A.L., T.L. and W.Z.; writing—original draft preparation, T.L.; writing—review and editing, T.L. and A.L.; visualization, Y.L. and T.L.; supervision, W.Z.; project administration, A.L. and W.Z. All authors have read and agreed to the published version of the manuscript.

Funding: This research was supported by the Third Xinjiang Scientific Expedition and Research Program (2021xjkk0203).

Data Availability Statement: This study's data is publicly available in this article.

Acknowledgments: We thank the editor and anonymous reviewers for their very helpful comments and suggestions for improving the quality of the article.

Conflicts of Interest: The authors declare no conflict of interest.

References

1. Shi, Y.; Shen, Y.; Li, D.; Zhang, G.; Ding, Y.; Hu, R.; Kang, E. Discussion on the present climate change from warm-dry to warm wet in northwest China. *Quat. Sci.* **2003**, *23*, 152–164.
2. Wu, P.; Ding, Y.; Liu, Y.; Li, X. The characteristics of moisture recycling and its impact on regional precipitation against the background of climate warming over Northwest China. *Int. J. Climatol.* **2019**, *39*, 5241–5255. [CrossRef]
3. Wang, C.; Zhang, S.; Zhang, F.; Li, K.; Yang, K. On the increase of precipitation in the northwestern China under the global warming. *Adv. Earth Sci.* **2021**, *36*, 980–989.
4. Lv, A.; Qu, B.; Jia, S.; Zhu, W. Influence of three phases of El Niño–Southern Oscillation on daily precipitation regimes in China. *Hydrol. Earth Syst. Sci.* **2019**, *23*, 883–896. [CrossRef]
5. Yao, J.Q.; Li, M.Y.; Dilinuer, T.; Chen, J.; Mao, W. The assessment on "warming-wetting" trend in Xinjiang at multi-scale during 1961–2019. *Arid Zone Res.* **2022**, *39*, 333–346.
6. Li, M.; Sun, H.Q.; Su, Z.C. Research progress in dry/wet climate variation in northwest China. *Geogr. Res.* **2021**, *40*, 1180–1194.
7. Hou, Y.; Chen, Y.; Ding, J.; Li, Z.; Li, Y.; Sun, F. Ecological Impacts of Land Use Change in the Arid Tarim River Basin of China. *Remote Sens.* **2022**, *14*, 1894. [CrossRef]
8. Bai, J.; Li, J.; Bao, A.; Chang, C. Spatial-temporal variations of ecological vulnerability in the Tarim River Basin, Northwest China. *J. Arid Land* **2021**, *13*, 814–834. [CrossRef]
9. Wang, G.Y.; Wang, Y.J.; Gui, D.W. A review on water resources research in Tarim river basin. *Arid Land Geogr.* **2018**, *41*, 1151–1159.
10. Ren, Q.; Long, A.H.; Yang, Y.M. Analysis on remote sensing monitoring of eco-environment variationof main stream basin of Tarim river in recent 20 years. *Water Resour. Hydropower Eng.* **2021**, *52*, 103–111.
11. Liu, Q.; Liu, Y.; Niu, J.; Gui, D.; Hu, B.X. Prediction of the Irrigation Area Carrying Capacity in the Tarim River Basin under Climate Change. *Agriculture* **2022**, *12*, 657. [CrossRef]
12. Li, H.; Wang, W.; Fu, J.; Chen, Z.; Ning, Z.; Liu, Y. Quantifying the relative contribution of climate variability and human activities impacts on baseflow dynamics in the Tarim River Basin, Northwest China. *J. Hydrol. Reg. Stud.* **2021**, *36*, 100853. [CrossRef]
13. Han, Y.; Zhang, S.; Lv, A.F.; Zeng, H. Risk Assessment of Water Resources Carrying Capacity in China. *JAWRA J. Am. Water Resour. Assoc.* **2021**, *57*, 539–551. [CrossRef]
14. Chen, Y.; Deng, H.; Li, B.; Li, Z.; Xu, C. Abrupt change of temperature and precipitation extremes in the arid region of Northwest China. *Quat. Int.* **2014**, *336*, 35–43. [CrossRef]
15. Li, B.; Chen, Y.; Shi, X.; Chen, Z.; Li, W. Temperature and precipitation changes in different environments in the arid region of northwest China. *Theor. Appl. Climatol.* **2013**, *112*, 589–596. [CrossRef]
16. Yang, H.; Xu, J.; Chen, Y.; Li, D.; Zuo, J.; Zhu, N.; Chen, Z. Has the Bosten Lake Basin been dry or wet during the climate transition in North -west China in the past 30 years? *Theor. Appl. Climatol.* **2020**, *141*, 627–644. [CrossRef]
17. Zhang, Q.; Zhu, B.; Yang, J.; Ma, P.; Liu, X.; Lu, G.; Wang, Y.; Yu, H.; Liu, W.; Wang, D. New characteristics about the climate humidification trend in northwest China. *Chin. Sci. Bull.* **2021**, *66*, 3757–3771. [CrossRef]
18. Gessner, U.; Naeimi, V.; Klein, I.; Kuenzer, C.; Klein, D.; Dech, S. The relationship between precipitation anomalies and satellite-derived vegetation activity in Central Asia. *Glob. Planet. Change* **2013**, *110*, 74–87. [CrossRef]

19. Botne, O.; Fraedrich, K.; Zhu, X. Precipitation climate of Central Asia and the large-scale atmospheric circulation. *Theor. Appl. Climatol.* **2012**, *108*, 345–354.
20. Jiang, J.; Zhou, T.; Chen, X.; Zhang, L. Future changes in precipitation over Central Asia based on CMIP6 projections. *Environ. Res. Lett.* **2020**, *15*, 054009. [CrossRef]
21. Wang, H.; He, M.; Yan, W.; Ai, J.; Chu, J. Ecosystem vulnerability in the tianshan mountains and Tarim oasis based on remote sensed gross primary productivity. *Acta Ecol. Sin.* **2021**, *41*, 9729–9737.
22. Fan, L.; Lv, A.F.; Zhang, W.X. Temporal-spatial variation characteristics of drought and its relationship with atmospheric circulation in qinghai province. *J. Arid Land Resour. Environ.* **2021**, *35*, 60–65.
23. Ma, Z.G.; Fu, Z.B. The facts of global aridity in the second half of the 20th Century and its connection to the large-scale context. *Sci. Sin. (Terrae)* **2007**, *37*, 222–233.
24. Guan, X.D.; Huang, J.P.; Guo, R.X. Changes in Aridity in Response to the Global Warming Hiatus. *J. Meteorol. Res.* **2017**, *31*, 117–125. [CrossRef]
25. Zhou, H.; Shen, M.; Chen, J.; Xia, J.; Hong, S. Trends of natural runoffs in the Tarim river basin during the last 60 years. *Arid. Land Geogr.* **2018**, *41*, 221–229.
26. Zhang, Y.L.; Chen, Y.N.; Yang, Y.H.; Hao, X.M. Analysis of historical change and future trends Tarim river. *Arid Land Geography.* **2016**, *39*, 582–589.
27. Song, C.H. Research on adjusting effect for different mathe matical model of precipitation series. *J. Appl. Meteorol. Sci.* **1998**, *9*, 86–91.
28. Li, X.Y.; Huang, S.P.; Ye, H.; Xiong, Y.Z.; Chen, F.; Wang, K.; Pan, L.Y. Comparison characteristics andinhomogeneity adjustments of mean, maximum and minimum surface air temperature for Xiamen meteorological station. *Sci. Geogr. Sin.* **2010**, *30*, 796–801.
29. Zhu, J.J.; Zhang, C.Y.; Wang, W.; Li, C. Analysis of correlation between groundwater, air pressure and precipitation in plain areas of Hubei and Hunan province. *Sci. Surv. Mapp.* **2015**, *40*, 7–11.
30. Li, Z.Q.; Wu, Y.X. Inverse Distance Weighted Interpolation Involving Position Shading. *Acta Geod. Et Cartogr. Sin.* **2015**, *44*, 91–98.
31. Souza Echer, M.P.; Echer, E.; Nordemann, D.J.; Rigozo, N.R.; Prestes, A. Wavelet analysis of a centennial (1895–1994) southern Brazil rainfall series (Pelotas, 31°46′19″ S 52°20′33″ W). *Clim. Chang.* **2008**, *87*, 489–497. [CrossRef]
32. Kumar, P. Foufoula-Georgiou, E. Wavelet analysis in geophysics: An introduction. *Wavelets Geophys.* **1994**, *4*, 1–43.
33. Xu, J.H. *Mathematical Methods in Modern Geography*; Higher Education Press: Beijing, China, 1996.
34. Huang, J.Y. *Meteorological Statistical Analysis and Forecasting Methods*; China Meteorological Press: Beijing, China, 2004; pp. 28–30.
35. Zhang, W.; Bao, W.M.; Feng, X.Q.; Lu, J.L.; Luo, Q.F. Study on the change of meteorological elements in Tarim river basin in recent 50 years. *Water Power* **2020**, *46*, 28–34.
36. Fu, A.H.; Chen, Y.N.; Li, W.H. Assessment on ecosystem health in the Tarim river basin. *Acta Ecol. Sin.* **2009**, *29*, 2418–2426.
37. Kong, Z.J.; Deng, M.J.; Ling, H.B.; Wang, G.Y.; Xu, S.W.; Wang, Z.R. Ecological security assessment and ecological restoration countermeasures in the dry-uparea of the lower Tarim river. *Arid Zone Res.* **2021**, *38*, 1128–1139.
38. Chen, Y.N.; Li, Z.; Fan, Y.T.; Wang, H.J.; Fang, G.H. Research progress on the impact of climate change on water resources in the arid region of northwest China. *Acta Geogr. Sin.* **2014**, *69*, 1295–1304.
39. Wei, G.H.; Yang, P.; Zhou, H.Y.; Xia, J.; Chen, J.; Gui, D.W.; Bayindala. An analysis of the drought characteristics and driving factors of Tarim river basin based on downscaled terrestrial water storage from GRACE. *China Rural Water Hydropower* **2020**, *7*, 12–19.
40. Tao, H.; Bai, Y.G.; Mao, W.Y. Observed and projected climate changes in Tarim river basin. *J. Glaciol. Geocryol.* **2011**, *33*, 738–743.
41. Chen, L.; Wang, S.G.; Shang, K.Z.; Yang, D.B. Atmospheric circulation anomalies of large-scale extreme high temperature events in northwest China. *J. Desert Res.* **2011**, *31*, 1052–1058.
42. Gao, M.; Yang, J.; Gong, D.; Shi, P.; Han, Z.; Kim, S.J. Footprints of Atlantic Multidecadal Oscillation in the Low-Frequency Variation of Extreme High Temperature in the Northern Hemisphere. *J. Clim.* **2019**, *32*, 791–802. [CrossRef]
43. Liu, Y.; Wu, C.; Jia, R.; Huang, J. An overview of the influence of circulation on the climate in arid and semi-arid region of central and east asia. *Sci. Sin. (Terrae)* **2018**, *48*, 1141–1152. [CrossRef]
44. Huang, J.; Guan, X.; Ji, F. Enhanced cold-season warming in semi-arid regions. *Atmos. Chem. Phys.* **2012**, *12*, 5391–5398. [CrossRef]

atmosphere

MDPI

Article

Mapping Potential Soil Water Erosion and Flood Hazard Zones in the Yarlung Tsangpo River Basin, China

Shan Chen [1], Shaocheng Zhu [2], Xin Wen [3], Huaiyong Shao [3,*], Chengjin He [3], Jiaguo Qi [4], Lingfeng Lv [3], Longbin Han [3] and Shuhan Liu [3]

1 School of Mechanical and Electrical Engineering, Chengdu University of Technology, Chengdu 610059, China
2 Southwest Jiaotong University Chengdu Design Institute Co., Ltd., Chengdu 610059, China
3 College of Earth Sciences, Chengdu University of Technology, Chengdu 610059, China
4 Center for Global Change and Earth Observations, Michigan State University, East Lansing, MI 48824, USA
* Correspondence: shaohuaiyong@cdut.edu.cn

Abstract: The ubiquity of soil water erosion in the Yarlung Tsangpo River Basin leads to a series of natural hazards, including landslides, debris flows and floods. In this study, the Revised Universal Soil Loss Equation model (RUSLE) was used to quantify potential soil water erosion, while the Height Above Nearest Drainage model (HAND) was used to delimit potential flood hazard zones. Remote sensing and geographic information system technologies were employed to spatialize the results, which showed that the annual soil loss from water erosion was less than 1239 t ha^{-1} y^{-1}. The total soil loss was estimated to be over 108 $\times$ 10^6 tons, of which about 13 $\times$ 10^6 tons (12.04% of the total) occurred from the agricultural land in the downstream valley. Soil erosion mapping was performed using six levels of soil erosion intensity and the effects of precipitation, land use/land cover and topography on soil erosion were revealed. Increases in precipitation and slope gradient significantly increased the soil loss rate, while the maximum rate of soil loss occurred from densely vegetated land, reaching 9.41 t ha^{-1} y^{-1}, which was inconsistent with erosion preconceptions for this land type. This may be due to a combination of the region's unique climate of high intensity rainfall and steep slopes. Flood hazard mapping showed that all regional cities were located in a flood hazard zone and that, within the total basin area (~258 $\times$ 10^5 ha), 9.84% (2,537,622 ha) was in a high flood occurrence area, with an additional 1.04% in aa vulnerable to moderate flood hazard area. Approximately 1.54% of the area was in a low flood risk area and 4.15% was in a very low flood risk area. The results of this study provide an initial identification of high-risk soil water erosion and flood hazard locations in the basin and provide a foundation upon which decision-makers can develop water and soil conservation and flood prevention policies.

Keywords: soil loss; potential flood risk; Revised Universal Soil Loss Equation (RUSLE); Height Above Nearest Drainage (HAND)

Citation: Chen, S.; Zhu, S.; Wen, X.; Shao, H.; He, C.; Qi, J.; Lv, L.; Han, L.; Liu, S. Mapping Potential Soil Water Erosion and Flood Hazard Zones in the Yarlung Tsangpo River Basin, China. *Atmosphere* **2023**, *14*, 49. https://doi.org/10.3390/atmos14010049

Academic Editors: Jinping Liu, Quoc Bao Pham, Arfan Arshad and Masoud Jafari Shalamzari

Received: 3 December 2022
Revised: 20 December 2022
Accepted: 21 December 2022
Published: 27 December 2022

1. Introduction

Soil erosion is an important environmental issue related to global ecology, environment, economy and security. Soil erosion reduces soil productivity and water quality, threatens food security and the future development of agriculture, increases sediment accumulation and leads to the possibility of floods [1]. Severe soil erosion eventually leads to frequent natural disasters such as siltation in rivers, lakes, weirs, ponds, floods, landslides and debris flows [2,3]. The accumulation of sediments caused by soil erosion is an important factor causing flood disasters. Since the 21st century, about 74% of natural disasters were related to water disasters [4] and the frequency of global floods has increased significantly, for example India is a region prone to chronic floods [5]. Ethiopia, one of the countries with the highest erosion risk in the world and is also prone to flooding. Developing countries, with more fragile soils and often very sensitive to climate change, often face more serious human and financial consequences and post-disaster reconstruction problems brought about by floods and related land disasters. Therefore, the control of soil erosion and avoidance of geological disasters caused by soil erosion is the focus of current research.

Against the background of global encouragement of preventing soil erosion and protecting the soil environment, the Yarlung Tsangpo River Basin (YTRB) [6], as a soil environment sensitive area and an area with strong physical freeze–thaw erosion, is located in the southern part of the Qinghai-Tibet Plateau, with high altitude and complex terrain. Its unique geographical features such as loose soil, high precipitation intensity and obvious climate change lead to various forms of soil erosion that are prone to occur widely [3]. Among them, the main type of soil erosion is freeze–thaw erosion, although water erosion cannot be ignored [7]. Therefore, quantitative evaluation of soil erosion in the Yarlung Zangbo River Basin, identification and analysis of soil erosion sensitive areas and their spatial and temporal patterns will bring important scientific guidance to the protection of soil environment and the reduction of natural disasters.

At present, soil erosion research methods have been widely investigated, mainly in the form of control simulation experiments, field fixed-point observations, remote sensing image analysis, soil erosion model simulation and sediment element tracer methods [2,8]. At present, many scholars are conducting research on different spatial and temporal scales based on different soil erosion models. The widely used models include the Revised Universal Soil Loss Equation (RUSLE), the US Universal Soil Loss Model (USLE), the Chinese Soil Loss Model (CSLE), Soil and Water Assessment Tool (SWAT), etc. [9]. Compared with the methods that rely on long-term soil survey data and parameters that are difficult to collect and calculate, the RUSLE model has more obvious advantages in data operability, high precision and a wide application range [10]. Combining remote sensing (RS) and geographic information system (GIS) technologies, the RUSLE model is widely used in complex areas of various terrain types and in the study of spatial pattern characteristics of soil erosion at different scales [11,12]. However, the application of the model has obvious regional characteristics, so it is necessary to comprehensively consider the regional characteristics and select a reasonable calculation method for the parameter localization calculation.

The Relative Altitude to Nearest Neighbor Channel (HAND) model is a quantitative terrain model based on the Shuttle Radar Terrain Mission-Digital Elevation Model (SRTM-DEM) [13]. The local relative terrain with reference to the river network water system is defined mathematically as the elevation difference between any position on the surface and the point where it enters the river network from the hillside along the flow of water and describes the relative height of any position on the surface compared to the local water system [14,15]. Compared with the classic DEM method, HAND can provide supplementary local terrain information. The flood mapping method based on HAND is feasible and reasonable; it can highlight local terrain features and reflect the water flow path of the hillside-valley, which is more conducive to establishing the relationship between local topography and hydrological response [15]. Many scholars have carried out research based on the HAND model. Zheng et al. developed a synthetic rating curve based on flow and water level [16] and Chow et al. created an exclusion mask that can improve the accuracy of flood mapping [17]. Liu et al. and Speckhann et al. demonstrated the applicability of the HAND model for flood mapping over large areas [18,19]. Therefore, this paper will use the DEM-based HAND model to draw flood hazard maps. However, it cannot dynamically show the inundation changes in the process of flood evolution and the influence of river network density on the accuracy of flood mapping cannot be ignored.

Soil erosion is caused by multiple factors, including rainfall intensity, soil characteristics, topography, runoff, land use type, vegetation coverage and human activities [20,21]. Although precipitation and topography are the dominant factors of soil erosion, human activities such as rapid population growth, deforestation, land plowing and overgrazing have also been reported to accelerate soil erosion globally [22]. In terms of factors affecting soil erosion, scholars have carried out much research mainly from two aspects: natural factors and human factors [23–25]. Studies have shown that climate change and human activities have a dual impact on soil erosion, which will lead to significant changes in land use/cover, causing drastic fluctuations in the soil erosion modulus [8].

Based on the research scale/region aspect of the RUSLE model, its feasibility has been effectively verified in different regions and research areas at multiple scales around the world. Global research involves areas such as Rondonia in Brazil, Ethiopia and so on [26]. However, scholars in China often conduct research on the influencing factors of soil erosion spatio-temporal differentiation feature sets, involving various spatial scales. These include Chengde City, Taihang Mountains, Dianchi Lake Basin, Maotiao River Basin, typical small watersheds in the middle reaches of the Yellow River, small watersheds in sugarcane fields, etc. [2,4,6,8,9,27–36]. The research on soil erosion in the source area of the Yarlung Tsangpo River based on the RUSLE model usually focuses on a small watershed, such as the Maquan River Basin [37–39] and the assessment of sudden soil erosion for the entire source area has not yet been carried out. Studies have shown that freeze–thaw erosion is the main form of soil erosion in this watershed. Overgrazing is common in the upper and middle reaches of the watershed and the middle and lower reaches are the main planting areas. Long-term grazing and farming activities will accelerate soil erosion. Furthermore, estimating the scale and distribution of soil erosion and flood hazards at different spatio-temporal scales should be undertaken to focus on environmental management and flood control in larger river basins [40,41].

Although research on soil erosion has received considerable attention, the exploration of the relationship between soil erosion and flood disasters has been neglected. In view of this, based on the RUSLE and HAND models, this paper uses RS and GIS spatial information analysis techniques to quantitatively analyze the spatial distribution patterns of soil erosion and potential flood disasters in the YTRB, aiming to provide planners and decision makers with the data to implement soil and water conservation and flood control policies. Another aim is to provide reliable information in order to assist the area's soil and water conservation projects, disaster prevention tasks and ecological restoration.

2. Study Area

The YTRB is located in Tibet, SW China, between 27°49 and 31°16 N and 81°57 and 7°6 E (Figure 1). The basin lies at the northern foot of the Himalayas, with an average altitude of 4500 m and an area of ~ 25.8 × 10^6 ha. The Yarlung Tsangpo (YT) river originates in the Angsi Glacier in the SW of the Tibetan Plateau; it has an annual flow of 1.359 × 10^{11} m^3 and flows for 2057 km into Assam, India [42].

The plateau valley in the upper reaches has the typical semiarid climate of a plateau cold temperate zone, with an annual average precipitation of 300 mm. The floodplain in the middle is approximately 1200 × 300 km in extent and there are numerous tributaries here—including the Lhasa, Parlung Tsangpo, Nyang, Nyang Qu and Dkzhung Tsangpo as the main five—which provide water for crop cultivation. The middle reaches, including 153,300 ha of agricultural land, with an annual precipitation of 300–600 mm, enjoy a plateau temperate climate and constitute the most developed agricultural region in Tibet. The downstream YT Grand Canyon is the deepest valley in the world, where the river flows around the Namcha Barwa peak and arrives at Pasighat (India) through a large, horseshoe bend. In this region, large altitudinal variations provide adequate hydraulic resources. The YTRB reaches its lowest temperatures in January and its highest in June and July, with annual average temperatures ranging from 4.3–8.3 °C. Precipitation in the basin gradually increases from the NW to the SE, reaching 600–800 mm y^{-1} in the lower reaches [43].

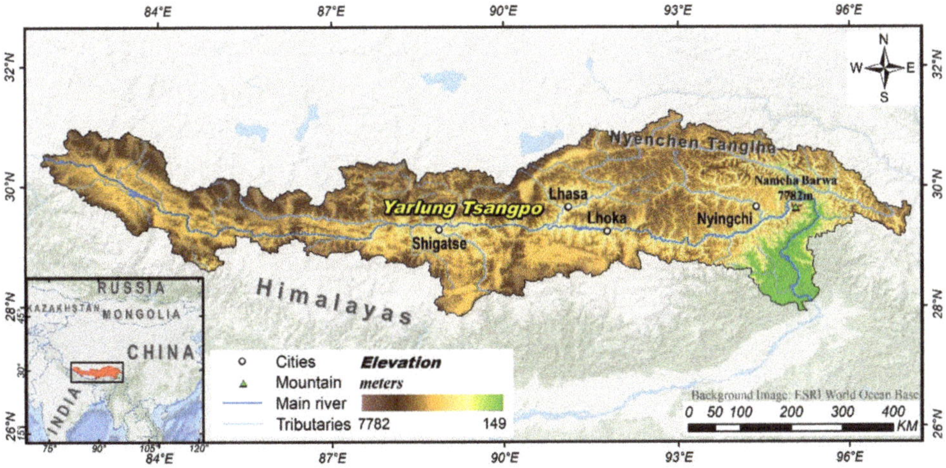

Figure 1. Yarlung Tsangpo river basin location, including major cities, mountains and rivers.

3. Materials and Methods

3.1. Data Sources

Weather stations are sparsely distributed in the YTRB and in their absence, widely used interpolation methods could obscure regional-level precipitation characteristics. For this reason, Chinese monthly precipitation data for the period 1901–2017, at a spatial resolution of 1 km and using 0.1 mm precipitation units were used in this study. These data were provided by the National Tibetan Plateau Data Center (http://www.data.tpdc.ac.cn (accessed on 8 March 2022)), based on the global 0.5° climate dataset published by CRU and the global high-resolution climate dataset published by WorldClim and obtained by the Delta spatial downscaling scheme [44–48]. Soil map information (at a 1 km spatial resolution) was extracted from the China Soil Map, which originated from data held in the Harmonized World Soil Database (HWSD) (version 1.1) [49,50]. Imagery of 30 × 30 m from Landsat 8 OLI was downloaded from the USGS (http://www.earthexplorer.usgs.gov (accessed on 13 March 2022)) to generate land use/land cover (LULC) study area mapping and 90 × 90 m SRTM3 DEM datasets from the USGS were also applied. The Chinese Vegetation Dataset and the YTRB outer cadaster were both downloaded from the Data Center for Resources and Environmental Sciences (RESDC) of the Chinese Academy of Sciences (http://www.resdc.cn (accessed on 15 March 2022)).

3.2. Revised Universal Soil Loss Equation

Soil water erosion was estimated using the RUSLE model. RS technology was used for LULC classification and GIS technology was employed to compute various factors (Figure 2).

The RUSLE model is empirically expressed as shown in Equation (1):

$$A = R \times K \times LS \times C \times P \quad (1)$$

where A indicates the average annual soil erosion per unit area (t ha^{-1} y^{-1}); R represents the rainfall erosivity factor (MJ mm ha^{-1} h^{-1} y^{-1}); K denotes the soil erodibility factor (t ha h MJ^{-1} ha^{-1} mm^{-1}); LS stands for the combination of slope length and slope steepness (dimensionless); C represents the cover management factor (dimensionless); and P indicates the support practice factor (dimensionless).

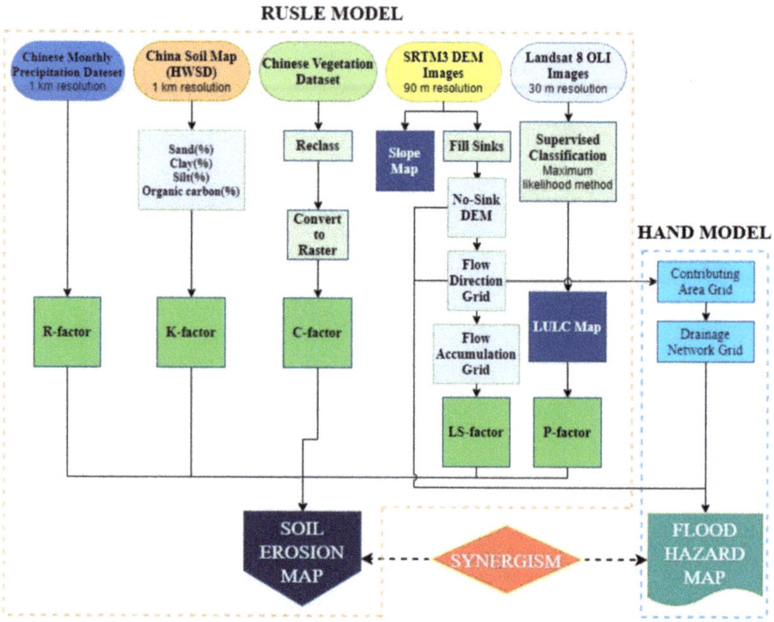

Figure 2. RUSLE model methodology flow chart, as used in this study.

3.2.1. Precipitation Erosivity (R) Factor

Precipitation is the direct driver of soil erosion because splashing raindrops separate soil particles and runoff from rainfall can further wash away and carry soil, causing soil erosion [11]. Thus, the precipitation erosion force factor reflects the ability of raindrops to separate and transport soil particles. Primarily, the R factor consists of the multiplication of total rainfall kinetic energy and maximum rainfall intensity for 30 consecutive minutes. Since intensity data were not available for the YTRB, a regression equation based on monthly precipitation data was applied [51], as shown in Equation (2):

$$R = \sum_{i=1}^{12} 1.735 \times 10^{(1.5\log_{10}(p_i^2/P) - 0.08188)} \quad (2)$$

where R indicates the precipitation erosivity factor (MJ mm ha^{-1} h^{-1} y^{-1}); p_i represents monthly precipitation (mm month^{-1}); and p shows annual precipitation (mm y^{-1}). R-factor mapping was generated with the ArcGIS (version 10.2, ESRI, Redlands, CA, USA) software raster calculator, using Equation (2).

3.2.2. Soil Erodibility (K) Factor

Soil erodibility is a measure of the inherent sensitivity of a standard plot to precipitation erosivity. The K factor demonstrates the inherent resistance of soil particles to the separation and transport capacity of precipitation and runoff [11]. In this study, K-factor estimation was carried out using the Erosion Productivity Impact Calculator (EPIC) empirical model, as developed by Williams [52] and represented by Equation (3):

$$\begin{gathered} K = \left\{0.2 + 0.3\exp\left[-0.0256SAN1 - \frac{SIL}{100}\right]\right\}\left(\frac{SIL}{CLA + SIL}\right)^{0.3} \\ \left[1 - \frac{0.25C}{C + \exp(3.72 - 2.95C)}\right]\left[1 - \frac{0.7SN1}{SN1 + \exp(-5.51 + 22.9SN1)}\right] \end{gathered} \quad (3)$$

where K indicates the soil erodibility factor (t ha h MJ^{-1} ha^{-1} mm^{-1}); SAN represents the percentage of sand (0.1–2 mm diameter) (%); CLA represents the percentage of clay (0.002–0.1 mm diameter) (%); SIL denotes the percentage of silt (diameter < 0.002 mm) (%); C shows the percentage of organic carbon (%); and SN1 = 1 − SAN/100. The soil types and respective characteristic values in the YTRB were acquired from the China Soil Map, while K-factor mapping was generated according to Equation (3), in the spatial raster layer (Figure 3).

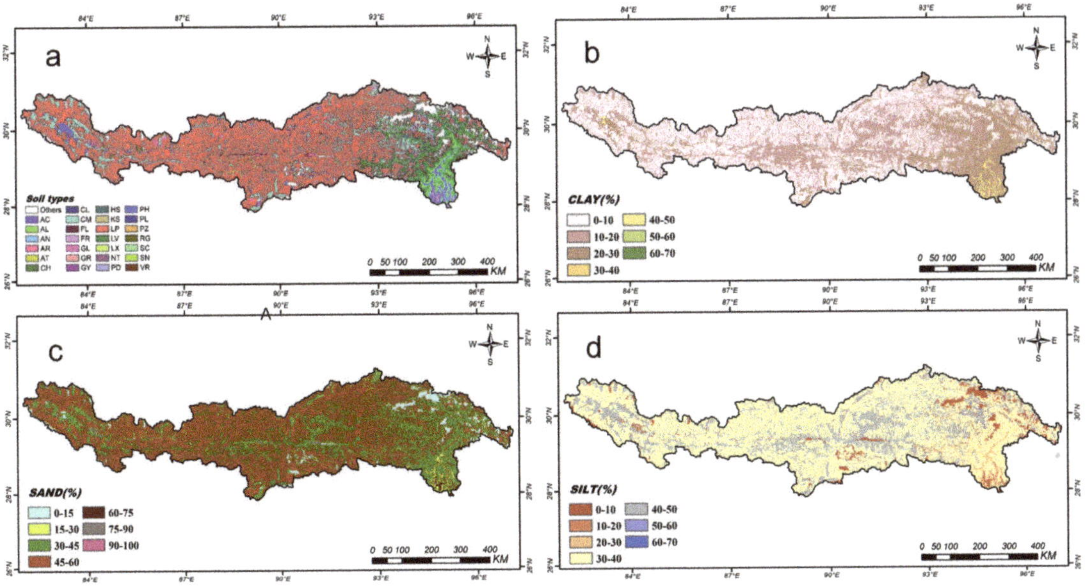

Figure 3. Yarlung Tsangpo soil data. (**a**) soil types; (**b**) percentage of clay; (**c**) percentage of sand; (**d**) percentage of silt.

3.2.3. Slope Length and Slope Steepness (LS) Factor

The LS factor consists of the ratio of soil loss per unit area of a field slope to that of the RUSLE standard slope, with a slope length of 22.13 m and a slope steepness of 9°, under the same conditions [11]. The effect of topography on soil erosion is mainly reflected in the movement and flow of sediment and water, as driven by gravity. The LS factor is made up of the slope length (L) and slope steepness (S) factors. Slope length refers to the distance from the initiation point of overland flow to the designated receiving channel along the flow route, while slope steepness is the ratio of the vertical height of the slope surface to the horizontal distance. Therefore, Equations (4) and (5) were used to calculate the LS factor [53,54]:

$$LS = \left(\frac{\lambda}{22.1}\right)^{m} \times \left(0.065 + 0.045S_g + 0.0065S_g^2\right) \tag{4}$$

$$\lambda = [flow\ accumulation * cell\ size]\ ; S_g = \frac{\sin(0.01745 \times \theta)}{0.09} \tag{5}$$

where LS indicates the slope length and steepness factor; λ represents slope length (m); S_g shows the grid slope in percentage; and θ denotes the slope (°). According to the above paragraph, m is taken as 0.5. SRTM-DEM imagery, with a 90 m spatial resolution, was used to identify filled sinks, which allowed flow direction and accumulation grids to be deduced. LS-factor mapping was computed using ArcGIS 10.2 software.

3.2.4. Cover Management Factor

The cover management factor is the ratio of soil loss from a specific land area to that from continuous fallow land under certain environmental conditions [42]. The C factor, which indicates the influence of surface vegetation, land use and agricultural activities on soil erosion, is an important factor in soil erosion assessment and can be estimated using several methods. In this study, the authors determined C-factor values for different vegetation types in the Chinese vegetation dataset (Figure 4), which, when combined with previous research results (Table 1), facilitated C-factor map development.

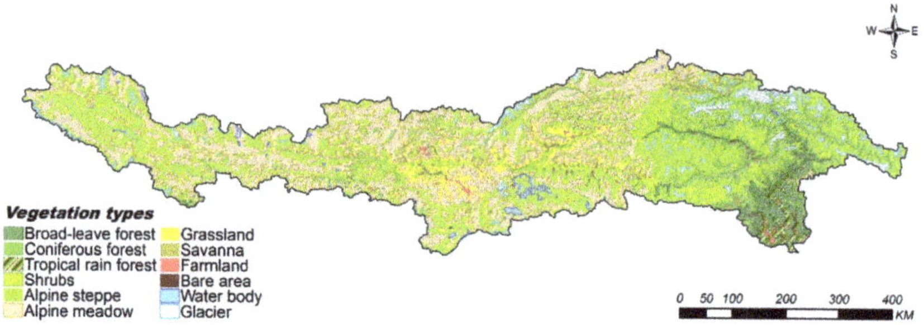

Figure 4. Yarlung Tsangpo river basin vegetation types.

Table 1. Yarlung Tsangpo river basin vegetation types and their C values.

Vegetation Type	Area (ha)	Proportion (%)	C Value	Source
Broad-leaved forest	840,050	3.26	0.06	Xiao et al. [55]
Coniferous forest	1,654,860	6.41	0.09	Xiao et al. [55]
Tropical rain forest	268,494	1.04	0.004	Xiao et al. [55]
Shrubs	3,810,622	14.77	0.09	Xiao et al. [55]
Alpine steppe	7,920,449	30.69	0.15	Wang & Jiao [56]
Alpine meadow	8,813,183	34.15	0.15	Wang & Jiao [56]
Grassland	910,231	3.53	0.11	Wang & Jiao [56]
Savanna	22,289	0.09	0.04	Xiao et al. [55]
Farmland	341,261	1.32	0.55	Yu et al. [55]
Bare area	2347	0.01	0.55	Yu et al. [55]
Water body	174,376	0.68	1	Zhou et al. [1]
Glacier	1,043,678	4.05	1	Zhou et al. [1]

3.2.5. Support Practice Factor

The support practice factor is the ratio of soil loss to soil erosion on upslopes and downslopes under the influence of specific support measures [11]. Large-scale P-factor mapping was performed using the method provided by the USDA handbook for LULC types, as classified using satellite images (Table 2) [57]. LULC mapping was generated using supervised classification of Landsat 8 OLI imagery and applying maximum likelihood settings in ENVI (version 5.1, Exelis Visual Information Solutions Inc., Boulder, CO, USA) software (Figure 5). For the classification accuracy of LULC data, a 2m GF image was used to evenly select 30 points for each of the ten land use types for verification and the results showed that the accuracy rate was about 88%.

Table 2. Yarlung Tsangpo river basin land use types and their P values.

Land Use Type	Area (ha)	Proportion (%)	P
Agricultural	405,554	1.57	0.5
Degraded forest	1,282,931	4.97	0.8
Densely vegetated	2,508,603	9.72	1.0
Grassland	12,322,344	47.76	0.9
Open forest	1,516,450	5.89	0.8
Rocky areas	5,508,307	21.35	1.0
Sandy areas	932,961	3.62	1.0
Settlements	25,155	0.10	0.1
Snow areas	402,839	1.56	1.0
Water bodies	892,381	3.46	0.0

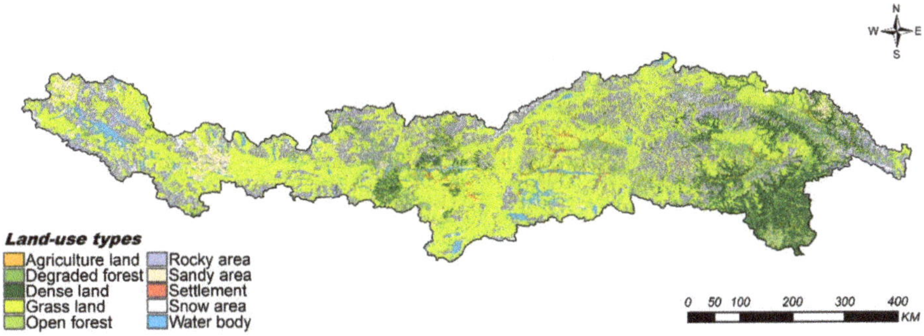

Figure 5. Yarlung Tsangpo river basin land use/land-cover map.

3.3. Using the Height above Nearest Drainage Model

The elevation from each raster cell above the nearest catchment unit can be described using the HAND model [13], which can separate flood-prone areas from areas with lower flood occurrence probability [17]. Topography is an influencing factor on hydrology, which determines the water flow direction and rate; thus, in order to establish a drainage network, it is necessary to maintain flow direction topological continuity.

The actual topography represented in SRTM data approximates the upper canopy [58] so that, for areas covered by dense or tall vegetation, a variable degree of relief masking occurs in SRTM data, producing depressions and pits. This causes inconsistencies between local drainage directions calculated using topography data (LDD) and actual flow paths. For this reason, in order to calculate a corrected LDD for this study, the original SRTM-DEM data were filled by raising the heights of pits to those of their pour points, in a procedure performed using ArcGIS 10.2. The contribution area grid was then determined using the D8 method proposed by Mark [59]. Then, since normalized terrain heights had been calculated using drainage network pixel elevations, the drainage network needed to be correctly defined by the contribution area threshold [19]. The HAND model, which was integrated into TerraHidro software, was applied by inputting the no-sinks DEM, the D8 flow grid and the drainage network grid (Figure 2), which allowed the HAND model nearest drainage mapping to be generated.

4. Results

4.1. Soil Loss Assessment Using the RUSLE Model

The method of combining GIS with the RUSLE model was used to determine the amount and spatial distribution of soil loss in the study area, including the following five erosion risk factors: rainfall erosivity, soil erosivity, slope length and steepness, soil cover management and soil conservation.

High-intensity precipitation is more likely to cause severe soil erosion and flood hazards. The R factors estimated for this study ranged from 38–4857 MJ mm ha^{-1} h^{-1} y^{-1}, with an average of 658 MJ mm ha^{-1} h^{-1} y^{-1} (Figure 6a). Actual weather station rainfall data were used to verify the accuracy of the R-factor estimation process. Data from five weather stations were compared with that of the Chinese Monthly Precipitation Dataset (Table 3), with the results showing that the latter were generally lower, with relative errors ranging between −17.88% and 5.03%.

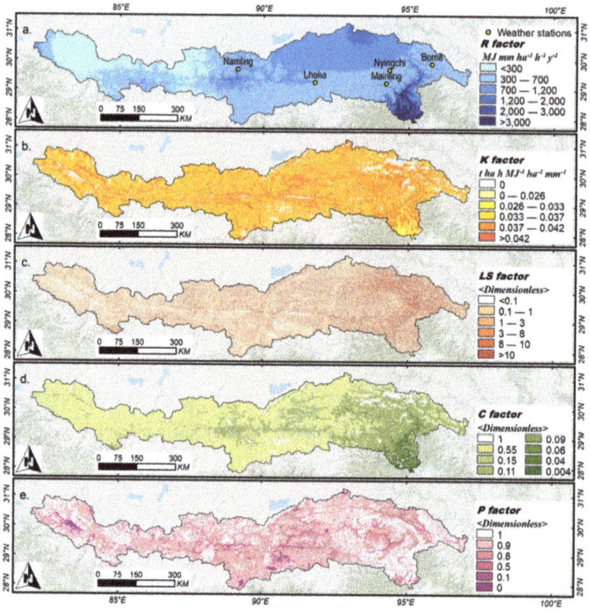

Figure 6. The RUSLE model factor maps: (**a**). R factor (precipitation erosivity), including locations of five weather stations used for verification; (**b**). K factor (soil erodibility); (**c**). LS factor (slope length and slope steepness); (**d**). C factor (cover management); (**e**). P factor (support practices).

Table 3. Validation of R-factor estimation using the Chinese Monthly Precipitation Dataset.

Station	Average Annual Precipitation (mm)			R Factor (MJ mm ha^{-1} h^{-1} y^{-1})		
	Station Value	In This Study	Relative Error (%)	Station Value	In This Study	Relative Error (%)
Namling	576	473	−17.88	1391	1230	−11.57
Lhoka	378	397	5.03	895	722	−19.33
Mainling	707	663	−6.22	638	698	9.40
Nyingchi	709	663	−6.49	709	721	1.69
Bomê	929	878	−5.49	992	920	−7.26

The relative errors for the R-factor estimates ranged from −19.33% to 9.40%, which was very consistent. The YTRB is very broad and also has elevation differences of > 7000 m (149–7782 m). These unique geographical conditions give rise to very large precipitation differences, with rainfall increasing from upstream to downstream, reaching a maximum of 2357 mm y^{-1}. The 490 mm month^{-1} peak in the downstream monthly average occurs in June (Figure 7). It was found that the R value was higher in areas with large rainfall and intensity and vice versa. However, a single R value cannot satisfactorily prove the change of rainfall distribution in the study area. Rainfall erosivity is an important factor in assessing soil erosion risk for future land use and climate change.

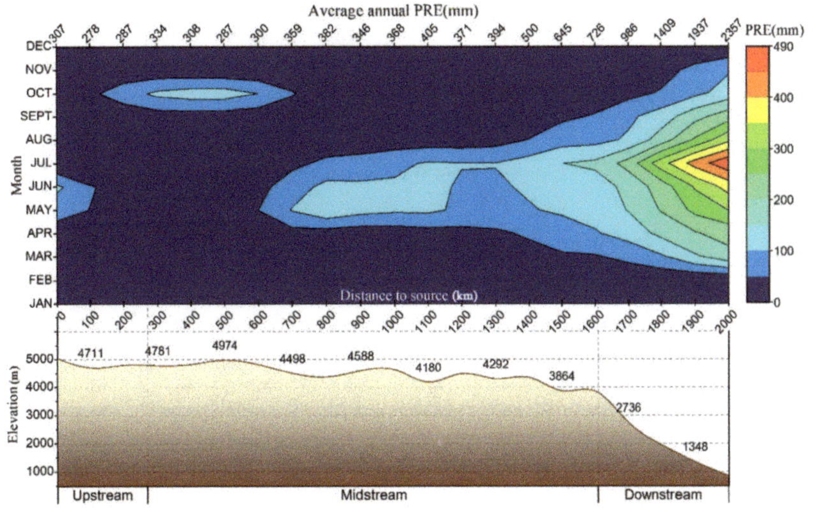

Figure 7. Spatial distribution of average monthly and annual precipitation. The bottom axis shows the distance to the river source and the top axis shows the average annual precipitation in each distance range. Blue (red) indicates low (high) precipitation. The lower panel indicates topography and elevation along the river.

The K factor shows the resistance of different soil types to runoff erosion and raindrop impact, with higher K values indicating lower resistance to soil erosion and vice versa [60]. According to the percentages of sand, silt, clay and organic carbon in different soils, the K values in the YTRB varied from 0 to 0.049 t ha h MJ^{-1} ha^{-1} mm^{-1}. The average K value was 0.034 t ha h MJ^{-1} ha^{-1} mm^{-1} (Figure 6b). Since settlements, glaciers and water bodies were not considered, they were assigned zero values.

The LS factor reveals the influence of topography on surface runoff and soil particle transport and in this study, its values increased gradually from the NW of the YTRB to the SE, with an average of 1.54 (Figure 6c). In several individual mountain areas, the values grew gradually from the valley to the peak, generally reaching > 5 at the ridges. In the Nyenchen Tanglha Mountains, for example, LS factor values reached 43.78 at Namcha Barwa inside the horseshoe bend in the SE Nyenchen Tanglha, which constituted the largest elevation difference in the lower reaches. In such areas, the steep terrain makes the region vulnerable to soil erosion.

Vegetation coverage and the depth of plant roots affect resistance to soil erosion. Vegetation types in the Chinese vegetation dataset were reclassified to obtain the C factor map; here, higher C factor values indicate a higher susceptibility to soil erosion and vice versa. Most of the YTRB is covered by alpine steppe and alpine meadow (64.84% of the total basin), with C-factor values here being 0.15. In contrast, values of 0.11 were observed in grasslands on hillsides (Figure 6d), while the middle and lower YTRB reaches were mostly covered by shrubs and coniferous forests (21.18% of the total basin), where C values were 0.09. Values of 0.06 were noted for those hillsides where the land coverage was characterized by broad-leaved forests. Tropical rain forests in the valley regions have high soil and water conservation, while agricultural land in the middle and lower reaches exhibited C-factor values of 0.55, showing that they are highly susceptible to soil erosion.

P factor values range from zero to one, where zero indicates effective manmade erosion resistance and one indicates that there is no resistance [12]. P values of 0.9 were observed in grassland and values of 0.5 were noted for riparian areas, where a large portion of the land was characterized as farmland (Figure 6e). Rocky and sandy areas were found to be widely distributed on high altitude mountains and forest coverage in the downstream canyons reached > 79.2%, with P values of 1. P values of 0.1 were observed for settled areas, reflecting their good manmade erosion resistance. Manmade erosion resistance was not considered to be a factor for water bodies, so they were assigned P factor values of zero.

4.2. Soil Erosion in the YTRB

Estimates for the annual YTRB soil erosion ranged from 0–1239 t ha^{-1} y^{-1} (Figure 8) and these results are consistent with the soil erosion spatial distribution estimates reported by Yang et al. The soil erosion average was estimated to be 4.21 t ha^{-1} y^{-1}, which was lower than the 5.43 t ha^{-1} y^{-1} estimated by Yang et al.; their higher figure was caused by an overestimation of the R value [61].

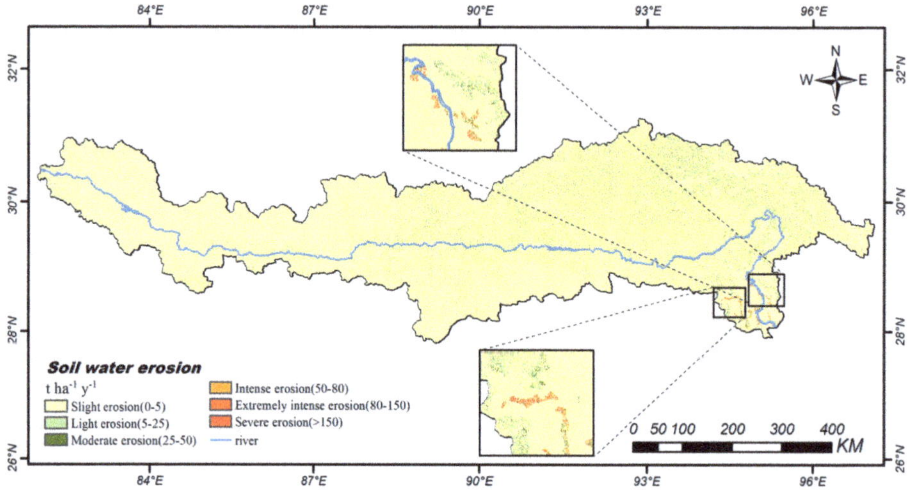

Figure 8. Yarlung Tsangpo river basin soil erosion; yellow (red) indicates slight (severe) soil erosion.

In order to map soil erosion susceptibility, the YTRB water soil erosion was mapped using six grades: slight (< 5 t ha^{-1} y^{-1}), light (5–25 t ha^{-1} y^{-1}), moderate (25–50 t ha^{-1} y^{-1}), intense (50–80 t ha^{-1} y^{-1}), extremely intense (80–150 t ha^{-1} y^{-1}) and severe (> 150 t ha^{-1} y^{-1}), as shown in Table 4. Erosion susceptibility rated as slight covered most of the study area (20,452,950 ha or 79.46% of the total basin), occurring mainly in the plateau of the upper and middle–upper reaches. Total soil loss in this region was estimated to be 28,272,583 t y^{-1}, accounting for 26.07% of the YTRB basin total. Due to the prevalence of low-intensity precipitation in these regions, it is difficult for soil particles to be transported by surface runoff, making local soil loss here easily sustainable.

Table 4. Yarlung Tsangpo river basin soil erosion rates and coverage.

Soil Erosion Range (t ha^{-1} y^{-1})	Soil Erosion Grade	Area (ha)	Area (%)	Annual Soil Erosion (t)	Total Soil Erosion (%)
<5	Slight	20,452,949.71	79.46	28,272,582.80	26.07
5–25	Light	4,543,178.91	17.65	46,992,121.22	43.34
25–50	Moderate	614,943.14	2.39	20,107,953.27	18.54
50–80	Intense	86,096.40	0.33	5,149,888.65	4.75
80–150	Extremely intense	23,608.96	0.09	2,428,166.16	2.24
>150	Severe	20,770.33	0.08	5,477,968.03	5.05

Erosion susceptibility rated as light covered 4,543,179 ha and accounted for 46,992,121 tons of soil loss. Much of this erosion susceptibility could be attributed to large terrain fluctuations, which amplify the effect of gravity on sediment. Moderate erosion susceptibility ratings, which occurred mostly on ridges, covered 614,943 ha and the soil loss from these areas was estimated to be 20,107,953 tons (18.54% of the total). This erosion was thought to be due mostly to the presence of Haplic Greyzems and Dystric Podzoluvisols, which are susceptible to soil erosion.

Farmland in the downstream valley had intense, extremely intense and severe water soil erosion susceptibility ratings, ranging from 50 to 1239 t ha^{-1} y^{-1}. In these regions, 13,056,023 tons (12.04% of the basin total) of soil erosion occurred over a land area of 130,476 ha (0.51% of the total) at a rate much greater than the sustainable soil loss rate. These figures showed that frequent agricultural activities have seriously damaged soil stability here, with surface runoff formed by high-intensity precipitation events upstream taking away large amounts of loose soil, resulting in severe soil loss.

4.3. Effects of Precipitation, LULC and Topography on Soil Erosion

The relationship between precipitation and soil erosion is illustrated in Figure 9a, with the results showing that soil loss rates were positively correlated with precipitation. The soil loss rate varied from 0.71 to 15.69 t ha^{-1} y^{-1}, with precipitation from < 200 mm to > 2000 mm, which indicated the predominant influence of precipitation on soil erosion. Precipitation in 60.23% of the basin ranged from 200 to 500 mm, especially in the middle-upper reaches, which accounted for 39.83% of the soil loss total. In contrast, regions with precipitation levels between 500 and 1000 mm (28.80% of the study area) were estimated to be responsible for 43.18% of the soil loss total. The lowest soil loss—1.11% of the annual total—was found in the arid plateau regions which experienced an annual average precipitation of <200 mm.

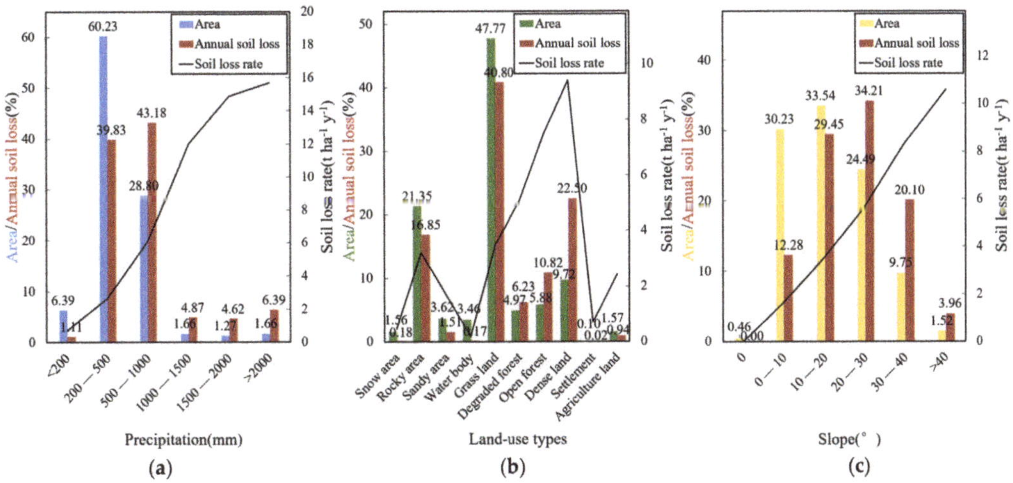

Figure 9. (**a**) Soil erosion associated with different precipitation ranges; (**b**) soil erosion in different land use types; and (**c**) soil erosion in different slope ranges. Blue, green and yellow columns represent area percentages in various precipitation ranges, land use types and slope ranges, respectively; red columns represent annual soil loss percentages.

Soil erosion rates and the annual soil losses associated with different land use types are mapped in Figure 9b. Open forests and densely vegetated land in the lower reaches had the highest soil erosion rates. The data show that although dense vegetation has a positive effect on soil conservation, the high precipitation downstream of the study area leads to its high erosion (33.32% of the total). The figures also showed that 16.85% of the soil erosion total occurred in rocky areas, which occupy 21.35% of the YTRB area. Grasslands constitute the main grazing areas in the basin; this is the land use that directly promotes sediment transfer and transportation and indirectly contributes to soil erosion by exposing fresh soil through grazing, trampling and digging. Thus, the soil erosion rate in these areas was higher than that estimated for agricultural land and as these areas are widespread throughout the study area, covering nearly half of the basin, they were a source of significant soil loss (40.80% of the total).

Overall, our estimates showed that 80% of the study area soil erosion occurred from grasslands, rocky areas and densely vegetated land, strongly indicating that these are the three land use types where soil loss protection measures should be focused. The influence of topography on soil erosion can be seen in Figure 9c, which shows that the soil loss rate estimate was significantly lower for flat terrain. The YTRB consists mostly of slopes between 10 and 20°, from which 29.45% of the soil erosion total was derived. It was also seen that, although the area with slopes between 20 and 30° was found to cover less area, its soil loss estimates were more greater, accounting for 34.21% of the total. Very steep regions, with 30–40° slopes, accounted for only 9.8% of the total area, but incurred 20.10% of the total soil loss. The significant effect of gravity on YTRB soil erosion was thus clearly evident.

4.4. Flood Hazard Mapping Using the HAND Model

The flood-risk map created using HAND model computations can be seen in Figure 10. In the absence of hydrological data, the model produced potential flood hazard zones with different flood risk levels, namely: none (>100 m depth), very low (15–100 m), low (10–15 m), moderate (5–10 m) and high (<5 m). The HAND model estimates suggested that approximately 9.84% (2,537,622 ha) of the area is vulnerable to high flood occurrence, with 1.04% located in a moderate hazard zone, as can be seen in Table 5. Approximately 1.54% of the area was found to have a low flood risk and 4.15% had a very low flood risk. With an area of 21,525,909 ha, the 'no flood risk' zone covered most of the basin, a total of 83.43%.

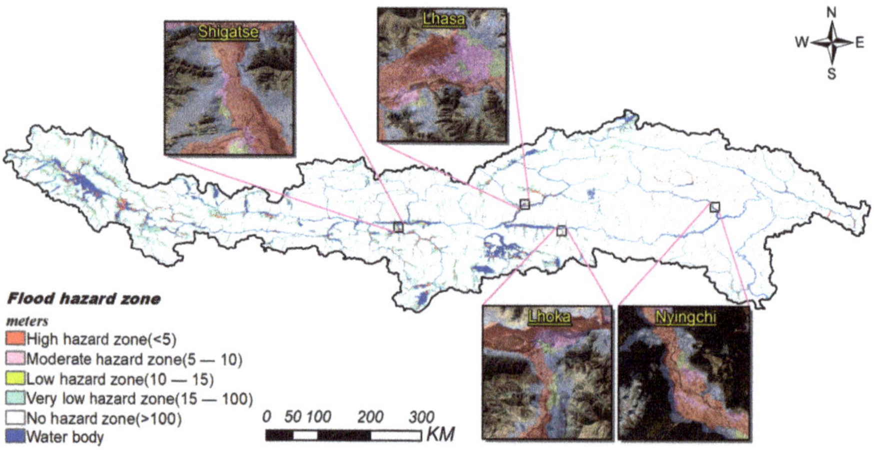

Figure 10. Yarlung Tsangpo river basin flood hazard zones. Satellite images of major cities are from Google Maps. White (red) indicates no (high) flood risk.

Table 5. Yarlung Tsangpo river basin flood hazard zones and coverage.

Flood Hazard Zone (m)	Area (ha)	Area (%)
High (<5)	2,537,622.95	9.84
Moderate (5–10)	268,151.88	1.04
Low (10–15)	397,292.32	1.54
Very low (15–100)	1,071,023.77	4.15
No hazard (>100)	21,525,909.08	83.43

4.5. Soil Erosion and Flood Hazard Synergies

Examining the potential soil erosion and flood hazard zones in areas with different slopes revealed contrasting scenarios. Most of the extreme and severe soil erosion estimates occurred in areas with 20–30° slopes, as shown in Table 6, with steeper regions indicating higher soil loss rates. Conversely, most moderate and high potential flood risk zones occurred in the areas with slopes of 0–10°, which only covered ~ 30% of the total basin, as can be seen in Table 7. It was also noted that moderate and high flood hazard zones were found to be prevalent in areas that included major basin cities and towns (including Lhasa, the Tibetan capital), with their highly concentrated populations, buildings and facilities. These cities are located in plain and terrace areas with slopes of 0-10°, with little topographic relief and are low-lying flood-prone areas within the watershed. Additionally, the increase in surface imperviousness due to urbanization makes surface runoff higher and increases the likelihood of flooding.

Table 6. Areas of soil erosion under different Yarlung Tsangpo river basin slope ranges.

Slope Degree (°)	Slight Soil Erosion		Light		Moderate		Intense		Extremely Intense		Severe	
	ha	%	ha	%	ha	%	ha	%	ha	%	ha	%
0	118,191.55	0.46	2.85	0.00	0.52	0.00	0	0.00	0	0.00	0	0.00
0–10	7,206,708.72	28.03	510,034.31	1.98	46,884.67	0.18	7264.24	0.03	2446.48	0.01	1365.86	0.01
0–20	6,982,718.36	27.16	1,476,670.55	5.74	134,854.66	0.52	19,621.67	0.08	7166.27	0.03	5675.01	0.02
0–30	4,409,361.76	17.15	1,637,668.97	6.37	217,520.93	0.85	28,982.35	0.11	8003.05	0.03	7963.62	0.03
0–40	1,498,273.05	5.83	809,853.16	3.15	172,528.87	0.67	23,121.38	0.09	4782.41	0.02	4742.85	0.02
>40	231,078.53	0.90	108,417.10	0.42	43,119.20	0.17	7098.06	0.03	1207.26	0.00	1022.98	0.00

Table 7. Flood hazard zone areas under different Yarlung Tsangpo river basin slope ranges.

Slope Degree (°)	No Hazard Zone		VERY LOW		Low		Moderate		High	
	ha	%	ha	%	ha	%	ha	%	ha	%
0	340.96	0.00	1611.24	0.01	655.84	0.00	1137.51	0.00	113,926.83	0.44
0–10	4,632,904.37	17.96	1,808,157.27	7.01	217,530.54	0.84	336,071.54	1.30	801,184.48	3.11
10–20	8,060,866.55	31.24	442,714.69	1.72	28,390.06	0.11	34,933.84	0.14	82,164.17	0.32
20–30	6,046,054.18	23.43	195,204.15	0.76	14,207.27	0.06	16,594.81	0.06	46,046.45	0.18
30–40	2,402,127.25	9.31	77,187.37	0.30	6208.52	0.02	6908.68	0.03	23,181.79	0.09
>40	373,117.74	1.45	12,582.55	0.05	1137.51	0.00	1090.94	0.00	4484.20	0.02

Various scenarios present a synergy between soil erosion and flood hazard. For example, the flood disaster areas are mainly distributed in the relatively low and gentle areas of the study area. The process of soil erosion mainly occurs on steep hillsides and a high slope gradient produces significant water impact forces, which in turn cause sediments to accumulate in sinks between slopes and floodplains. In spring and autumn, the temperature changes frequently, the freeze–thaw erosion is strong and the broken material erodes under the action of snow (ice) meltwater or gravity and accumulates in the gentle slope area to form an alluvial fan (alluvial skirt) [6]. Whenever a sink is filled, the material will overtop and be stored again at the next downslope opportunity, doing this repeatedly until it reaches the floodplain, where its presence results in a rise in the potential flood incidence level [62]. This indicates that the increase in sediment content in surface runoff and flooding during soil erosion in flood season may increase the possibility of a flood disaster. On the contrary, heavy rainfall and floods will lead to serious soil erosion. It is obvious that soil loss and flood incidence will have a synergistic impact on the environmental degradation of the catchment area. During the flood season, higher sediment concentrations in surface runoff prevent surface runoff from being used efficiently; the runoff cannot be taken from the river and is forced to discharge to the lower reaches of the river, increasing the flood risk [63,64].

5. Conclusions

Soil erosion prevention planning and decision making require potential soil loss to be quantified. In this study, RS and GIS technologies were employed in the RUSLE and HAND model calculation processes to describe the scale and spatial distribution of soil erosion and flood hazards in the YTRB. This resulted in soil erosion estimates ranging from 0 to 1239 t ha^{-1} y^{-1}, with an average of 4.21 t ha^{-1} y^{-1}. Very intense and severe soil erosion was found to occur in downstream valleys, where frequent and high precipitation combined with agricultural land uses aggravate erosion.

Slight levels of erosion, which covered most of the basin, were spatially distributed in the upper and middle–upper reaches, where there was low precipitation. Contrary to previous findings, this study found that dense forests had the highest soil loss rate, which may result from the combined effects of high precipitation and steep slopes in these regions. It was also shown that the soil loss rate increased with slope gradient, confirming a high correlation between soil erosion intensity and topographic relief.

Flood hazard prediction showed that approximately 17% of the YTRB was threatened by flood hazards, with the high flood hazard rating covering 2,537,622 ha of the basin, with an additional 268,151 ha vulnerable to moderate floods. Approximately 397,292 ha of the basin was found to be in a low hazard zone, while 1,071,023 ha was found to be in the very low hazard zone. The most flood-fragile areas were mostly distributed in riverside valleys and plains and in areas with high concentrations of residents and buildings. Thus, enhancing flood warnings and protection measures in these regions should be considered a priority.

6. Discussion

We calculated the soil erosion intensity of the Yarlung Tsangpo River basin using the RUSLE model, considering precipitation erodibility, different land use types, physical and chemical properties of soil and topographic factors. In the calculation of the R factor, we used the R factor based on monthly precipitation, pi and annual rainfall, p, due to the lack of data on rainfall intensity. This resulted in the same R factor for high-intensity rainfall over multiple consecutive days and scattered over a month, which is usually different from the actual situation [65]. Additionally, due to the large scale of the study, we assigned the same P value to the same LULC type in the calculation of the P factor. However, in areas where human interventions are present, P factor values for the same land use types should take into account for the effect of a slope [66–68]. At the same time, field-based soil loss measurements are not currently available for the YTRB and so we have been unable to verify the results of our study using field data. Therefore, it is suggested that government and policy makers conduct field surveys in the regions identified as being susceptible to soil erosion and establish soil loss measurement stations, so that accurate and real-time soil loss information can be captured.

Author Contributions: Conceptualization, S.C., S.Z, X.W., H.S. and J.Q.; formal analysis, S.C., S.Z., X.W. C.H. and S.L.; funding acquisition, S.Z. and H.S.; investigation, S.C., S.Z., X.W., C.H. and L.L.; methodology, S.C., S.Z. and X.W.; validation, S.C., X.W., C.H. and L.H.; writing—original draft, S.C., S.Z. and X.W.; writing—review and editing, H.S., J.Q., L.L., L.H. and S.L. All authors have read and agreed to the published version of the manuscript.

Funding: This research was funded by the National Natural Science Fund of China (Grant No.41302282; Grant No.41401659), the Science & Technology Department of Sichuan Province (Grant No. 2015JY0145), the Remote sensing monitoring of water conservation response to vegetation cover change in alpine wetlands in northwest Sichuan (Grant No.202110616001), the Quantitative Study on Evapotranspiration Emission in Zoige Wetland in the Upper Reaches of the Yellow River (Grant No. s202210616002).

Data Availability Statement: The data presented in this study are contained within the article.

Acknowledgments: We greatly appreciate the support of Key Lab of Earth Exploration & Information Techniques of Chinese Ministry Education and are thankful for the support of the Scientific Innovation Team of Remote Sensing Science and Technology of Chengdu University of Technology. We are also grateful for the anonymous reviewer for their insight and critical review of the manuscript.

Conflicts of Interest: The authors declare no conflict of interest.

References

1. Zhou, P.; Luukkanen, O.; Tokola, T.; Nieminen, J. Effect of vegetation cover on soil erosion in a mountainous watershed. *Catena* **2008**, *75*, 319–325. [CrossRef]
2. Xiaoping, Y.; Leixiang, W.; Jun, X.; Yongqian, W. Assessment of soil erosion sensitivity and response analysis of soil erosion sensitivity to land use change in Chengde City. *Water Resour. Hydropower Eng.* **2021**, *52*, 221–232. [CrossRef]
3. Zhang, X.Y.; Wang, J.Q.; Wang, T.Q. Water and soil loss in Tibet and measures for conservation. *Adv. Sci. Technol. Water Resour.* **2005**, 45–48.
4. Fang, L.; Huang, J.; Cai, J.; Nitivattananon, V. Hybrid approach for flood susceptibility assessment in a flood-prone mountainous catchment in China. *J. Hydrol.* **2022**, *612*, 128091. [CrossRef]
5. Gupta, S.; Javed, A.; Datt, D. Economics of flood protection in India. *Nat. Hazards* **2003**, *28*, 199–210. [CrossRef]
6. Haidong, L.; Weishou, S.; Changxin, Z.; Lei, Y. Soil Erosion in the Source Area of the Yarlung Zangbo in China. *J. Ecol. Rural. Environ.* **2010**, *26*, 25–30.
7. Liu, S.Z.; Zhang, J.G.; Gu, S.X. Study on the soil erosion types in Tibet. *Mt. Res.* **2006**, *24*, 592–596. [CrossRef]
8. Xingmin, M.; Pengfei, L.; Bintao, L.; Guangju, Z.; Peng, G.; Wenyi, S. Spatial-Temporal Development and Driving Mechanisms of Erosion on the Chinese Loess Plateau Between 1901 and 2016. *Yellow River* **2022**, *44*, 36–45. [CrossRef]
9. Weili, H.; Junqing, Z.; Zhibo, W.; Zhenting, W.; Gaowa, B. Research Progress of Domestic Soil Erosion Prediction Model. *J. Inn. Mong. For. Sci. Technol.* **2022**, *48*, 55–59.
10. Shuihong, S.; Hengkai, L.; Zhiwei, Z. Spatial-temporal variation of soil erosion in Dongjiang source area based on RUSLE model. *J. Earth Environ.* **2022**, *13*, 344–353. [CrossRef]
11. Wischmeier, W.; Smith, D. Predicting Rainfall Erosion Losses—A Guide to Conservation Planning. In *Agriculture Handbooks, USA, No. 537*; Department of Agriculture, Science and Education Administration: West Lafayette, IN, USA, 1978; Volume 537, 62p.
12. Renard, K.G.; Foster, G.R.; Weesies, G.A.; Mccool, D.K.; Yoder, D.C. Predicting soil erosion by water: A guide to conservation planning with the Revised Universal Soil Loss equation (RUSLE). In *Agricultural Handbook, USA, No. 703*; USDA, Agricultural Research Service: Beltsville, MD, USA, 1997
13. Rennó, C.; Nobre, A.; Cuartas, L.; Soares, J.; Hodnett, M.; Tomasella, J.; Waterloo, M.J. HAND, a new terrain descriptor using SRTM-DEM: Mapping terra-firme rainforest environments in Amazonia. *Remote Sens. Environ.* **2008**, *112*, 3469–3481. [CrossRef]
14. Nobre, A.D.; Cuartas, L.A.; Hodnett, M.; Renno, C.D.; Rodrigues, G.; Silveira, A.; Waterloo, M.; Saleska, S. Height Above the Nearest Drainage—A hydrologically relevant new terrain model. *J. Hydrol.* **2011**, *404*, 13–29. [CrossRef]
15. Tingting, L.; Peijun, S.; Chunying, M.; Fanya, S.; Juan, N.; Bo, C. Height above nearest drainage and application in flood inundation mapping in China. *J. Beijing Norm. Univ.* **2022**, *58*, 300–309. [CrossRef]
16. Zheng, X.; Tarboton, D.G.; Maidment, D.R.; Liu, Y.Y.; Passalacqua, P. River Channel Geometry and Rating Curve Estimation Using Height above the Nearest Drainage. *J. Am. Water Resour. Assoc.* **2018**, *54*, 785–806. [CrossRef]
17. Chow, C.; Twele, A.; Martinis, S. An assessment of the Height Above Nearest Drainage terrain descriptor for the thematic enhancement of automatic SAR-based flood monitoring services. In *Remote Sensing for Agriculture, Ecosystems and Hydrology XVIII*; SPIE: Bellingham, WA, USA, 2016; Volume 9998.
18. Liu, Y.; Maidment, D.; Tarboton, D.; Zheng, X.; Yıldırım, A.; Sazib, N.; Wang, S. A CyberGIS Approach to Generating High-resolution Height Above Nearest Drainage (HAND) Raster for National Flood Mapping. In Proceedings of the Third International Conference on CyberGIS and Geospatial Data Science, Urbana, IL, USA, 26–28 July 2016.
19. Speckhann, G.A.; Borges Chaffe, P.L.; Goerl, R.F.; de Abreu, J.J.; Altamirano Flores, J.A. Flood hazard mapping in Southern Brazil: A combination of flow frequency analysis and the HAND model. *Hydrol. Sci. J.* **2018**, *63*, 87–100. [CrossRef]
20. Nyssen, J.; Poesen, J.; Moeyersons, J.; Deckers, J.; Haile, M.; Lang, A. Human impact on the environment in the Ethiopian and Eritrean highlands—A state of the art. *Earth Sci. Rev.* **2004**, *64*, 273–320. [CrossRef]
21. Qin, W.; Zhu, Q.K.; Zhang, Y. Soil erosion assessment of small watershed in Loess Plateau based on GIS and RUSLE. *Trans. Chin. Soc. Agric. Eng.* **2009**, *25*, 157–163. [CrossRef]

22. Reusing, M.; Schneider, T.; Ammer, U. Modelling soil loss rates in the Ethiopian Highlands by integration of high resolution MOMS-02/D2-stereo-data in a GIS. *Int. J. Remote Sens.* **2010**, *21*, 1885–1896. [CrossRef]
23. Ren, Y.; Liu, J.; Liu, S.; Wang, Z.; Liu, T.; Shalamzari, M.J. Effects of Climate Change on Vegetation Growth in the Yellow River Basin from 2000 to 2019. *Remote Sens.* **2022**, *14*, 687. [CrossRef]
24. Liu, J.; Zhang, W.; Nie, N. Spatial Downscaling of TRMM Precipitation Data Using an Optimal Subset Regression Model with NDVI and Terrain Factors in the Yarlung Zangbo River Basin, China. *Adv. Meteorol.* **2018**, *2018*, 3491960. [CrossRef]
25. Liu, J.; Zhang, W.; Liu, T.; Li, Q. Runoff Dynamics and Associated Multi-Scale Responses to Climate Changes in the Middle Reach of the Yarlung Zangbo River Basin, China. *Water* **2018**, *10*, 295. [CrossRef]
26. Lu, D.S.; Li, G.Y.; Valladares, G.; Batistella, M. Mapping soil erosion risk in Rondônia, Brazilian Amazonia: Using RUSLE, remote sensing and GIS. *Land Degrad. Dev.* **2004**, *15*, 499–512. [CrossRef]
27. Xu, Y.Q.; Shao, X.M. Estimation of soil erosion supported by GIS and RUSLE: A case study of Maotiaohe Watershed, Guizhou Province. *J. Beijing For. Univ.* **2006**, *28*, 67–71.
28. Mengmei, W.; Suhua, F.; Baoyuan, L. Quantitative Research of Water Erosion on the Qinghai-Tibet Plateau. *Adv. Earth Sci.* **2021**, *36*, 740–752. [CrossRef]
29. Tongde, C.; Juying, J.; Haolin, W.; Chunjing, Z.; Hong, L. Progress in Research on Soil Erosion in Qinghai-Tibet Plateau. *Acta Pedol. Sin.* **2020**, *57*, 547–564. [CrossRef]
30. Feng, W.; Jintong, L.; Tonggang, F.; Hui, G.; Fei, Q. Spatio-temporal variations in soil erosion and its influence factors in Taihang Mountain area based on RUSLE modeling. *Chin. J. Eco-Agric.* **2022**, *30*, 1064–1076. [CrossRef]
31. Ruiyin, C.; Dongchun, Y.; Anbang, W.; Chenggang, L.; Zhonglin, S. Research on Soil Erosion in Key Prevention and Control Region of Soil and Water Loss Based on GIS/CSLE in Sichuan Province. *J. Soil Water Conserv.* **2020**, *34*, 17–26. [CrossRef]
32. Zhengfa, C.; Aimin, G.; Dongwei, N.; Liudong, Z.; Jianxiong, W.; Biao, X. Characteristics of Soil Erosion and Nutrient Lossin Yunnan Province Based on RUSLE Model. *J. Soil Water Conserv.* **2021**, *35*, 7–14. [CrossRef]
33. Cuiyan, L.; Xihua, S.; Yueqing, L. Quantitative attribution of soil erosion for different geomorphological types in mountainous area of Mid-Shandong. *Water Resour. Hydropower Eng.* **2020**, *51*, 209–219. [CrossRef]
34. Congjian, S.; Ruojing, L.; Zhenjing, Z.; Jiarui, W.; Jiulin, S. Characteristic analysis of soil and water loss in typical small watersheds of the Middle Yellow River based on RUSLE model. *Southwest China J. Agric. Sci.* **2022**, *35*, 200–208. [CrossRef]
35. Xiaomei, L.; Yong, L.; Zhigang, H.; Yusong, D.; Hao, G.; Qingheng, H. Soil Erosion Characteristics of Sugarcane-Growing Watershed Based on RUSLE. *Bull. Soil Water Conserv.* **2022**, *42*, 82–88+397. [CrossRef]
36. Shuangyun, P.; Kun, Y.; Liang, H.; Quanli, X.; Yajun, H. Spatio-temporal evolution analysis of soil erosion based on USLE model in Dianchi Basin. *Trans. Chin. Soc. Agric. Eng.* **2018**, *34*, 138–146+305. [CrossRef]
37. Jian, Z.; Rong, L. Soil Erosion and Subarea Characteristics in Yarlung Tsangpo RiverBasi. *J. Yangtze River Sci. Res. Inst.* **2008**, *25*, 42–45.
38. Xiaobu, C.; Cheng, Q.; Jie, H. Soil Erosion and Its Countermeasures in the Middle Reaches of Yarlung Zangbo River in Tibet Autnomous Region. *Bull. Soil Water Conserv.* **1996**, *5*, 48–53. [CrossRef]
39. Manyi, C.; Gang, Z.; Dahong, Z.; Shiqiang, Z. Global snowmelt flood disasters and their impact from 1900 to 2020. *J. Glaciol. Geocryol.* **2022**, *10*, 0361. [CrossRef]
40. Chen, T.; Niu, R.Q.; Li, P.X.; Zhang, L.P.; Du, B. Regional soil erosion risk mapping using RUSLE, GIS and remote sensing: A case study in Miyun Watershed, North China. *Environ. Earth Sci.* **2011**, *63*, 533–541. [CrossRef]
41. Dutta, S.; Ghosh, S. Impact of climate and land use changes on the flood hazard of the middle Brahmaputra Reach, India. *J. Disaster Res.* **2012**, *7*, 573–581. [CrossRef]
42. Zhao, W.W.; Fu, B.J.; Qiu, Y. An Upscaling Method for Cover-Management Factor and Its Application in the Loess Plateau of China. *Int. J. Environ. Res. Public Health* **2013**, *10*, 4752–4766. [CrossRef]
43. Song, M.H.; Ma, Y.M.; Zhang, Y.; Li, M.S.; Ma, W.Q.; Sun, F.L. Analyses of Characteristics and Trend of Air Temperature Variation along the Brahmaputra Valley. *Clim. Environ. Res.* **2011**, *16*, 760–766. [CrossRef]
44. Peng, S.Z. *1-km Monthly Precipitation Dataset for China (1901–2017)*; National Tibetan Plateau Data Center: Beijing, China, 2020.
45. Ding, Y.X.; Peng, S.Z. Spatiotemporal Trends and Attribution of Drought across China from 1901–2100. *Sustainability* **2020**, *12*, 477. [CrossRef]
46. Peng, S.Z.; Ding, Y.X.; Liu, W.Z. 1 km monthly temperature and precipitation dataset for China from 1901 to 2017. *Earth Syst. Sci. Data* **2019**, *11*, 1931–1946. [CrossRef]
47. Peng, S.Z.; Gang, C.C.; Cao, Y.; Chen, Y.M. Assessment of climate change trends over the Loess Plateau in China from 1901 to 2100. *Int. J. Climatol.* **2018**, *38*, 2250–2264. [CrossRef]
48. Peng, S.Z.; Ding, Y.X.; Wen, Z.M.; Chen, Y.M.; Cao, Y.; Ren, J.Y. Spatiotemporal change and trend analysis of potential evapotranspiration over the Loess Plateau of China during 2011–2100. *Agric. For. Meteorol.* **2017**, *233*, 183–194. [CrossRef]
49. Food and Agriculture Organization of the United Nations; International Institute for Applied Systems Analysis. *China Soil Map Based Harmonized World Soil Database (HWSD) (v1.1) (2009)*; Food and Agriculture Organization of the United Nations (FAO): Rome, Italy, 2019.
50. Fischer, G.; Nachtergaele, F.; Prieler, S.; van Velthuizen, H.T.; Verelst, L.; Wiberg, D. *Global Agro-Ecological Zones Assessment for Agriculture (GAEZ 2008)*; IIASA: Laxenburg, Austria; FAO: Rome, Italy, 2008.
51. Wischmeier, W.H.; Mannering, J.V. Relation of Soil Properties to its Erodibility. *Soil Sci. Soc. Am. J.* **1969**, *33*, 131–137. [CrossRef]

52. Williams, J.R. EPIC-erosion/productivity impact calculator: 1. Model documentation. *Tech. Bull. United States Dep. Agric.* **1990**, *4*, 206–207. [CrossRef]
53. Niyonsenga, J.D.; Mugabowindekwe, M.; Mupenzi, C. Spatial analysis of soil erosion sensitivity using the Revised Universal Soil Loss Equation model in Nyamasheke District, Western Province of Rwanda. *Trans. Gis* **2020**, *25*, 735–750. [CrossRef]
54. Ganasri, B.P.; Ramesh, H. Assessment of soil erosion by RUSLE model using remote sensing and GIS—A case study of Nethravathi Basin. *Geosci. Front.* **2016**, *7*, 953–961. [CrossRef]
55. Xiao, Y.; Xie, G.; An, K. The function and economic value of soil conservation of ecosystems in Qinghai-Tibet Plateau. *Acta Ecol. Sin.* **2003**, *23*, 2367–2378.
56. Wang, W.Z.; Jiao, J.Y. Quantitative evaluation on factors influencing soil erosion in China. *Bull. Soil Water Conserv.* **1996**.
57. Wischmeir, W.H. *Predicting Rainfall-Erosion Losses from Cropland East of the Rocky Mountain: Guide for Selection of Practices for Soil and Water Conservation*; USDA: Washington, DC, USA, 1965; Volume 282.
58. Valeriano, M.M.; Kuplich, T.M.; Storino, M.; Amaral, B.D.; Mendes, J.N.; Lima, D.J. Modeling small watersheds in Brazilian Amazonia with shuttle radar topographic mission-90m data. *Comput. Geosci.* **2006**, *32*, 1169–1181. [CrossRef]
59. Mark, D.M. Network models in geomorphology. In *Modelling Geomorphological Systems*; John Wiley: New York, NY, USA, 1988.
60. Kumar, S.; Kushwaha, S.P.S. Modelling soil erosion risk based on RUSLE-3D using GIS in a Shivalik sub-watershed. *J. Earth Syst. Sci.* **2013**, *122*, 389–398. [CrossRef]
61. Yang, Y.Y.; Zhao, R.Y.; Shi, Z.; Viscarra Rossel, R.A.; Wan, D.; Liang, Z.Z. Integrating multi-source data to improve water erosion mapping in Tibet, China. *Catena* **2018**, *169*, 31–45. [CrossRef]
62. Dotterweich, M. The history of human-induced soil erosion: Geomorphic legacies, early descriptions and research and the development of soil conservation—A global synopsis. *Geomorphology* **2013**, *201*, 1–34. [CrossRef]
63. Li, X.; Wei, X.; Wang, N.; Cheng, H. Maximum Grade Approach to Surplus Floodwater of Hyperconcentration Rivers in Flood Season and its Application. *Water Resour. Manag.* **2011**, *25*, 2575–2593. [CrossRef]
64. Li, X.; Wei, X. Analysis of the Relationship between Soil Erosion Risk and Surplus Floodwater during Flood Season. *J. Hydrol. Eng.* **2014**, *19*, 1294–1311. [CrossRef]
65. Xu, Z.; Zhang, S.; Zhou, Y.; Hou, X.; Yang, X. Characteristics of watershed dynamic sediment delivery based on improved RUSLE model. *Catena* **2022**, *219*, 106602. [CrossRef]
66. Xiong, M.; Sun, R.; Chen, L. Effects of soil conservation techniques on water erosion control: A global analysis. *Sci. Total Environ.* **2018**, *645*, 753–760. [CrossRef]
67. Xiong, M.; Sun, R.; Chen, L. Global analysis of support practices in USLE-based soil erosion modeling. *Prog. Phys. Geogr. Earth Environ.* **2019**, *43*, 391–409. [CrossRef]
68. Kumar, M.; Sahu, A.P.; Sahoo, N.; Dash, S.S.; Raul, S.K.; Panigrahi, B. Global-scale application of the RUSLE model: A comprehensive review. *Hydrol. Sci. J.* **2022**, *67*, 806–830. [CrossRef]

atmosphere

MDPI

Article

Integrated SWAT-MODFLOW Modeling-Based Groundwater Adaptation Policy Guidelines for Lahore, Pakistan under Projected Climate Change, and Human Development Scenarios

Rana Ammar Aslam [1,*], Sangam Shrestha [2], Muhammad Nabeel Usman [3], Shahbaz Nasir Khan [1], Sikandar Ali [4], Muhammad Shoaib Sharif [4], Muhammad Waqas Sarwar [5], Naeem Saddique [4], Abid Sarwar [4], Mohib Ullah Ali [6] and Arfan Arshad [7,*]

1 Department of Structures and Environmental Engineering, University of Agriculture Faisalabad, Jail Road, Faisalabad 38040, Pakistan
2 Water Engineering and Management, Asian Institute of Technology, Klong Luang, P.O. Box 4, Pathum Thani 12120, Thailand
3 Hydrology and River Basin Management, Technical University of Munich, 80331 Munich, Germany
4 Department of Irrigation and Drainage, University of Agriculture Faisalabad, Jail Road, Faisalabad 38040, Pakistan
5 School of Engineering, University of Waikato, Waikato 3216, New Zealand
6 Punjab Irrigation Department, Lahore 54000, Pakistan
7 Department of Biosystems and Agricultural Engineering, Oklahoma State University, Stillwater, OK 74078, USA
* Correspondence: ammar.aslam@uaf.edu.pk (R.A.A.); aarshad@okstate.edu (A.A.)

Citation: Aslam, R.A.; Shrestha, S.; Usman, M.N.; Khan, S.N.; Ali, S.; Sharif, M.S.; Sarwar, M.W.; Saddique, N.; Sarwar, A.; Ali, M.U.; et al. Integrated SWAT-MODFLOW Modeling-Based Groundwater Adaptation Policy Guidelines for Lahore, Pakistan under Projected Climate Change, and Human Development Scenarios. *Atmosphere* **2022**, *13*, 2001. https://doi.org/10.3390/atmos13122001

Academic Editor: Haishan Chen

Received: 26 October 2022
Accepted: 28 November 2022
Published: 29 November 2022

Abstract: Urban aquifers are experiencing increasing pressures from climate change, land-use change, and abstraction, consequently, altering groundwater levels and threatening sustainable water availability, consumption, and utilization. Sustainability in such areas requires the adaptation of groundwater resources to these stressors. Consequently, this research made projections about future climate, land use, and abstraction, examines how these drives will affect groundwater levels, and then proposes adaptation strategies to reduce the impact on Lahore's groundwater resources. The objectives are achieved using an integrated modeling framework involving applications of Soil Water Assessment Tool (SWAT) and MODFLOW models. The results indicated a projected rise in T_{min} by ~2.03 °C and T_{max} by ~1.13 °C by 2100 under medium (RCP 4.5) and high-end (RCP 8.5) scenarios, respectively. Future precipitation changes for mid, near and far periods are projected to be −1.0%, 25%, and 24.5% under RCP4.5, and −17.5%, 27.5%, and 29.0% under RCP8.5, respectively. The built-up area in the Lahore division will dominate agricultural land in the future with an expansion from 965 m^2 to 3716 km^2 by the year 2100 under R1S1 (R2S2) land-use change scenarios (significant at p = 5%). The future population of the Lahore division will increase from 6.4 M to 24.6 M (28.7 M) by the year 2100 under SSP1 (SSP3) scenarios (significant at p = 5%). Groundwater level in bult-up areas will be projected to decline from 185 m to 125 m by 2100 due to increasing groundwater abstraction and expansion in the impermeable surface under all scenarios. In contrast, agricultural areas show a fluctuating trend with a slight increase in groundwater level due to decreasing abstraction and multiple recharge sources under combined scenarios. The results of this study can be a way forward for groundwater experts and related institutions to understand the potential situation of groundwater resources in the Lahore division and implement adaptation strategies to counteract diminishing groundwater resources.

Keywords: groundwater sustainability; impact assessment; climate change; adaptation strategies; land use change; abstraction; SWAT; MODFLOW

1. Introduction

Groundwater resources in megacities of the developing world are susceptible to over-exploitation and other stresses [1]. The multiple interactive triggers such as climate change, population, agriculture, and industry increase pressure on groundwater resources [2–6]. These pressures alter water table elevation and its temporal variations. Established literature highlights some areas in different regions around the world where human-induced climate change, land-use change, and abstraction are altering the groundwater attributes up to a level at which local systems are unable to cope with their negative impacts [7–10] and thereby affecting the sustainability of groundwater reserves.

Lahore is a metropolitan city in Pakistan where groundwater is the dominant source for domestic, agricultural, and industrial use. Due to overexploitation, Lahore is observing a fast groundwater level recession in certain areas [4,11,12]. Historical analysis of the climate reveals a rise in temperature in the region [13–16]. The available literature on water use highlights that more than 75% of users in Lahore have access to the piped water supply system (WSS) through direct connections. The estimated abstraction by users from sources other than the Water and Sanitation Agency (WASA) represents ~30% of the water consumed. Historical population records in Lahore show increasing trends from 6.3 M in 1998 to over 10 M at present. The city's extensive growth has resulted in significant urban development [17], resulting in a rapid rise in tube wells, abstraction rates, and an increase in water table depth with time [11].

Although the groundwater system in Lahore is part of the groundwater reservoir of the Indus basin, exploitation has formed a depression in the local groundwater table, expanding east and southward. Ref. [4] studied the expansion of depression zones as a function of change in groundwater depth during five consecutive years (2007–2011). The results reveal an alarming situation in some areas where the water table has reached more than 38 m since 2007. The findings depict a gradual expansion in depression to the east and south directions due to an increase in the rate of abstraction and decreasing recharge. A study by [11] discovered the worst situation in a few areas of Lahore city where the water table lowered to about 45 m in depth. Synthesis of the existing literature highlights a gap between investigating studies and policy institutions and the lack of (modeling-based) studies on the integrated impact of multiple stresses (i.e., climate, land use, and abstraction). A handful of studies covering quality aspects of the local groundwater resources are readily available. However, quantity aspects are not fully covered [18–20]. Ref. [18] evaluated the impact of past climate change and abstraction on groundwater resources of Lahore with a focus on management issues. Ref. [20] used chemical, isotopes, and numerical techniques. They identified types and sources of recharge for the Lahore aquifer and simulated the impact of abstraction on groundwater levels up to 2019.

There are two basic techniques to groundwater modeling, the first of which is the volume-based approach that makes use of direct groundwater monitoring data from observational wells [21–23], while other one used physical-based hydrological models, e.g., 3D groundwater flow modeling [24], semi- and fully distributed hydrological models [25] and remotely assisted simple water balance models [2,25]. Volume based methods driven by observational data are generally more accurate, however, uneven, and sparse distribution of monitoring wells make this approach less useful particularly for Lahore study region. Physical-based hydrological models and remotely driven water balance approaches on other hand can be useful for data sparse regions to monitor the groundwater changes [24,25]. For instance, physical based models include the global scale hydrological model (GHMs: e.g., PCRaster Global Water Balance (PCR-GLOBWB)) [26], and regional scale models (e.g., Soil Water Assessment Tool (SWAT) [27], MODFLOW [28]). SWAT and MODFLOW applications in surrounding areas of study region have well documented in previous studies [2,29–33]. Most of the studies have employed modeling-based methodologies with a focus on groundwater use. However, almost all of them have focused on agricultural areas [2,25,33–37], while ignoring the impacts of Uran areas. A few other observation-based and quantitative studies are also part of the literature that points out the continuously deteriorating situation associated with management issues in Lahore city in context of historical records [4,38,39]. The literature is sufficient for understanding the severity of the problem for current conditions. However, future changes in groundwater levels, particularly in the urban and peri-urban areas of Lahore City, have not yet been investigated in the context of projected climate change and human activities. To achieve sustainability of the groundwater levels in an urban area, we need to develop a sound knowledge of the situation for the future. Consequently, knowledge development could help in proposing a concise groundwater management policy to guide structural and non-structural aspects.

Therefore, this study uses an integrated modeling framework consisted of SWAT and MODFLOW applications to investigate the groundwater changes in context of projected changes in climate change and anthropogenic drivers, e.g., land-use change, and abstraction on the groundwater resources in the Lahore division. We further provided a proposed formulation of a few adaptation options to counteract diminishing groundwater resources.

2. Materials and Methods

2.1. Study Area

The study uses a modeling approach involving SWAT for hydrologic modeling and MODFLOW for hydro-geologic modeling. Considering the different requirements of both models, two segments of the study area were formulated. The geographical distribution of the Ravi River basin (28,000 km^2) comprises a 20% upstream area having rugged topography. Mountains in the foothills of the Himalayas form the topography of the upstream basin (4300 m above mean sea level (amsl)) that mostly remains covered with snow, while 80% of the basin area comprising the middle and downstream is plain. The Middle and downstream areas of the basin are agricultural lands [14,40]. River Ravi is one of the major tributaries of the Indus River. The river Ravi covers a total stretch of 720 km through the basin from its origin in the Kailash Mountains and drains into the Chenab River in the South-west area. Peek flows seasonally occur during monsoon months (July to Sept), while the rest of the months observe low flows. The river Ravi basin has a semiarid-tropical climate. Upstream areas receive heavier rainfall than downstream plain areas. The rainfall season mostly shows bimodal patterns, November to January are the low rainfall months, and July to September is heavy rainfall. The annual rainfall in the basin varies from 300 to 1200 mm, and mean temperature changes between 8 to 40 °C.

Lahore division lies between latitude 31°15′ N and 31°42′ N, longitude 74°01′ E and 74°39′ E and comprises of an area around 6800 km^2, covers four districts such as Lahore, Sheikhupura, Kasur, and Nankana Sahib. Geographically, the Lahore division has a plane topography that changes from 200 to 210 m amsl and a general slope of 1:3000 towards the south and southwest. Based on the latest census Lahore division is home to over 19.4 million population. Groundwater is the dominant source of water supply for domestic (95%) and agriculture (39%) consumption. The Lahore division aquifer majorly comprises unconsolidated alluvial complexes. Silt, sand, and clay in varying proportions are the main constituents of the alluvial complex. These constituents consolidate to form a thick sedimentary complex of more than 400 m thickness. Clay is the dominant constituent of the alluvial complex with small quantities near the Ravi River and gradually increases with distance from the river [41] (Figure 1).

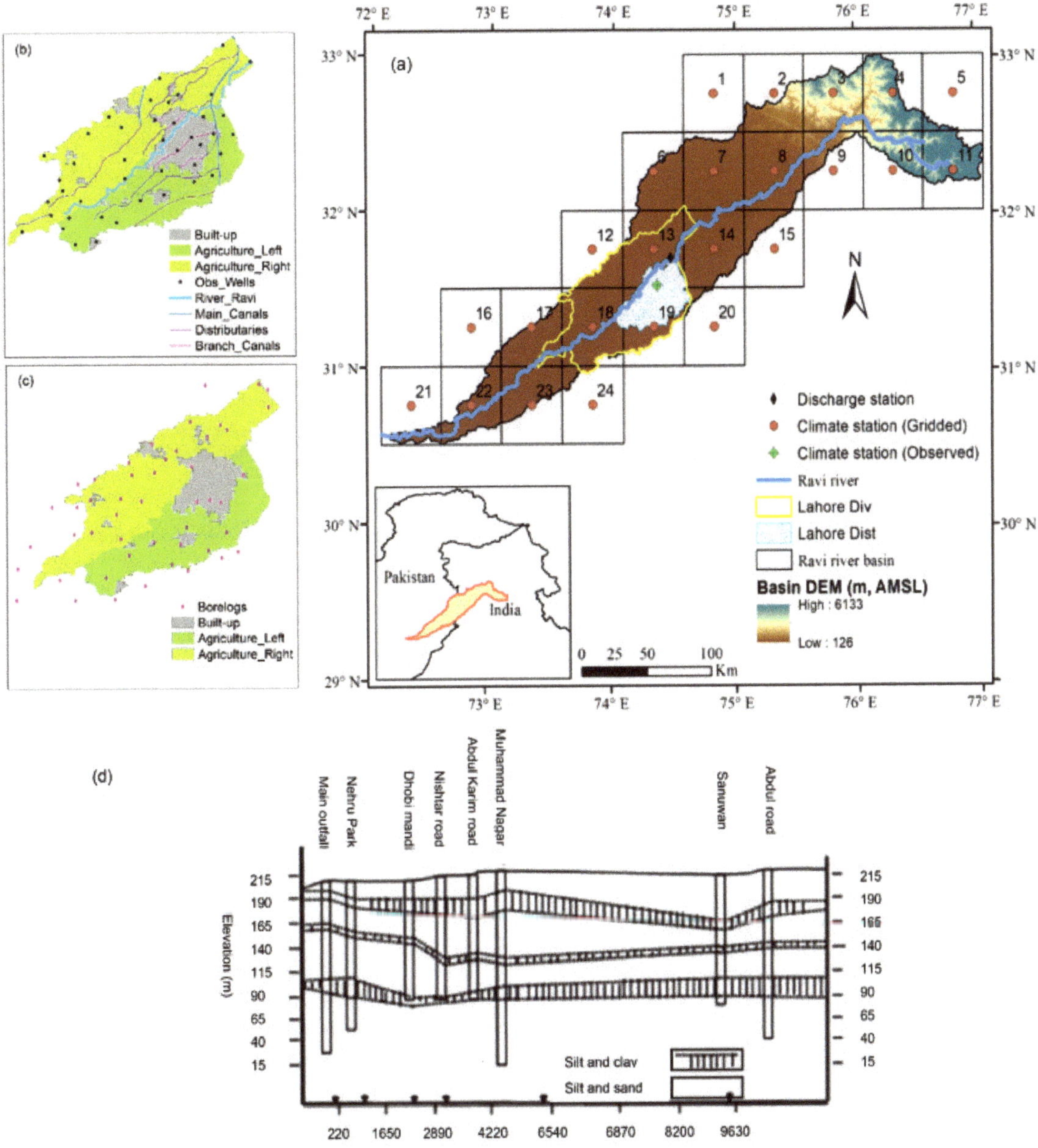

Figure 1. The geographic location of RRB. Topography is based on ASTER data. Climate stations (observed and virtual) are shown together with the 0.5° × 0.5° grid points (**a**). Three zones of Lahore division (Urban and Agriculture), river, canals, and observation wells are shown in (**b**). Bore-logs and their locations are shown in (**c**). Lithological classification of Lahore division aquifer in (**d**) [14].

2.2. Climate Data Collection and Evaluation

Scanty observed data is an issue of the study area, therefore reanalysis and interpolated data on three major components; Tmax, Tmin, and rainfall were downloaded from Princeton University (PU)website: (http://hydrology.princeton.edu/data/; access date: January 2022), APHRODITE (APH): (http://www.chikyu.ac.jp; access date: January 2022) and Climate change Prediction Center-National Oceanic and Atmospheric Administration (CPCNOAA) (https://www.ncdc.noaa.gov/cdo-web/; access date: January 2022). Likewise, future climate data on Tmax, Tmin, and rainfall for two representative concentration pathways (RCP4.5&8.5) scenarios and seven regional climate models (RCM) were obtained from the CORDEX South Asia website: (http://www.cordex.org/; access date: January 2022) (Table 1). The downloaded data was performance checked. Two GCD and four RCM with outstanding performance were shortlisted for climate change analysis and used as input to the SWAT model. The detailed methodology and performance results are reported in [14].

Table 1. Data required for assessment of the impact of future climate, land use, and abstraction on groundwater resources in Lahore.

Data Type	Duration	Resolution	Sources(s)
Meteorological Data			
Observed Climate Data Rainfall, Tmax, and Tmin	1982–2015	Daily	Pakistan Meteorological Department (PMD) Chandigarh Meteorological Station (CMS)
Gridded Climate Data Princeton University forcing APHRODITE dataset Rainfall	1982–2015	0.5° × 0.5° (daily)	Princeton University (http://hydrology.princeton.edu/data/; access date: January 2022) Research Institute for Humanity and Nature (http://www.chikyu.ac.jp; access date: January 2022)
NOAA climate dataset Tmax and Tmin	1982–2015	Daily	NOAA's National Centers for Environmental Information (NCEI) (https://www.ncdc.noaa.gov/cdo-web/; access date: January 2022)
Hydrological Data			
River discharge and Canal discharge	2000–2014	Daily	Punjab Irrigation Department (PID)
Regional Climate Models (RCMs)	1982–2100	Daily (0.5° × 0.5°)	CORDEX (http://www.cordex.org/; access date: January 2022)
Spatial Data			
Land use data Soil date	2007	1 km	Ref. [40] World Map (https://worldmap.harvard.edu/data/geonode:DSMW_RdY; access date: January 2022)
Digital Elevation Model (DEM)		30 m	(https://earthexplorer.usgs.gov/; access date: January 2022)
Aquifer lithology and hydraulic data			Groundwater division of Water and Power Development Authority (WAPDA)
Population data (Counts, density)	1998–2017		Pakistan Bureau of Statistics (PBS) (http://www.pbs.gov.pk/; access date: January 2022)

2.3. *Hydrological and Hydrogeological Data Collection and Processing*

The daily hydrological data on river flows and canal discharge for fifteen years (2000–2014) were obtained from the Punjab irrigation department (PID) which is passed through quality check before use. The hydro-geological data on groundwater levels measured in Jun and Sept every year was obtained from PID, aquifer lithology from CSIRO, hydraulic conductivity, specific storage, and specific yield was extracted from literature [37].

2.4. *Spatial Data Collection and Processing*

The land use map of the Indus basin developed for 2007 was provided by [40]. This map is one of the several inputs to the hydrological model (SWAT). Two other land use maps, one developed by the European space agency (ESA), cover the global domain from 1990 to 2015. The other map prepared by [42] (covers the Lahore district from 1999 to

2021 (1999, 2011, 2013, 2021, and 2035). Both maps were obtained and used for land-use change analysis. The ASTER based digital elevation model (DEM) data with spatial resolution of 30 m was obtained from the United States Geological Survey (USGS) website: (https://earthexplorer.usgs.gov/; access date: January 2022). The global soil data were obtained from the Harvard University data archive: (https://worldmap.harvard.edu/data/geonode:DSMW_RdY; access date: January 2022). The population data of four censuses from 1951 to 2017 was obtained from the Pakistan bureau of statistics.

2.5. Future Climate Change Projections

The three most relevant parameters of climate (Tmax, Tmin, and rainfall) were used, because of their significance in the climate system, their most often use in climate change studies, and their relevance to the current work. The temporal changes in Tmax, Tmin, and rainfall in the base period (1982–2014) were calculated using PU and CPC-NOAA datasets. Future projections (2020–2095) were made using four shortlisted RCMs under RCP4.5 and RCP8.5 scenarios. Annual and seasonal scale projections were made where the wet season covers months (Apr to Sept) that typically receive heavy rainfall. The dry season covers months (Oct to Mar) that receive low or no rainfall at all. Their significance was tested using a student t-test, and linear trends were fitted using a linear regression model [43].

2.6. Future Land Use Change Projections

Two land-use maps were used; one developed by the [44] hereafter referred to as map-1 and the second by [45], referred to as map-2. Map-1 classifies the area into eleven classes and map-2 into five classes. Considering the scope of the study, both maps were reclassified into three broad categories: Agriculture, built-up, and water. The land-use changes estimated using map-1 and map-2 were compared with each other. These land-use changes were used to develop linear regression models and future land-use change projections made using map-2 under two transition scenarios: R1S1-Business as usual and R2S2-Conservation. The R2S2 somehow includes the effect of policies by the current federal government in Pakistan, such as the construction of high-rise buildings and tree plantation campaigns. The study area is lacking in research on land-use change projections, with the available literature instead focusing more on historical changes. In this situation, although land use projections are based on simplified assumptions, they are still informative for interested individuals or institutions.

2.7. Projected Groundwater Projections

The groundwater abstraction in urban and agricultural areas of Lahore was estimated independently. For urban areas, the population projections were made beforehand by fitting multiple regression models on population data projected by [46] for some random years (2025, 2050, 2075, and 2100) under SSP1 and SSP3 scenarios developed by [47]. Based on population density, the study area was divided into different zones. Groundwater abstraction was calculated as a multiplication product of per capita water demand ($m^3{\cdot}c^{-1}{\cdot}d^{-1}$) and total population. For agricultural areas, the groundwater abstraction was calculated using the water balance approach of SWAT [25]. This study assumes that seasonal abstraction is half that of annual and will remain constant during each season. Therefore, seasonal scale analysis is not part of this study.

2.8. Hydrological Modeling Using SWAT

The Soil and Water Assessment Tool (SWAT) model was used to calculate the recharge and abstraction. This model was developed by the USDA-ARS. It can simulate hydrological and biogeochemical cycles and impacts of stimuli such as climate and land-use changes on these processes at the watershed scale [48–51]. The SWAT model offers to include a wide range of components such as weather, hydrology, soil characteristics, crop growth characteristics, land management operations, and nutrients load and flows in the runoff.

An insightful description of this model is available in [52]. SWAT simulates water balance using Equation (1).

$$SW_t = SW_O + \sum_{i=1}^{t} (R_{day} - Q_{surf} - E_a - W_{seep} \times Q_{gw}) \quad (1)$$

where:

SW_t final soil water content at time t in (mm H_2O)
SW_o initial soil water content on day i in (mm H_2O)
t time in (days)
R_{day} precipitation on day i in (mm)
Q_{surf} surface runoff on day i in (mm)
E_a evaporation on day i in (mm)
W_{seep} percolation and bypass flow leaving the bottom of soil strata (mm)
Q_{gw} return flow on day i (mm)

SWAT calculates the surface runoff using the SCS curve number method. It is a function of land use, soil characteristics, and previous soil moisture conditions. The mechanism of the SCS curve number method has been given in Equation (2).

$$Q_{surf} = \frac{(R_{day} - I_s)^2}{\left(R_{day} - I_a + S\right)} \quad (2)$$

where:

Q_{surf} = runoff or excess rainfall (mm).
R_{day} = precipitation on a given day (mm).
I_a = initial abstractions such as surface storage, interception, and infiltration prior to runoff (mm) and
S = retention parameter (mm).

$$S = 25.4\left(\frac{1000}{CN} - 10\right) \quad (3)$$

CN = curve number a function of land characteristics, varies between 0 to 100; 0 shows easy conversion of water to direct runoff and 100 shows difficult conversion.

SWAT Model Calibration, Validation, and Performance Evaluation

The model was manually calibrated for eight years (2000–2007) and validated for seven years (2008–2014) at Ravi Syphon and Shahdara gauges. Sensitivity analysis of the SWAT model parameters is required to obtain their effect on simulated components and characterize uncertainties in those components [53]. In total, 32 parameters were considered in the sensitivity analysis. The parameter values changed up to four levels (±10 and ±25%) to check their influence on flow and groundwater components. The three most widely used statistical parameters: the coefficient of determination (R^2), the percentage bias (PBIAS), and the Nash-Sutcliffe efficiency (NSE), were considered in this study to evaluate the performance of the SWAT model. An insightful description of these parameters and performance rating for SWAT model evaluation is reported in [54].

2.9. Hydrogeologic Modeling Using MODFLOW

In this study, MODFLOW was used to simulate the impact of climate change, land-use change, and abstraction on groundwater levels. MODFLOW is a three-dimensional finite-difference groundwater flow model. Groundwater flow within the aquifer is simulated in MODFLOW using a block-centered finite-difference approach. The partial differential equation (Equation (4)) describes the groundwater flow in each grid.

$$\frac{\partial}{\partial_x}\left(K_{xx}\frac{\partial h}{\partial x}\right)+\frac{\partial}{\partial y}\left(K_{yy}\frac{\partial h}{\partial y}\right)+\frac{\partial}{\partial z}\left(K_{zz}\frac{\partial h}{\partial z}\right)-W=S_s\frac{\partial h}{\partial t} \quad (4)$$

In Equation (4), h is the hydraulic head (m) and an independent variable. The K_{xx}, K_{yy}, and K_{zz} represent the hydraulic conductivity (m.day^{-1}) in x, y, and z directions, respectively. The S_S shows specific storage that is a dimensionless quantity, while the letter W represents the source or sink. It is measured in day^{-1}. The W has two signs a positive sign shows the recharge/injection of water into the aquifer, a negative sign means abstraction. In this study, the groundwater flow model developed for Lahore also includes the surrounding agricultural areas. The geographical boundaries of the model area are presented in Figure 1. The model area was divided into uniform grids of 500 m spatial resolution in horizontal (X, Y) directions.

Based on the lithological data available at 149 locations, a total depth of 400 m was modeled with four layers of dominant materials. The model was calibrated and validated for transient conditions using the wet seasons from 2003–2014 with two stress periods each year: dry and wet having a duration of 182 and 183 days, respectively. The stress periods were further subdivided into 10-time steps with a time step multiplier of 1.2 to characterize the temporal variation in piezometric heads. The flow components were simulated using four MODFLOW modules.

MODFLOW Model Calibration and Performance Evaluation

In the study, a built-in Parameter Estimation (PEST) module was used to calibrate and validate the transient model for nine years (2003–2009). The PEST takes control of the MODFLOW and performs iterations as many times as necessary to determine the optimum parameter values. Annual piezometric water levels recorded by WASA were used as a reference for calibration. The aquifer parameters: specific yield, hydraulic conductivity, specific storage, and groundwater abstraction were adjusted for four aquifer layers in calibration to obtain the best-simulated heads. The coefficient of determination (R^2) was used for performance evaluation.

2.10. Conceptualization of Scenarios Combination for Future Groundwater Level Projections

The study projects future climate, land use, and abstraction (population) under six combined scenarios (RCP4.5 and RCP8.5; R1S1 and R2S2; SSP1 and SSP3). The RCP and SSP are global-scale scenarios. These scenarios project the future evolution of climate and socio-economic development based on multiple variables such as socio-economic, technological, land use, energy, greenhouse gas emissions, and air pollutants. Unlike other studies [47,55] that combine RCP and SSP scenarios based on CO_2 emission projections, we used similar assumption to combine those scenarios based on the global population projections. Comparison of global population projections made under the SSP scenarios and reported in [56]. Background population projections for the RCP scenarios are reported in [57], revealing close agreements between SSP1 and RCP4.5 and SSP3 and RCP8.5 scenarios. The land-use change scenarios are applicable on the local scale. They were combined simultaneously with the climate and population scenario for the future groundwater level projections (Table 2).

Table 2. Scenarios combinations for projected groundwater levels.

RCP Scenarios	Land Use Scenarios	SSP Scenarios	Scenario Combinations
RCP4.5	R1S1, R2S2	SSP1	RCP4.5-R1S1-SSP1 RCP4.5-R2S2-SSP1
RCP8.5	R1S1, R2S2	SSP3	RCP8.5-R1S1-SSP3 RCP8.5-R2S2-SSP3

3. Results and Discussion

3.1. Climate Change Projections

The mean climate will evolve in the future (2020–2095) as both the annual temperature and rainfall will increase under RCP4.5 and RCP8.5 scenarios, as shown in Figure 2a–c. Compared to the base period (1982–2005), the T_{max} will increase by 0.38 °C, 0.33 °C, and 0.64 °C in near, mid, and far future, respectively, under the RCP4.5, and 0.21 °C, 0.96 °C, and 2.04 °C in near, middle, and far future, respectively, under RCP8.5. Future Tma_x in the wet season (April to September) will decrease, except in the far future under RCP8.5, the expected absolute change will be −0.36 °C, −0.28 °C, and −0.07 °C in near, mid, and far future, respectively, under the RCP4.5 scenario, and −0.41 °C, −0.29 °C, and +0.45 °C in near, middle, and far future, respectively, under RCP8.5. Future Tma_x in the dry season (April to September) will increase, except near futures under RCP4.5, the expected absolute change will be −1.91 °C, +0.95 °C, and +1.36 °C in near, mid, and far future, respectively, under the RCP4.5 scenario, and +0.84 °C, +2.21 °C, and +3.65 °C in near, middle, and far future, respectively, under RCP8.5 (Figure 2a). The Tmin will increase. The absolute change in future T_{min} will be 0.66 °C, 0.94 °C, and 1.23 °C in near, mid, and far future, respectively, under the RCP4.5, and 0.85 °C, 2.05 °C, and 3.41 °C in near, middle, and far future, respectively, under RCP8.5. Future T_{min} in the wet season (April to September) will decrease, except in the far future under RCP8.5, the expected absolute change will be +0.73 °C, +0.70 °C, and +0.86 °C in near, mid, and far future, respectively, under the RCP4.5 scenario, and +0.81 °C, +1.73 °C, and +2.86 °C in near, middle and far future, respectively, under RCP8.5. Future T_{min} in the dry season (October to March) will increase, except near future under RCP4.5, the expected absolute change will be +0.59 °C, +1.19 °C, and +1.61 °C in near, mid, and far future, respectively, under the RCP4.5 scenario, and +0.89 °C, +2.39 °C, and +3.97 °C in near, middle, and far future, respectively, under RCP8.5 (Figure 2b).

Results show an increase in temperature in the future under the climate change scenarios. A comparison of seasonal scale projections shows that the dry season will be warmer than the wet season except for the near future, during which the T_{max} will decrease under the RCP4.5 scenario. Both Tmax and Tmin will increase while Tmin will rise more than the Tmax. Previous studies report similar results for the study area and other parts of the world [16,58–61]. Rising future temperatures will increase agricultural water demand due to an increase in evapotranspiration [62–64]. All-purpose domestic and industrial water consumption will also increase [63,65]. Consequently, rising demand will exacerbate pressure on groundwater resources.

Compared to the base period (1982–2005), the mean annual rainfall will increase in the future (2020–2095), except near future, under RCP4.5. Changes in rainfall in the near, mid, and far future will be −1%, 25%, and 24% under the RCP4.5, 17%, 27%, and 29% under RCP8.5. Future rainfall in the wet season will increase under RCP4.5 and RCP8.5 scenarios. The dry season will become drier under both climate change scenarios (Figure 2c). A comparison of the climate change scenarios shows agreement for the future projections, especially in the mid and far future. Rainfall projections are in close agreement with the previous studies [66]. Increasing future annual rainfall depicts more water availability which can offset negative pressure on groundwater resources. Since the increase in rainfall will be in the wet season, which is shorter in length than the dry season, dry seasons, due to longer duration and decreasing, rainfall will further exacerbate the pressure on groundwater resources in Lahore.

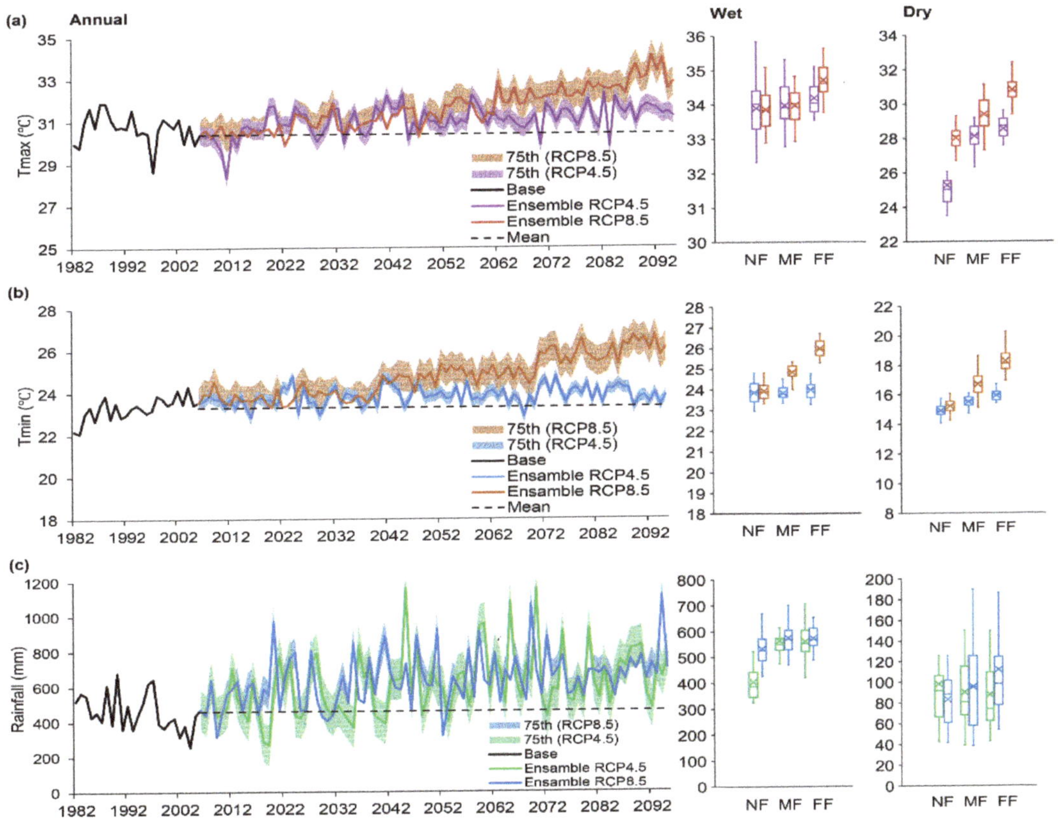

Figure 2. Temporal projections of (**a**) T_{max}, (**b**) T_{min}, and (**c**) rainfall in Ravi River Basin for climate change (RCP4.5 and RCP8.5) scenarios in the twenty-first century.

3.2. Land Use Change Projections

This section presents an analysis of the land-use change during the base period (2000–2014) and future period (2020–2095) under two scenarios: R1S1-Business as usual and R2S2-conservation in Lahore. The future projections for three broad urban land-use types are shown in Figure 3. Compared to the base period, the dominant land use (agriculture) will decrease from 5392 km^2 to 2862 km^2 under R1S1 and 3247 km^2 under the R2S2 scenario, while the built-up area will expand. The expansion in built-up will be from 1184 km^2 to 3716 km^2 and 3329 km^2 under R1S1 and R2S2, respectively. The change in agriculture and built-up areas will be significant across the entire century (Table 3). The third land-use type, water, will remain the same with minor changes throughout the twenty-first century. An earlier study accomplished in the study area projected expansion in the urban area (built-up) [67]. The future contraction in Agricultural areas and expanding built-up will have implications for groundwater resources [6,33,68]. Besides putting pressure on groundwater, the agricultural (cultivated) area also plays a role in replenishing groundwater resources. Because a portion of the water received from rainfall and irrigation supplies percolates to groundwater, with the future contraction of agricultural areas, both the demand for irrigation and recharge will decrease [33]. The effect of the future expansion of built-up on groundwater resources will be more severe than the decrease in the agricultural area. It is due to a decrease in recharge and an increase in population-triggered water demand. Since the land use projections are driven by simplified assumptions (land use change (expansion or contraction) will only occur on the outskirts of existing landscapes). This assumption might overestimate land use change and its impact on groundwater resources.

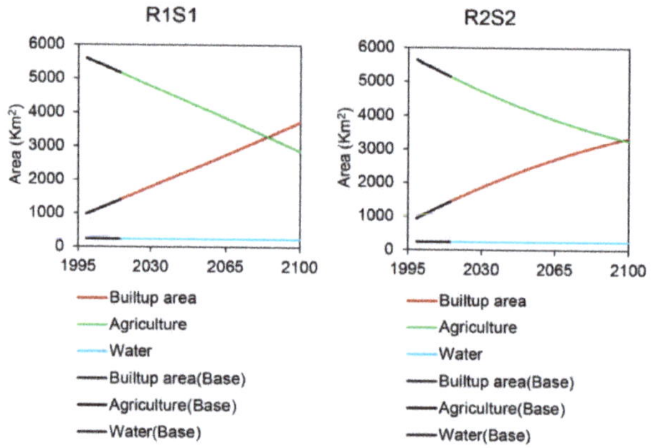

Figure 3. Future projections of three dominant land use and land cover types under R1S1 and R2S2 scenarios using Map-1.

Table 3. Area under three broad land use and land cover types for the NF, MF and FF, projected under R1S1 and R2S2 scenarios.

Year	Land Use and Land Cover Type (km^2)					
	R1S1			R2S2		
	Agriculture	Built-Up	Water	Agriculture	Built-Up	Water
Base period	5392	1184	240	5392	1184	240
2020	5048 *	1528 *	240	5001 *	1575 *	240
2043	4325 *	2251 *	240	4287 *	2289 *	240
2072	3642 *	2935 *	240	3736 *	2840 *	240
2100	2860 *	3716 *	240	3247 *	3329 *	240

Note. Each future time window was tested for significance relative to the base period. The p value for all future time windows was less than 0.05 significance level. * Shows a significant change (at 5%) in land use relative to the base period.

3.3. SWAT Model Sensitivity Analysis

Nine parameters are found to be the most influential (Table 2). The CN2.mgt has the most influence and LAT_TIME.hru the least while the other seven parameters fall between the two. In terms of their relationship with simulated parameters, CANMX.hru, TLAPS.sub, ESCO.hru, and CH_K2.rte have a direct relationship with the surface flow components, while CN2.mgt, SOL_AWC.sol, and LAT_TIME.hru have an inverse relationship (Table 2). Only ESCO.hru has a direct relationship with groundwater components. CN2.mgt, TLAPS.sub, CH_K2.rte, and PLAPS.sub have no relationship at all with the groundwater components. Some parameters such as GW_DELAY.gw, SOL_K.sol and PLAPS.sub have no proper relationship with flow components. Besides their relationships, the effects of influential parameters were also observed during sensitivity analysis (Table S1) (supplementary material). For example, CN2.mgt and LAT_TIME.hru reduce instant peaks and smooth the hydrograph. TLAPS.sub and SOL_K.sol affect the occurrence (onset and withdrawal) of hydrograph peaks. Similarly, CANMX.hru and ESCO.hru affect the base flow and discharge while PLAPS.sub influences discharge only and SOL_AWC.sol affects soil moisture storage.

3.4. *SWAT Model Calibration and Validation—For Discharge*

Figure 4a–c shows a close matchup between observed and simulated hydrographs. Based on the statistics, the performance of the SWAT model varies from good to very good (Table 4). The coefficient of determination (R^2) has very good values; 0.77 for calibration and 0.72 for validation at Ravi Syphon, with 0.75 for calibration and 0.81 for the validation period at Shahdara gauge. The NSE also varies from good to very good; 0.76 for calibration and 0.72 for validation at Ravi Syphon with 0.74 and 0.75 at Shahdara gauge. The PBIAS only has a high value (−10.46) at Ravi Syphon for calibration, while it falls in a good range (1.68 and 0.79) at both stations for the validation period. The statistics for calibration and validation show that the model performance at both stations falls in a range from good to very good. The model can be used for further hydrological analysis.

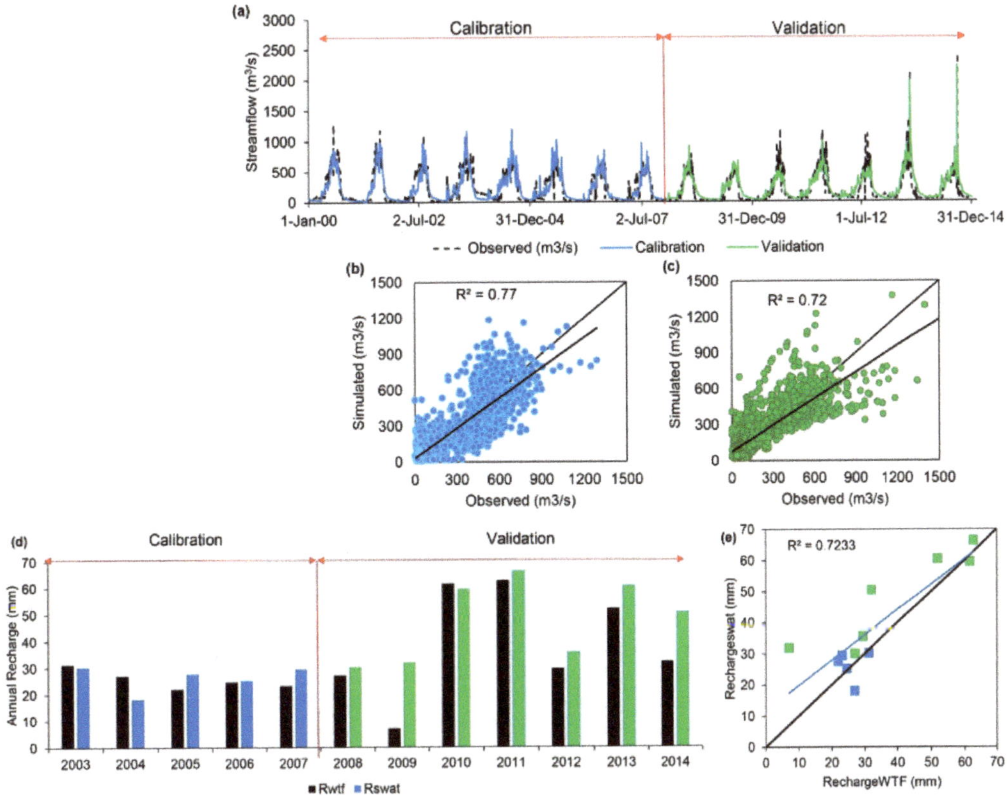

Figure 4. (**a**–**c**) Comparison between simulated and observed discharge for calibration and validation periods at Ravi siphon gauge based on daily time series and scatter plots; and (**d**,**e**) comparison between annual simulated and reference groundwater recharge for calibration and validation periods.

Table 4. SWAT model performance for calibration and validation at Ravi syphon and Shahdara gauges.

Performance Statistics	Calibration	Validation
Ravi Syphon gauge (upstream)		
Coefficient of determination (R^2)	0.77 (very good)	0.72 (good)
Nash-Sutcliffe efficiency (NSE)	0.76 (very good)	0.72 (good)
Percentage bias in volume (PBIAS)	−10.46 (good)	1.68 (very good)
Shahdara gauge (downstream)		
Coefficient of determination (R^2)	0.75 (very good)	0.81 (very good)
Nash-Sutcliffe efficiency (NSE)	0.74 (good)	0.75 (very good)
Percentage bias in volume (PBIAS)	−0.20 (very good)	0.79 (very good)

3.5. SWAT Model Calibration and Validation—For Recharge

The SWAT model was calibrated for groundwater recharge as well. The calibration period spans five years (2003–2007) and the validation period seven years (2008–2014). Groundwater recharge, calculated using the water table fluctuation (WTF) method, was used as a reference. The optimized parameters are enlisted in Table 5. The performance of calibrated SWAT model falls in the range of good to very good, with a minimum difference between annual simulated and reference groundwater recharge (Figure 4d,e). The calibrated model mostly overestimates groundwater recharge during validation. There is a considerable difference between the years 2009 and 2014 (25 mm and 18 mm, respectively). For the remaining years, the model matches well with the reference groundwater recharge. The coefficient of determination (R^2) falls in a very good range for calibration (0.84) and validation (0.82), and the NSE falls in the class of satisfactory (0.60) for both calibration and validation periods. The PBIAS shows a very good value (2.0) for calibration and a high value (16) for validation. The SWAT model's satisfactory performance has also been addressed by previous studies in the study area [2,10,25,30,69]. The success of the calibration of SWAT model depends on the choice of parameters and sensitivity [70]. Both factors largely depend on watershed characteristics and are determined by sensitivity analysis. Discharge in the Ravi River is controlled by a set of parameters that were determined during sensitivity analysis. The influence of selected parameters on discharge is described in Table S3. For the calibrated SWAT model, the CN2, CANMX, ESCO, SOL_AWC, and SOL_K were altered, depicting the dominant control of land surface characteristics and subsurface soil characteristics on river discharge. The calibrated CN2 value 47.63 depicts high infiltration. Small quantities of infiltered water are stored in the soil root zone, as indicated by low SOL_AWC (0.11). The possible reason is that much of the soil water is removed through evaporation, as shown by ESCO (0.35), and recharged to shallow groundwater, as depicted by GW_DELAY (19).

Table 5. List of sensitive parameters and their calibrated values for the swat model.

Parameter	Description of Parameters	Parameter Range	Calibrated Value	Sensitivity Rank
CN2.mgt	SCS runoff curve number	(35, 98)	47.63	1
CANMX.hru	Maximum canopy storage (mm H_2O)	(0, 100)	6.02	2
TLAPS.sub	Temperature lapse rate (°C/km)	(−50, 50)	−3.80	3
ESCO.hru	Soil evaporation compensation factor	(0, 1)	0.35	4
SOL_AWC.sol	Available water capacity of the soil layer (mm H_2O/mm soil)	(0, 1)	0.11	5
CH_K2.rte	Effective hydraulic conductivity in main channel alluvium	(−0.01, 500)	9.50	6
GW_DELAY.gw	Threshold depth of water in the shallow aquifer required for return flow to occur	(0, 500)	19.00	7
SOL_K.sol	Saturated hydraulic conductivity (mm/h)	(0, 2000)	42.22	8
PLAPS.sub	Precipitation lapse rate (mm H_2O/km)	(−500, 500)	305.10	9
LAT_TIME.hru	Horizontal flow travel time (days)	(0, 180)	8.00	10

3.6. Groundwater Abstraction Projections

Groundwater abstraction in urban areas is dominated by domestic water use while in agricultural (rural) areas by water demand for agricultural consumption. This study considers only the dominant factors for each sector. Compared to the base period (2000–2014), annual groundwater abstraction will decrease in the 21st century under RCP4.5 and RCP8.5 scenarios (Figure 5). The reduction will be highest in the near future under RCP scenarios. At an annual scale, future abstraction will decrease by −17, −7, and −13% in the near future, mid future, and far future, respectively, under RCP4.5, and by −11, −9, and −10% in the near future, mid future, and far future, respectively, under RCP8.5. Future changes in abstraction will be significant under all scenarios. The groundwater abstraction follows the same pattern of change in both wet and dry seasons under RCP4.5 and RCP8.5 scenarios. Urban area abstraction will increase in the twenty-first century under SSP1 and SSP3 scenarios, as shown in Figure 6. Compared to the base period, groundwater abstraction will increase by 80% and 61% in the near future, by 158 and 135% in the mid future, and by 193 and 203% in the far future under SSP1 and SSP3, respectively. Groundwater abstraction increases during the twenty-first century, stabilizing in the far future under SSP1, while in contrast, under SSP3, it shows a continuous increase. The projected decrease in groundwater abstraction in agricultural areas is associated with increasing future rainfall that will enhance water availability for agriculture, and groundwater recharge and pressure on groundwater resources will decrease. Unlike agricultural areas, the likely rise in future abstraction in urban areas can be associated with increasing population. Previous studies [2,71] suggest pressure on groundwater resources is dominated by anthropogenic pumping. Rising future temperatures and expansion in the built-up area due to imperviousness will exacerbate the pressure on groundwater resources.

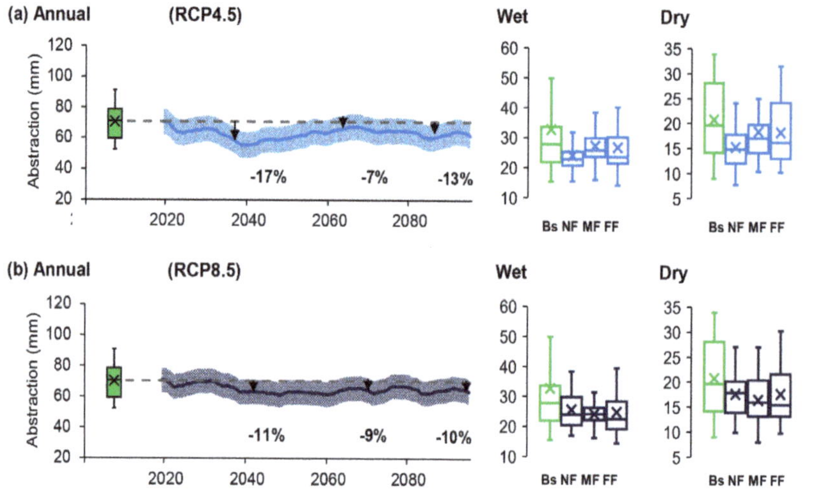

Figure 5. Annual and seasonal projections of groundwater abstraction in the agricultural areas of Lahore, under the (**a**) RCP4.5, and (**b**) RCP8.5 climate change scenarios for the 21st century. The shaded area in annual graphs represents the 25th and 75th percentile, and smooth lines are twenty years moving averages. All abstraction changes are significant at $p = 0.05$.

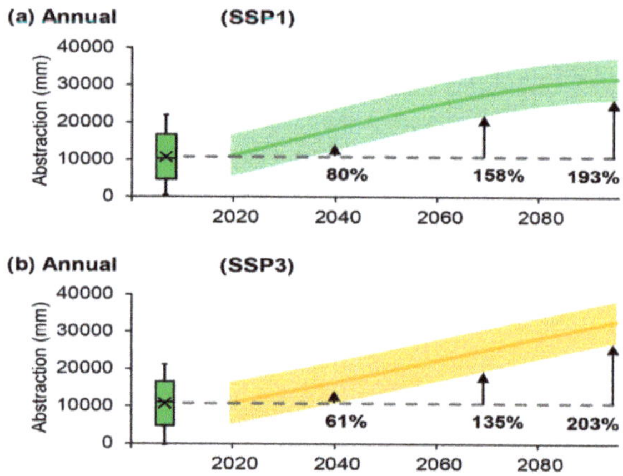

Figure 6. Annual projections of groundwater abstraction in the urban area of Lahore, under (**a**) SSP1, and (**b**) SSP3 population scenarios for the 21st century. The shaded area in annual graphs represents the 25th and 75th percentile, and smooth lines are twenty years moving averages. Change in abstraction will be significant at $p = 0.05$.

3.7. MODFLOW Sensitivity Analysis

The study used a parameter estimation package (PEST) for model calibration and sensitivity analysis. Sensitivity analysis of the parameters such as; specific storage, specific yield, vertical and horizontal hydraulic conductivities, and recharge was performed to check their influence on groundwater level. The groundwater model was most sensitive to horizontal hydraulic conductivity, followed by recharge, specific storage, vertical hydraulic conductivity, and specific yield.

3.8. MODFLOW Calibration—Steady State

The study used the parameter estimation (PEST) module to calibrate the steady-state MODFLOW model. The observed piezometric water levels for the wet season in 2003 (Dec-03) were used as reference targets. The specific yield (Sy), horizontal (Kh), and vertical (Kv) hydraulic conductivities, specific storage (Ss), and recharge were used as calibration parameters. The optimized hydraulic conductivity of the total formation varies from 66.53 to 89.52 mm.day-1 and vertical hydraulic conductivity from 3.47 to 6.02 mm.day-1. Specific storage and specific yield vary from 1.09E-04 to 2.47E-04 and 0.17 to 0.22, respectively. The calculated heads are given in Figure 7. Results depict that the plotted points are close to the best fit line with a small degree of scattering (less than 2 m), as shown in Figure 7. The degree of scattering varies between 2 and −2 m with an average value of −0.077 m (0.25 ft), thereby depicting overall good agreement between simulated water levels. The coefficient of determination is very high (R^2 = 0.9869), showing a high level of correspondence between observed and simulated water levels.

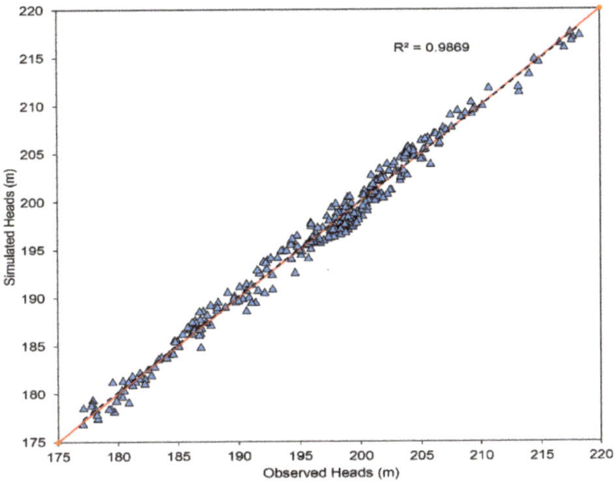

Figure 7. Scatter plot of the observed and simulated heads in Lahore.

3.9. MODFLOW Calibration—Transient

The transient model was calibrated for seven years; December 2003 to June 2009. The horizontal and vertical hydraulic conductivities were not considered because hydraulic conductivity is a non-time variant parameter. Therefore, recharge, specific yield, and specific storage were optimized using the PEST model. The optimized parameters are listed in Table S2 (supplementary material). The average values of specific storage and yield for the total formation vary from 1.09E-06 to 2.47E-03 and 0.17 to 0.22, respectively. The calibrated groundwater flow model performed well in simulating the water levels, as shown in Figure 8. The simulated hydrographs showed well match with observed hydrographs by effectively capturing the seasonal variations and overall trends in urban and agricultural zones. The average bias of model-simulated water levels in urban zones varies between 0.59 m and 1.23 m, and in the agriculture zones between 0.52 m and 0.81 m. The falling water levels during the calibration period show increasing pressure on groundwater resources in urban zones.

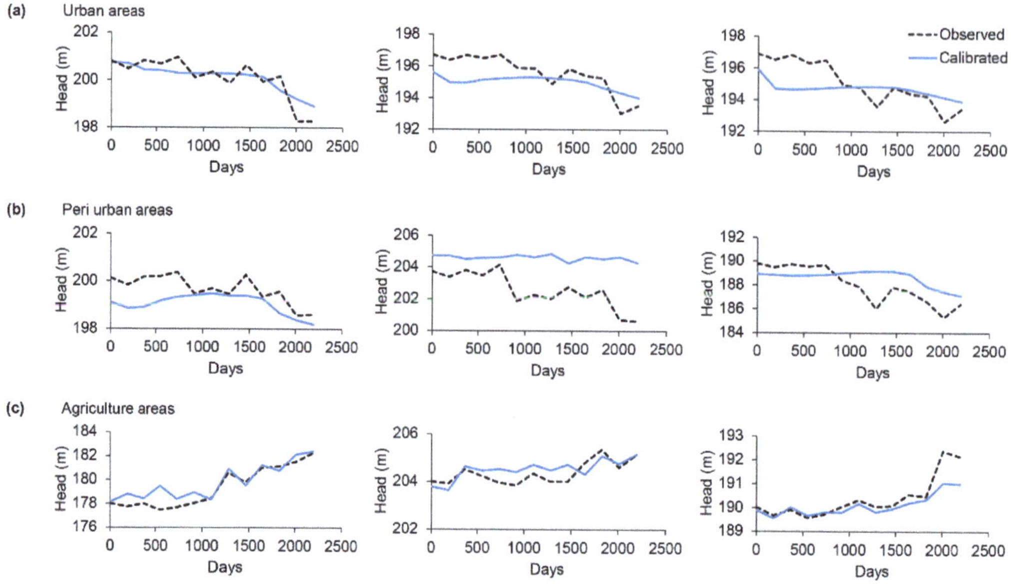

Figure 8. Calibration results of a transient GMS-MODFLOW model for groundwater levels in Lahore across (**a**) urban areas, (**b**) peri urban areas, and (**c**) agriculture areas. Note: Peri-urban areas are referred to as areas situated on the outskirts of urban areas and comprise mixed and fragmented landscapes of rural and urban areas [72].

3.10. Groundwater Level Projections

This section presents the future evolution of groundwater levels projected using the MODFLOW model for four scenarios. The simulations were performed to predict the behavior of groundwater resources up to the year 2100. All three factors, climate, land use, and abstraction, pose a combined effect on groundwater resources. Therefore, the following sections present the results, projected under combined climate change, land-use change, and abstraction scenarios. Changes in groundwater levels are a measure of groundwater storage, and therefore, this study focuses on groundwater levels only. Groundwater levels of urban and agricultural zones were projected using the MODFLOW up to 2100, as shown in Figure 9a–c. The results elucidate that groundwater levels will decrease at a much faster rate in urban areas. The drawdown in urban areas will vary from 45 to 55 m under all scenarios by 2100. The surroundings of urban areas (peri-urban) will observe a drawdown of about 9 to 10 m. In agricultural zones, there will be a drawdown of more than 2 m under the RCP4.5R1S1 scenario.

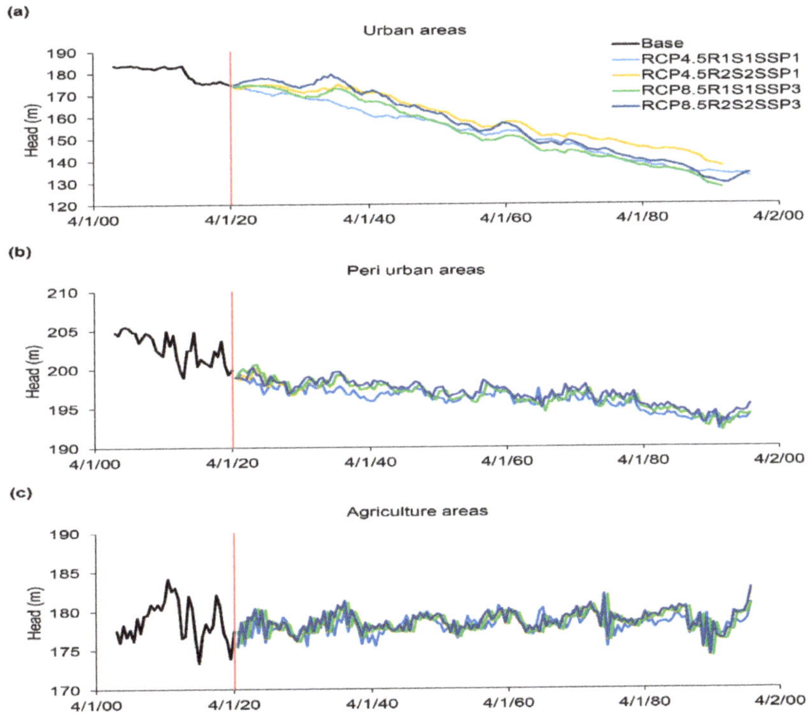

Figure 9. Projections of groundwater levels in Lahore division, for combined climate change, land-use change, and abstraction scenarios for the 21st century, (**a**) urban areas, (**b**) per urban areas, and (**c**) agricultural areas.

The changes in groundwater levels for urban and peri-urban areas are significant at $p = 0.05$ under both scenarios. Changes in groundwater levels for the left and right zones will be insignificant. The fast decline in groundwater levels for urban areas is due to urbanization, causing a rise in abstraction and a fall in recharge. Unlike the increase in total annual precipitation under climate change scenarios, groundwater levels will decrease. The decrease in water levels is due to the dominant effect of land-use change and abstraction, both of which marginalize the climate change effect, causing groundwater levels to reduce at a much faster rate. In peri-urban areas, the rate of decline is relatively slower. Low population density and recharge from rainfall and irrigation water supply slows the rate of decline. In agricultural zones, the fluctuation in groundwater levels is apparent. Unlike urban and peri-urban areas, groundwater levels will increase. Besides rainfall recharging, the dense network of irrigation, link canals, and irrigation return flows contribute to the likely increase in water levels.

3.11. Future Impact on Groundwater Resources

The impact of climate change, land-use change, and abstraction on groundwater levels was estimated under combined scenarios by subtracting the groundwater levels for the base period and future. An increase in depth was assigned a negative sign and vice versa. Results show that compared to the base period, the degree of impact will be much higher in the future, as shown in Figures 10 and 11. Unlike north and west, most areas will observe negative impacts under combined scenarios in the near future.

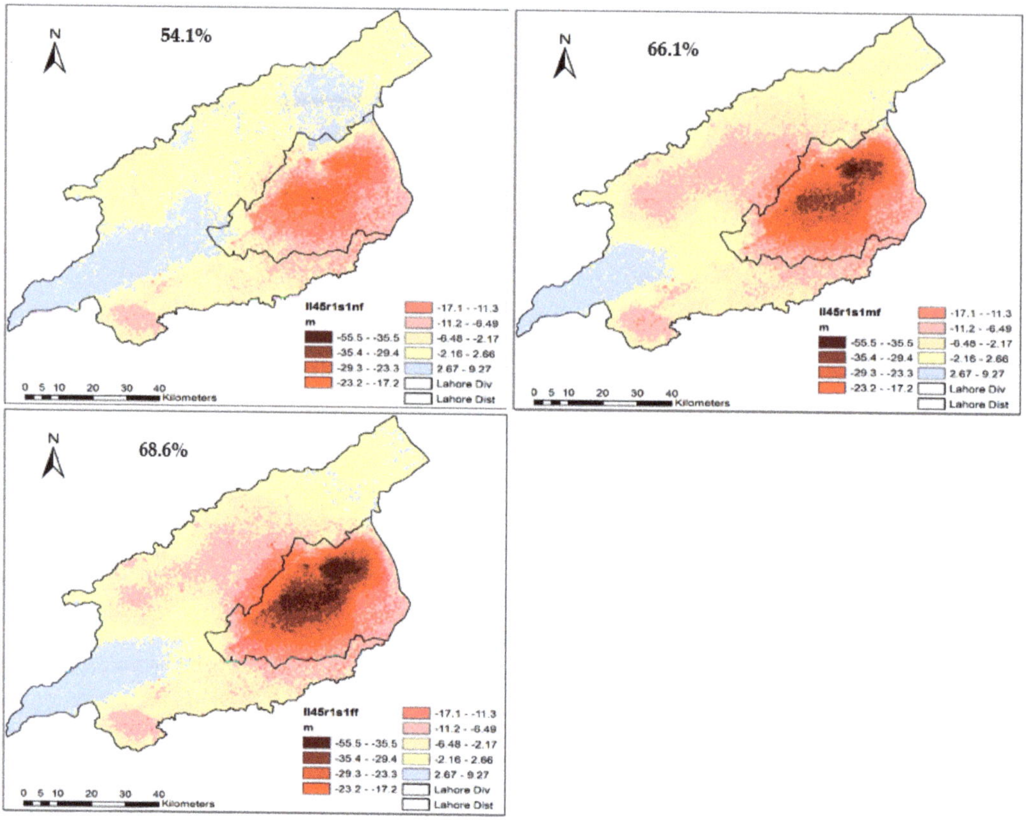

Figure 10. Spatial distribution of groundwater levels in Lahore under RCP4.5R1S1SSP1 scenarios. Near future (**top left**), mid future (**top right**) and far future (**bottom**). Note: Numbers in the top right corner show percentage of the total area with a significant change in impact.

In the north, the effect will change from positive to negative in the mid and far future. Urban and southeastern areas will observe the highest negative impact due to groundwater levels, decreasing at a much faster rate from 23 m in the near future to over 55 m in the far future. The surrounding areas being positively impacted in the base period will observe negative impacts due to the fast-outward expansion of the depression front. The groundwater levels in these areas are likely to decrease from 2 m in the near future to over 23 m by 2100. The agriculture areas, already facing a negative impact due to a decrease in groundwater levels during the base period, are likely to observe a further increase in the magnitude of negative impact. The likely decrease in groundwater levels in agriculture areas varies from 2 m in the near future to over 11 m in the far future. The degree of impact will be higher under RCP8.5-R1S1-SSP3 scenarios than RCP4.5-R1S1-SSP1. Similarly, the combined R1S1 scenario project higher changes than the R2S2 scenarios. The areas of a high negative impact and low adaptive capacity are likely to be more vulnerable in the future under all scenarios, as shown in Figures 10 and 11, including urban areas. The highly vulnerable urban areas will expand outward and triple spatially in the far future. Besides experiencing a negative impact during the base period and increasing in the future, agricultural areas will be less vulnerable in the future, under all scenarios. The surrounding areas of urban settlements are also likely to be less vulnerable in the future, although margins will decrease due to their proximity to the urban area. The degree and range of vulnerability will be higher and wider under RCP8.5-R1S1-SSP3 scenarios than

RCP4.5-R1S1-SSP1. Similarly, the combined R1S1 scenarios project a larger change than R2S2 scenarios.

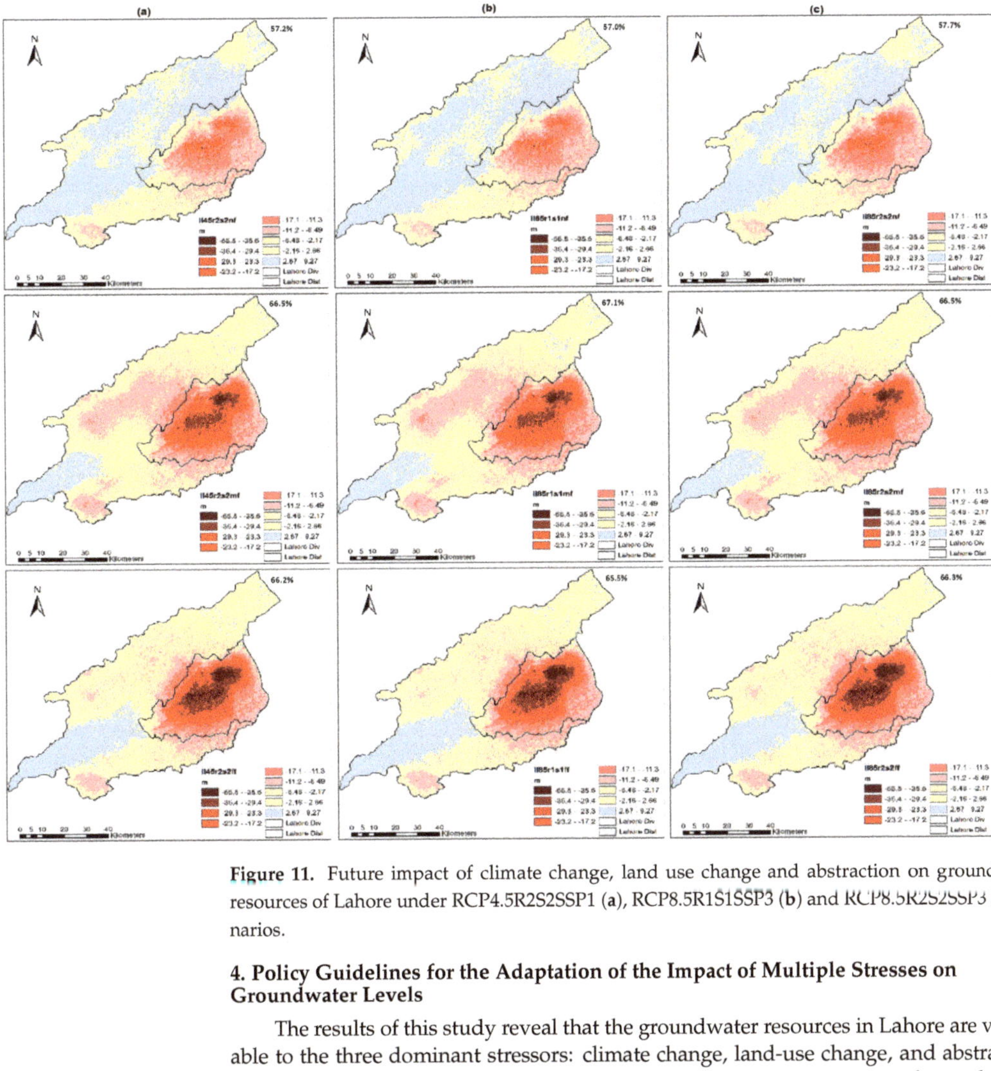

Figure 11. Future impact of climate change, land use change and abstraction on groundwater resources of Lahore under RCP4.5R2S2SSP1 (**a**), RCP8.5R1S1SSP3 (**b**) and RCP8.5R2S2SSP3 (**c**) scenarios.

4. Policy Guidelines for the Adaptation of the Impact of Multiple Stresses on Groundwater Levels

The results of this study reveal that the groundwater resources in Lahore are vulnerable to the three dominant stressors: climate change, land-use change, and abstraction. The main reason is the increased deficit over time due to rising water demand and a decreasing supply (recharge) to the Lahore aquifer. The situation will become severe in the future if current groundwater practices continue. Therefore, groundwater resources need to be protected against the impact of climate change (SDG13), land-use change, and abstraction to ensure the continuous supply of water in ample quantities (SDG6). Protection means complying with the United Nations Sustainable Development Goals: SDG6 ("clean water and sanitation") and SDG 13 ("climate action"). The said goal can be achieved by formulating practical adaptation measures.

The indicators used in this study can be grouped into physical and climatic. The results also reveal that population (density) has a dominant influence on groundwater, followed by recharge and water table depth, respectively. Irrigation supply and impervious areas have minimum effect compared to other factors.

The projected results show quantitively that the population (in urban areas) of Lahore will increase many-fold in the future, ultimately increasing abstraction. At present, the annual groundwater abstraction from the Lahore aquifer is dominated by domestic consumption (52.9%) [39] and is likely to exacerbate due to the future increase in population. An increase in population is associated with local birth rates and migration from other parts of the country due to socioeconomic factors [73]. Expansion in built-up (urban) areas involves many factors, defining the rate at which it overruns other land-use types [74,75]. These factors are biophysical, infrastructural, and socioeconomic. The socioeconomic (66.67%) and infrastructural (64.10%) factors contribute equally and twice the number of biophysical factors (33.33%) in Lahore [45]. The current study projects that built-up area expansion will be consequent to the increase in the future population. An increase in built-up areas depicts the contraction of other land use types, especially agriculture areas, thus reducing the recharge from rainfall and agricultural return flows. In order to control the rapid increase in abstraction and decrease in recharge, some workable adaptation policy actions should be implemented, supported by the results of this study, especially those presented in Table 6. The detailed description of adaptation options is provided in Table S3 (Supplementary material).

Table 6. Proposed adaptation options to counter the negative impacts on groundwater resources in Lahore.

Sr. No.	Improve	Adaptation Options	Time-Based Effectiveness	Approximate Time to Observe Outcome
1		Population control	Slow	10–15 years
2	A	New economic zones	Slow	10–20 years
3		Regulation of abstraction and zoning	Fast	3–5 years
4	A/R	Supplemental supply of treated sewage	Fast	1–2 years
5		Building development laws	Slow	5–10 years
6	R	Rainwater and storm water harvesting	Fast	4–5 years
7		River ponding	Fast	3–4 years

Note. A = Abstraction; R = Recharge. The time-based effectiveness of each adaption measure is defined based on experts' experience and judgment.

5. Assumptions, Limitations, and Future Work

Land use change, especially in the built-up category, is identified to be posing a severe impact on groundwater in Lahore. However, this study relies only on land use projections, developed based on simplified assumptions (future built-up will extend only around the existing built-up areas). Future studies can prepare future land-use change maps based on robust methodologies and replicate the results of this study for Lahore. Our study proposes some adaptation options that could improve the quantitative situation of groundwater in Lahore. Further research may evaluate the potential for groundwater replenishment of these adaptation options: river ponding, rainwater, stormwater harvesting, and reuse of treated wastewater using modeling-based tools. Because of the lack of observation-based data on groundwater quality, this study focuses only on the quantitative aspect of groundwater resources without the quality component. The groundwater quantity, as well as quality aspects, may be considered in future work and management policy formulation as well.

6. Conclusions

This study uses a multi-model integrated approach to investigate the impacts of climate change, land-use change, and abstraction on groundwater resources in Lahore. Following are the conclusions.

Future annual temperatures will rise, with the Tmin increasing more than the Tmax. The dry season will be warmer than the wet season under climate change scenarios. Future annual rainfall will increase, while patterns of annual and seasonal rainfall will remain the same.

The annual rainfall will increase by 24% and 29% under RCP4.5 and RCP8.5 scenarios, respectively, in the far future. The changes will be more apparent under the RCP8.5 compared to the RCP4.5. The built-up area will increase in the future and dominate the agricultural area under land-use change scenarios. An increase in the built-up area will at the expense of the agriculture area. Future annual abstraction in urban areas will increase under shared socioeconomic scenarios. Future annual groundwater abstraction in agricultural areas will decrease but, seasonal groundwater abstraction in the wet and dry seasons will follow the patterns of annual groundwater abstraction under climate change scenarios. Annual groundwater abstraction in urban areas will increase with the highest increase under SSP3 and lowest under SSP1. Annual groundwater abstraction in the agricultural areas will decrease. The decrease will be fast under RCP4.5 than RCP8.5 scenarios.

Future groundwater levels in urban areas will decrease fast due to an increase in abstraction and built-up expansion. Surrounding areas will observe a decline in groundwater levels due to the outward expansion of the water level depression front. Groundwater levels in urban areas will decrease by 51 m and 56 m under RCP4.5R1S1SSP1 and RCP8.5R1S1SSP1 combined scenarios, respectively, in the far future. The groundwater levels in the surroundings of urban areas will decrease by 10.7 m and 11 m under RCP4.5R1S1SSP1 and RCP8.5R1S1SSP1 combined scenarios, respectively, in the far future. Agricultural areas will experience an insignificant change in groundwater levels due to rainfall and irrigation under combined scenarios in the future. Seven adaptation options to offset the negative effect of climate change and human development on groundwater resources have been proposed based on consultation with experts. However, their actual potential has yet to be determined.

Supplementary Materials: The following supporting information can be downloaded at: https://www.mdpi.com/article/10.3390/atmos13122001/s1, Table S1: Functional relationship of the most influential SWAT parameters with the discharge and groundwater components, Table S2: List of sensitive parameters with their calibrated values for GMS-MODFLOW, Table S3: Description of proposed adaptation options to counter the negative impacts on groundwater resources in Lahore.

Author Contributions: Conceptualization, R.A.A.; methodology, R.A.A.; software R.A.A.; validation, R.A.A.; formal analysis, .A.S.; investigation, S.S.; resources R.A.A.; data curation, R.A.A.; writing—original draft preparation, R.A.A.; writing—review and editing, R.A.A., S.N.K. and A.A.; visualization, S.N.K., M.N.U., S.A., M.S.S., M.W.S., A.S., M.U.A. and N.S.; supervision, S.S.; project administration, S.S.; funding acquisition, A.A. All authors have read and agreed to the published version of the manuscript.

Funding: This research received no external funding.

Data Availability Statement: Not applicable.

Acknowledgments: We greatly appreciate Punjab Irrigation Department (PID) and Pakistan Meteorological Department (PMD) for proving data to carry out analysis for this study.

Conflicts of Interest: The authors declare no conflict of interest.

References

1. Ramnarong, V. Evaluation of groundwater management in Bangkok: Positive and negative. In *Groundwater in the Urban Environment: Selected City Profiles*; 1999; pp. 51–62.
2. Arshad, A.; Mirchi, A.; Samimi, M.; Ahmad, B. Combining downscaled-GRACE data with SWAT to improve the estimation of groundwater storage and depletion variations in the irrigated Indus basin. *Sci. Total Environ.* **2022**, *838*, 156044. [CrossRef]
3. Arshad, A.; Zhang, W.; Zaman, M.A.; Dilawar, A.; Sajid, Z. Monitoring the impacts of spatio-temporal land-use changes on the regional climate of city Faisalabad, Pakistan. *Ann. GIS* **2019**, *25*, 57–70. [CrossRef]
4. Mahmood, K.; Rana, A.D.; Tariq, S.; Kanwal, S.; Ali, R.; Haidar, A. Groundwater levels susceptibility to degradation in Lahore metropolitan. *Depression* **2011**, *150*, 8-01.

5. Saleem, F.; Arshad, A.; Mirchi, A.; Khaliq, T.; Zeng, X.; Rahman, M.; Dilawar, A.; Pham, Q.B.; Mahmood, K. Observed changes in crop yield associated with droughts propagation via natural and human-disturbed agro-ecological zones of Pakistan. *Remote Sens.* **2022**, *14*, 2152. [CrossRef]
6. Sajjad, M.M.; Wang, J.; Abbas, H.; Ullah, I.; Khan, R.; Ali, F. Impact of Climate and Land-Use Change on Groundwater Resources, Study of Faisalabad District, Pakistan. *Atmosphere* **2022**, *13*, 1097. [CrossRef]
7. Clifton, C.; Evans, R.; Hayes, S.; Hirji, R.; Puz, G.; Pizarro, C. Water and climate change. 2010.
8. Nyakundi, R.; Makokha, M.; Mwangi, J.; Obiero, C. Impact of rainfall variability on groundwater levels in Ruiru municipality, Kenya. *Afr. J. Sci. Technol. Innov. Dev.* **2015**, *7*, 329–335. [CrossRef]
9. Sishodia, R.P.; Shukla, S.; Graham, W.; Wani, S.P.; Garg, K.K. Bi-decadal groundwater level trends in a semi-arid south indian region: Declines, causes and management. *J. Hydrol. Reg. Stud.* **2016**, *8*, 43–58. [CrossRef]
10. Umar, M.; Khan, S.N.; Arshad, A.; Aslam, R.A.; Khan, H.M.S.; Rashid, H.; Pham, Q.B.; Nasir, A.; Noor, R.; Khedher, K.M.; et al. A modified approach to quantify aquifer vulnerability to pollution towards sustainable groundwater management in Irrigated Indus Basin. *Environ. Sci. Pollut. Res.* **2022**, *29*, 1–22. [CrossRef]
11. Basharat, M. Groundwater Environment in Lahore, Pakistan. In *Groundwater Environment in Asian Cities*; Elsevier: Amsterdam, The Netherlands, 2016; pp. 147–184.
12. Basharat, M.; Rizvi, S.A. Groundwater extraction and waste water disposal regulation. Is Lahore Aquifer at stake with as usual approach. In *Pakistan Engineering Congress, World Water Day, Lahore*; 2011.
13. Heureux, A.M.C.; Alvar-Beltrán, J.; Manzanas, R.; Ali, M.; Wahaj, R.; Dowlatchahi, M.; Afzaal, M.; Kazmi, D.; Ahmed, B.; Salehnia, N.; et al. Climate Trends and Extremes in the Indus River Basin, Pakistan: Implications for Agricultural Production. *Atmosphere* **2022**, *13*, 378. [CrossRef]
14. Aslam, R.A.; Shrestha, S.; Pal, I.; Ninsawat, S.; Shanmugam, M.; Anwar, S. Projections of climatic extremes in a data poor transboundary river basin of India and Pakistan. *Int. J. Climatol.* **2020**, *40*, 4992–5010. [CrossRef]
15. Hina, S.; Saleem, F.; Arshad, A.; Hina, A.; Ullah, I. Droughts over Pakistan: Possible cycles, precursors and associated mechanisms. *Geomat. Nat. Hazards Risk* **2021**, *12*, 1638–1668. [CrossRef]
16. Saleem, F.; Zeng, X.; Hina, S.; Omer, A. Regional changes in extreme temperature records over Pakistan and their relation to Pacific variability. *Atmos. Res.* **2021**, *250*, 105407. [CrossRef]
17. Ahmad, A.; Gilani, H.; Shirazi, S.A.; Pourghasemi, H.R.; Shaukat, I. Spatiotemporal urban sprawl and land resource assessment using Google Earth Engine platform in Lahore district, Pakistan. In *Computers in Earth and Environmental Sciences*; Elsevier: Amsterdam, The Netherlands, 2022; pp. 137–150.
18. Gabriel, H.; Khan, S. Climate responsive urban groundwater management options in a stressed aquifer system. *IAHS-AISH Publ.* **2010**, *338*, 166–168.
19. Ahmad, A.; Chao, W.; Yixian, T.; Sultan, M.; Falak, A.; Wei, D.; Jing, W. SAR-based Subsidence Monitoring and Assessment of the Factors Involved in the Occurrence of Subsidence, Lahore City. *J. Resour. Ecol.* **2022**, *13*, 826–841. [CrossRef]
20. Ahmad, N.; Ahmad, M.; Rafiq, M.; Iqbal, N.; Ali, M.; Sajjad, M.I. Hydrological modeling of the Lahore aquifer using isotopic chemical and numerical techniques. *Back Issues J. Sci. Vis.* **2002**, *7*, 16.
21. Scanlon, B.R.; Faunt, C.C.; Longuevergne, L.; Reedy, R.C.; Alley, W.M.; McGuire, V.L.; McMahon, P.B. Groundwater depletion and sustainability of irrigation in the US High Plains and Central Valley. *Proc. Natl. Acad. Sci. USA* **2012**, *109*, 9320–9325. [CrossRef]
22. Wang, S.; Liu, H.; Yu, Y.; Zhao, W.; Yang, Q.; Liu, J. Evaluation of groundwater sustainability in the arid Hexi Corridor of Northwestern China, using GRACE, GLDAS and measured groundwater data products. *Sci. Total Environ.* **2020**, *705*, 135829. [CrossRef]
23. MacAllister, D.; Krishan, G.; Basharat, M.; Cuba, D.; MacDonald, A.M. A century of groundwater accumulation in Pakistan and northwest India. *Nat. Geosci.* **2022**, *15*, 390–396. [CrossRef]
24. Liu, W.; Bailey, R.T.; Andersen, H.E.; Jeppesen, E.; Park, S.; Thodsen, H.; Nielsen, A.; Molina-Navarro, E.; Trolle, D. Assessing the impacts of groundwater abstractions on flow regime and stream biota: Combining SWAT-MODFLOW with flow-biota empirical models. *Sci. Total Environ.* **2020**, *706*, 135702. [CrossRef]
25. Cheema, M.; Immerzeel, W.; Bastiaanssen, W. Spatial quantification of groundwater abstraction in the irrigated Indus basin. *Groundwater* **2014**, *52*, 25–36. [CrossRef]
26. de Graaf, I.E.M.; Sutanudjaja, E.H.; van Beek, L.P.H.; Bierkens, M.F.P. A high-resolution global-scale groundwater model. *Hydrol. Earth Syst. Sci.* **2015**, *19*, 823–837. [CrossRef]
27. Melaku, N.D.; Wang, J. A modified SWAT module for estimating groundwater table at Lethbridge and Barons, Alberta, Canada. *J. Hydrol.* **2019**, *575*, 420–431. [CrossRef]
28. Behera, A.K.; Pradhan, R.M.; Kumar, S.; Chakrapani, G.J.; Kumar, P. Assessment of groundwater flow dynamics using MODFLOW in shallow aquifer system of mahanadi delta (east coast), India. *Water* **2022**, *14*, 611. [CrossRef]
29. Awan, U.K.; Ismaeel, A. A new technique to map groundwater recharge in irrigated areas using a SWAT model under changing climate. *J. Hydrol.* **2014**, *519*, 1368–1382. [CrossRef]
30. Awan, U.K.; Liaqat, U.W.; Choi, M.; Ismaeel, A. A SWAT modeling approach to assess the impact of climate change on consumptive water use in Lower Chenab Canal area of Indus basin. *Hydrol. Res.* **2016**, *47*, 1025–1037. [CrossRef]
31. Aslam, M.; Arshad, M.; Singh, V.P.; Shahid, M.A. Hydrological Modeling of Aquifer's Recharge and Discharge Potential by Coupling WetSpass and MODFLOW for the Chaj Doab, Pakistan. *Sustainability* **2022**, *14*, 4421. [CrossRef]

32. Aslam, M.; Shehzad, M.U.; Ali, A.; Ali, N.; Chaiyasan, K.; Tahir, H.; Joyklad, P.; Hussain, Q. Seepage and Groundwater Numerical Modelling for Managing Waterlogging in the Vicinity of the Trimmu–Sidhnai Link Canal. *Infrastructures* **2022**, *7*, 144. [CrossRef]
33. Ashraf, S.; Ali, M.; Shrestha, S.; Hafeez, M.A.; Moiz, A.; Sheikh, Z.A. Impacts of climate and land-use change on groundwater recharge in the semi-arid lower Ravi River basin, Pakistan. *Groundw. Sustain. Dev.* **2022**, *17*, 100743. [CrossRef]
34. Ali, S.; Liu, D.; Fu, Q.; Cheema, M.J.M.; Pal, S.C.; Arshad, A.; Pham, Q.B.; Zhang, L. Constructing high-resolution groundwater drought at spatio-temporal scale using GRACE satellite data based on machine learning in the Indus Basin. *J. Hydrol.* **2022**, *612*, 128295. [CrossRef]
35. Arshad, A.; Zhang, Z.; Zhang, W.; Dilawar, A. Mapping favorable groundwater potential recharge zones using a GIS-based analytical hierarchical process and probability frequency ratio model: A case study from an agro-urban region of Pakistan. *Geosci. Front.* **2020**, *11*, 1805–1819. [CrossRef]
36. Basharat, M.; Basharat, M. Developing Sukh-Beas as a potential recharge site during wet years for Bari Doab. *Appl. Water Sci.* **2019**, *9*, 1–15. [CrossRef]
37. Khan, S.; Rana, T.; Ullah, K.; Christen, E.; Nafees, M. Investigating Conjunctive Water Management Options Using a Dynamic Surface-Groundwater Modelling Approach: A Case Study of Rechna Doab. CSIRO Land and Water Technical Report 2003. Available online: http://hdl.handle.net/102.100.100/193016?index=1 (accessed on 1 January 2022).
38. Hussain, M.A.; Chen, Z.; Zheng, Y.; Shoaib, M.; Ma, J.; Ahmad, I.; Asghar, A.; Khan, J. PS-InSAR Based Monitoring of Land Subsidence by Groundwater Extraction for Lahore Metropolitan City, Pakistan. *Remote Sens.* **2022**, *14*, 3950. [CrossRef]
39. Qureshi, A.; Sayed, A.H. *Situation Analysis of the Water Resources of Lahore Establishing a Case for Water Stewardship*; WWF-Pakistan and Cleaner Production Institute (CPI): Lahore, Pakistan, 2014; pp. 1–45.
40. Cheema, M.; Bastiaanssen, W.G. Land use and land cover classification in the irrigated Indus Basin using growth phenology information from satellite data to support water management analysis. *Agric. Water Manag.* **2010**, *97*, 1541–1552. [CrossRef]
41. Khan, S.; Rana, T.; Gabriel, H.F.; Ullah, M.K. Hydrogeologic assessment of escalating groundwater exploitation in the Indus Basin, Pakistan. *Hydrogeol. J.* **2008**, *16*, 1635–1654. [CrossRef]
42. Basharat, M.; Ali, S.U.; Azhar, A.H. Spatial variation in irrigation demand and supply across canal commands in Punjab: A real integrated water resources management challenge. *Water Policy* **2014**, *16*, 397–421. [CrossRef]
43. Anandhi, A.; Hutchinson, S.; Harrington, J.; Rahmani, V.; Kirkham, M.B.; Rice, C.W. Changes in spatial and temporal trends in wet, dry, warm and cold spell length or duration indices in Kansas, USA. *Int. J. Climatol.* **2016**, *36*, 4085–4101. [CrossRef]
44. European Space Agency (ESA). *Land Cover CCI Product User Guide Version 2*; Tech. Rep.; 2017.
45. Bhatti, S.S.; Tripathi, N.K.; Nitivattananon, V.; Rana, I.A.; Mozumder, C. A multi-scale modeling approach for simulating urbanization in a metropolitan region. *Habitat Int.* **2015**, *50*, 354–365. [CrossRef]
46. Hoornweg, D.; Pope, K. Population predictions for the world's largest cities in the 21st century. *Environ. Urban.* **2017**, *29*, 195–216. [CrossRef]
47. O'Neill, B.C.; Kriegler, E.; Riahi, K.; Ebi, K.L.; Hallegatte, S.; Carter, T.R.; Mathur, R.; van Vuuren, D.P. A new scenario framework for climate change research: The concept of shared socioeconomic pathways. *Clim. Change* **2014**, *122*, 387–400. [CrossRef]
48. Arnold, J.G.; Srinivasan, R.; Muttiah, R.S.; Williams, J.R. Large area hydrologic modeling and assessment part I: Model development 1. *JAWRA J. Am. Water Resour. Assoc.* **1998**, *34*, 73–89. [CrossRef]
49. Gujree, I.; Zhang, F.; Meraj, G.; Farooq, M.; Muslim, M.; Arshad, A. Soil and Water Assessment Tool for Simulating the Sediment and Water Yield of Alpine Catchments: A Brief Review. In *Geospatial Modeling for Environmental Management*; CRC Press: Boca Raton, FL, USA, 2022; pp. 37–57.
50. Srinivasan, R.; Arnold, J.; Jones, C. Hydrologic modelling of the United States with the soil and water assessment tool. *Int. J. Water Resour. Dev.* **1998**, *14*, 315–325. [CrossRef]
51. Samimi, M.; Mirchi, A.; Moriasi, D.; Ahn, S.; Alian, S.; Taghvaeian, S.; Sheng, Z. Modeling arid/semi-arid irrigated agricultural watersheds with SWAT: Applications, challenges, and solution strategies. *J. Hydrol.* **2020**, *590*, 125418. [CrossRef]
52. Neitsch, S.L.; Arnold, J.G.; Kiniry, J.R.; Williams, J.R. *Soil and Water Assessment tool Theoretical Documentation Version 2009*; Texas Water Resources Institute: College Station, TX, USA, 2011.
53. Guo, J.; Su, X. Parameter sensitivity analysis of SWAT model for streamflow simulation with multisource precipitation datasets. *Hydrol. Res.* **2019**, *50*, 861–877. [CrossRef]
54. Leta, O.T.; El-Kadi, A.I.; Dulai, H.; Ghazal, K.A. Assessment of SWAT model performance in simulating daily streamflow under rainfall data scarcity in Pacific island watersheds. *Water* **2018**, *10*, 1533. [CrossRef]
55. Rogelj, J.; Popp, A.; Calvin, K.; Luderer, G.; Emmerling, J.; Gernaat, D. Scenarios towards limiting climate change below 1.5 °C Nat. *Clim. Change* **2018**, *8*, 325.
56. Lutz, W.; KC, S. Dimensions of global population projections: What do we know about future population trends and structures? *Philos. Trans. R. Soc. B Biol. Sci.* **2010**, *365*, 2779–2791. [CrossRef]
57. Van Vuuren, D.P.; Edmonds, J.; Kainuma, M.; Riahi, K.; Thomson, A.; Hibbard, K.; Hurtt, G.C.; Kram, T.; Krey, V.; Lamarque, J.-F.; et al. The representative concentration pathways: An overview. *Clim. Change* **2011**, *109*, 5–31. [CrossRef]
58. Arshad, A.; Zhang, Z.; Zhang, W.; Gujree, I. Long-term perspective changes in crop irrigation requirement caused by climate and agriculture land use changes in Rechna Doab, Pakistan. *Water* **2019**, *11*, 1567. [CrossRef]
59. Islam, A.; Sultan, S. Climate Change and South Asia: What Makes the Region Most Vulnerable? 2009. Available online: https://mpra.ub.uni-muenchen.de/21875/ (accessed on 2 October 2022).

60. Pomee, M.S.; Hertig, E. Precipitation projections over the Indus River Basin of Pakistan for the 21st century using a statistical downscaling framework. *Int. J. Climatol.* **2022**, *42*, 289–314. [CrossRef]
61. Ullah, I.; Saleem, F.; Iyakaremye, V.; Yin, J.; Ma, X.; Syed, S.; Hina, S.; Asfaw, T.G.; Omer, A. Projected changes in socioeconomic exposure to heatwaves in South Asia under changing climate. *Earth's Future* **2022**, *10*, e2021EF002240. [CrossRef]
62. Gondim, R.; Silveira, C.; Filho, F.D.S.; Vasconcelos, F.; Cid, D. Climate change impacts on water demand and availability using CMIP5 models in the Jaguaribe basin, semi-arid Brazil. *Environ. Earth Sci.* **2018**, *77*, 1–14. [CrossRef]
63. Jampanil, D.; Suttinon, P.; Seigo, N.; Koontanakulvong, S. Application of input-output table for future water resources management under policy and climate change in Thailand: Rayong province case study. *Change* **2012**, *27*, 29.
64. Khan, I.; Chowdhury, H.; Alam, F.; Alam, Q.; Afrin, S. An investigation into the potential impacts of climate change on power generation in Bangladesh. *J. Sustain. Energy Environ.* **2012**, *3*, 103–110.
65. Price, J.I.; Chermak, J.M.; Felardo, J. Low-flow appliances and household water demand: An evaluation of demand-side management policy in Albuquerque, New Mexico. *J. Environ. Manag.* **2014**, *133*, 37–44. [CrossRef] [PubMed]
66. Abbas, A.; Ullah, S.; Ullah, W.; Waseem, M.; Dou, X.; Zhao, C.; Karim, A.; Zhu, J.; Hagan, D.F.T.; Bhatti, A.S.; et al. Evaluation and projection of precipitation in Pakistan using the Coupled Model Intercomparison Project Phase 6 model simulations. *Int. J. Climatol.* **2022**. [CrossRef]
67. Ashraf, S.; Nazemi, A.; AghaKouchak, A. Anthropogenic drought dominates groundwater depletion in Iran. *Sci. Rep.* **2021**, *11*, 1–10. [CrossRef]
68. Wu, W.-Y.; Lo, M.-H.; Wada, Y.; Famiglietti, J.S.; Reager, J.T.; Yeh, P.J.-F.; Ducharne, A.; Yang, Z.-L. Divergent effects of climate change on future groundwater availability in key mid-latitude aquifers. *Nat. Commun.* **2020**, *11*, 1–9. [CrossRef]
69. Becker, R.; Koppa, A.; Schulz, S.; Usman, M.; der Beek, T.A.; Schüth, C. Spatially distributed model calibration of a highly managed hydrological system using remote sensing-derived ET data. *J. Hydrol.* **2019**, *577*, 123944. [CrossRef]
70. Arnold, J.G.; Moriasi, D.N.; Gassman, P.W.; Abbaspour, K.C.; White, M.J.; Srinivasan, R.; Santhi, C.; Harmel, R.D.; Van Griensven, A.; Van Liew, M.W.; et al. SWAT: Model use, calibration, and validation. *Trans. ASABE* **2012**, *55*, 1491–1508. [CrossRef]
71. Rodell, M.; Velicogna, I.; Famiglietti, J.S. Satellite-based estimates of groundwater depletion in India. *Nature* **2009**, *460*, 999–1002. [CrossRef]
72. Mbuligwe, S. *Physical Infrastructure Service and Environmental Health Deficiencies in Urban and Peri-Urban Areas*; Elsevier: Amsterdam, The Netherlands, 2011; pp. 189–198.
73. Muhammad, W.S. RURAL-URBAN MIGRATION (A CASE STUDY OF LAHORE DISTRICT). Ph.D. Dissertation, University of the Punjab, Lahore, Pakistan, 2004.
74. Dilawar, A.; Chen, B.; Trisurat, Y.; Tuankrua, V.; Arshad, A.; Hussain, Y.; Measho, S.; Guo, L.; Kayiranga, A.; Zhang, H.; et al. Spatiotemporal shifts in thermal climate in responses to urban cover changes: A-case analysis of major cities in Punjab, Pakistan. *Geomat. Nat. Hazards Risk* **2021**, *12*, 763–793. [CrossRef]
75. Mumtaz, F.; Tao, Y.; de Leeuw, G.; Zhao, L.; Fan, C.; Elnashar, A.; Bashir, B.; Wang, G.; Li, L.; Naeem, S.; et al. Modeling spatio-temporal land transformation and its associated impacts on land surface temperature (LST). *Remote Sens.* **2020**, *12*, 2987. [CrossRef]

atmosphere

MDPI

Article

Analysis of Spatio-Temporal Evolution Characteristics of Drought and Its Driving Factors in Yangtze River Basin Based on SPEI

Jieru Wei [1,†], Zhixiao Wang [2,†], Lin Han [1,*], Jiandong Shang [1,3,*] and Bei Zhao [3]

1 National Supercomputing Center in Zhengzhou, Zhengzhou University, Zhengzhou 450001, China
2 The School of Geo-Science & Technology, Zhengzhou University, Zhengzhou 450001, China
3 The School of Computer and Artificial Intelligence, Zhengzhou University, Zhengzhou 450001, China
* Correspondence: hanlin@zzu.edu.cn (L.H.); sjd@zzu.edu.cn (J.S.)
† These authors contributed equally to this work.

Abstract: Using a dataset of 114 meteorological stations in the Yangtze River Basin from 1980–2019, the standardized precipitation evapotranspiration index (SPEI) was calculated based on the Penman-Monteith evapotranspiration model for multiple time scales, and the spatial and temporal evolution characteristics and driving factors of drought in the Yangtze River Basin were analyzed by combining spatial and temporal analysis methods as well as geodetector. The main results obtained are as follows: (1) The climate of the Yangtze River Basin is an overall wet trend, and the trend of summer drought is more similar to the annual scale trend. (2) Most areas in the Yangtze River Basin showed mild drought or no drought, and there is little difference in drought condition among the Yangtze River Basin regions. The areas with drought conditions are mainly distributed in the southwest and east of the Yangtze River Basin. (3) There are significant seasonal differences in drought conditions in all regions, and the drought condition is more different in autumn compared to spring, summer and winter. (4) The average annual precipitation and elevation factors are the dominant driving factors of drought in the Yangtze River Basin, and the double-factor interaction has a greater influence on the drought variation in the Yangtze River Basin than the single-factor effect, indicating that the difference of drought condition in the Yangtze River Basin is the result of the combination of multiple factors.

Keywords: drought evolution characteristics; SPEI; space-time cube; geodetector; Yangtze River Basin; driving factors

Citation: Wei, J.; Wang, Z.; Han, L.; Shang, J.; Zhao, B. Analysis of Spatio-Temporal Evolution Characteristics of Drought and Its Driving Factors in Yangtze River Basin Based on SPEI. *Atmosphere* **2022**, *13*, 1986. https://doi.org/10.3390/atmos13121986

Academic Editors: Jinping Liu, Quoc Bao Pham, Arfan Arshad and Masoud Jafari Shalamzari

Received: 9 October 2022
Accepted: 22 November 2022
Published: 28 November 2022

1. Introduction

Drought is one of the most costly natural disasters, which has a very important impact on agricultural production [1], biodiversity [2], human health [3], hydrology [4] and other important fields related to human production and life. Droughts can be classified into four main types according to their causes [5]: meteorological drought, agricultural drought, hydrological drought and socio-economic drought. The frequency of drought events has become more frequent [6] because of the superposition of natural and anthropogenic factors [7] such as climate change and human activities. However, due to the complexity and variability of the many factors involved in drought, the identification and analysis of drought events pose a huge challenge.

The Yangtze River Basin is the largest basin in China, and it straddles the Qinghai-Tibet alpine region, the southwest tropical monsoon region and the central China subtropical monsoon region, with complex climatic conditions. The Yangtze River Basin, as a typical wet-semi-humid zone, has obvious alternation between wet and dry, and the Yangtze River Basin droughts are characterized by short-term fluctuations and the coexistence of droughts and floods, which makes the study of drought scenarios in the Yangtze River Basin particularly complex.

The drought index is an important method for quantitatively calculating the severity and impact of drought [8]. Drought indices are vital to objectively quantify and compare drought severity, duration, and extent across regions with varied climatic and hydrologic regimes [9]. In the past decades, a series of meteorological drought indexes have been developed, such as Standardized Precipitation Index (SPI) [10], Standardized Precipitation Evapotranspiration Index (SPEI) [11], Palmer Drought Severity Index (PDSI) [12] and Soil Moisture Deficit Index (SMDI) [13], which are widely used in different spatial scales globally, regionally, nationally and in different river basins [14]. The SPEI was proposed by Vicente Serrano et al. [11], which retains the core algorithms of the PDSI and SPI, and can combine multi-scale features with the ability to assess the impact of temperature change on drought [15]. At the same time, the temperature factor was considered, and the concept of potential evapotranspiration was introduced [16]. The SPEI is an important and useful tool for comparing meteorological drought [9]. Evapotranspiration is the major component of the water cycle, so a correct estimate of this variable is fundamental [17]. At present, there are two potential evapotranspiration models commonly used in the SPEI calculation process in China, which are Thornthwaite and Penman-Monteith. Temperature is the only meteorological element required in the Thornthwaite model. In contrast, the elements involved in the calculation based on the Penman-Monteith model, in addition to temperature, also take into account solar radiation, air pressure, wind speed, relative humidity and the geographical location of the meteorological station site [16]. If data permits, the Penman-Monteith model strikes a useful balance between consistency and minimal data requirements, requiring only the addition of minimum/maximum temperature and wind speed [9]. Liu et al. [18] calculated the SPEI (abbreviated as SPEI_TH and SPEI_PM, respectively) for the Chinese region using the Thornthwaite and Penman-Monteith models, respectively, and showed that SPEI_PM can describe the dry and wet variation characteristics of the study area relatively more reasonably.

SPEI has been widely used in drought research. Ling et al. [19] used SPEI to analyze the spatio-temporal evolution characteristics of drought in the Haihe River Basin from 1960 to 2020, and found that the frequency of drought was on the rise, with mild drought and moderate drought occurring frequently. Men et al. [20] analyzed the spatio-temporal characteristics of meteorological drought in the Chaobai River Basin, and the results showed that the variation trends of dry and wet conditions were not exactly the same at different time scales, but they were all mainly dominated by mild and moderate droughts. Wang et al. [21] used SPEI to analyze the effects of multi-temporal scale drought on vegetation dynamics in Inner Mongolia from 1982 to 2015, and found that the probability of vegetation productivity loss increased with increasing drought levels under different drought levels. Chen et al. [14] showed that SPEI_PM performed better than SPEI_TH in the results of drought monitoring in China, and that temperature changes in recent decades had the greatest weight in the natural factors causing drought. Li et al. [22] found that the SPEI_PM results for the Yangtze River Basin were better than SPI and SPEI_TH, but the study only used SPEI_PM to analyze the annual-scale drought evolution characteristics of the Yangtze River Basin without multi-scale analysis and analysis of drought drivers. Tian et al. [23] divided the Yangtze River Basin according to each sub-basin and used soil moisture data to study agricultural drought, but did not use SPEI for drought analysis. Huang et al. [24] analyzed the temporal evolution characteristics of drought area, spatial and temporal distribution characteristics of dry and wet scenarios, and change trends in the Yangtze River Basin based on PDSI; however, the PDSI used in this study lacked multi-scale characteristics and did not effectively analyze the multi-scale drought characteristics of the Yangtze River Basin, and the study also lacked the analysis of drought drivers in the Yangtze River Basin. However, when analyzing the conditions for the occurrence of drought events, previous studies often simply attributed them to average or extreme temperatures and precipitation, while ignoring the internal factors and exploring the patterns in the long-term changes of drought events, thus leading to a failure to effectively break through the core causes of regional drought phenomenon. Therefore, understanding the characteristics of drought and its influencing factors in the study area plays an important role in natural disasters and the pressure of production and life in the area.

At present, most studies on droughts in the Yangtze River Basin were conducted separately in time and space, without analyzing the characteristics and evolution of droughts at the overall spatial and temporal scales, and without analyzing the driving factors of droughts in the Yangtze River Basin. Exploring the distribution pattern, formation process and impact mechanism of meteorological drought in the Yangtze River Basin using spatial and temporal data models has important practical and realistic significance. Since the space-time cube model can ensure the continuity of spatio-temporal data, when compared with traditional spatio-temporal analysis, the space-time cube can show the spatio-temporal characteristics of the data as a whole, instead of only selecting individual years for analysis and presentation as in traditional spatio-temporal analysis. In this paper, the multi-scale SPEI of the Yangtze River Basin was visualized and analyzed by using the space-time cube model, and the clustering areas of the drought at each scale in the Yangtze River Basin were obtained by combining the time series clustering method. The trend of drought conditions in the Yangtze River Basin over the past 40 years was determined by using emerging hot spot analysis. Finally, the drought driving factors in the Yangtze River Basin were studied based on geodetector.

2. Materials and Methods

2.1. Study Region

The Yangtze River Basin covers a total area of 1.8 million square kilometers (Figure 1), accounting for 18.8% of China's territory, making it the largest basin in Asia. The Yangtze River Basin spans the Qinghai-Tibet alpine region, the southwest tropical monsoon region and the central China subtropical monsoon region [22]. The vegetation in the upper reaches is dominated by alpine meadow and natural grassland, the forest vegetation in the middle reaches is dominated, and farmland is widely distributed in the middle and lower reaches and Sichuan Basin [23]. With rich resources, large population clusters, and rapid industrial development [25], it plays an important role in ecological integrity and ecosystem services.

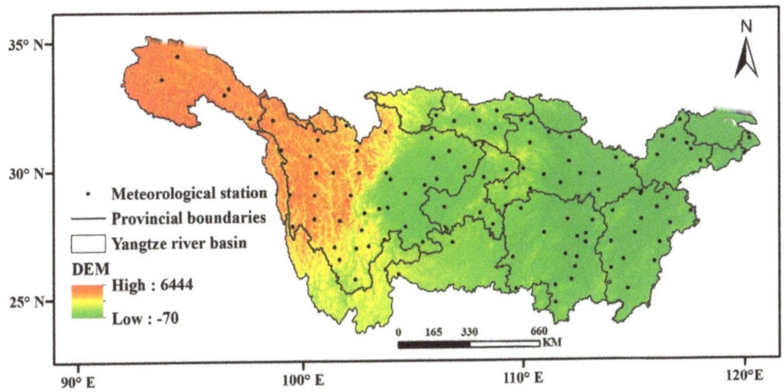

Figure 1. Study area.

2.2. Data Source

In this paper, 114 meteorological stations within the Yangtze River Basin were selected from 1980 to 2019, and data such as mean temperature and latitude were obtained from the China Meteorological Science Data Sharing Service (https://www.data.cma.cn/ accessed on 2 March 2020). In order to ensure the integrity of data in time series, the missing data of a few stations are interpolated by neighboring stations. The driving factor data (Table 1), provincial boundaries in the Yangtze River Basin and the boundary of the Yangtze River Basin were obtained from the Resource and Environmental Science and Data Center of Chinese Academy of Sciences (https://www.resdc.cn/ accessed on 26 March 2022) and Geospatial Data Cloud (https://www.gscloud.cn/ accessed on 20 March 2022).

Table 1. Yangtze River Basin SPEI driving factors.

Category	Factor
Topography	Elevation (X_1)
	Slop (X_2)
	Soil type (X_3)
Meteorology	Average annual temperature (X_4)
	Average annual precipitation (X_5)
Socio-economic	population density (X_6)
	GDP (X_7)
	Night light (X_8)
	Human footprints (X_9)
Traffic location	Distance to water system (X_{10})
	Distance to provincial road (X_{11})
	Distance to railroad (X_{12})

2.3. Methods

2.3.1. Standardized Precipitation-Evapotranspiration Index

SPEI is the result of standardized difference between precipitation and potential evapotranspiration [24]. In this paper, the Penman-Monteith model was selected as the potential evapotranspiration model to calculate the multiscale SPEI values for the period 1980–2019 at 114 meteorological stations in the study area, which provided a more accurate method for calculating the potential evapotranspiration and can better reflect the regional dry and wet conditions [19]. The specific calculation process of SPEI_PM is as follows [11].

(1) The calculation of the reference crop evapotranspiration (ET_0) was computed using the Penman-Monteith model with the following equation.

$$ET_0 = \frac{0.408\Delta(R_n - G) + \gamma\frac{900}{T+273}U_2(e_s - e_a)}{\Delta + \gamma(1 + 0.34U_2)} \tag{1}$$

where ET_0 is the evapotranspiration of the reference crop (mm/d), Δ is the slope of the saturated water pressure curve (kPa/°C), γ is the hygrometry constant (kPa/°C), R_n is net solar radiation (MJ·m^{-2}·d^{-1}), G is the heat flux of soil (MJ·m^{-2}·d^{-1}), T is the average temperature during the calculation period (°C), U_2 is the average wind speed at 2 m above the ground, e_s is the saturated water pressure (kPa), and e_a is the actual water pressure (kPa).

(2) Calculate the difference between month-by-month precipitation and evapotranspiration.

$$D_i = P - ET_0 \tag{2}$$

where D_i is the difference between precipitation and evapotranspiration, P is the monthly precipitation, ET_0 is the actual monthly evapotranspiration.

(3) Normalization of D_i data series. The log-logistic probability distribution F(x) is used to fit D_i, and the SPEI value corresponding to each D_i value is calculated.

$$w = \sqrt{-2\ln P} \tag{3}$$

when the cumulative probability $p \leq 0.5$:

$$SPEI = w - \frac{c_0 + c_1w + c_2w^2}{1 + d_2w + d_1w^2 + d_3w^3} \tag{4}$$

when the cumulative probability $p \geq 0.5$:

$$SPEI = -(w - \frac{c_0 + c_1 w + c_2 w^2}{1 + d_2 w + d_1 w^2 + d_3 w^3}) \tag{5}$$

where d_1 = 1.432788, d_2 = 0.189269, d_3 = 0.001308, c_0 = 2.515517, c_1 = 0.802853, c_2 = 0.010328.

2.3.2. Space-Time Cube

The space-time cube model is a method to aggregate sample points into space-time bars [26]. By creating space-time cube (Figure 2), spatio-temporal data can be visualized in the form of time series analysis, integrated spatial analysis and temporal analysis models [27]. In Figure 2, X and Y represent the spatial location of the geographic entity, Z represents time. The bottom layer is the starting time and the top layer is the latest time, and each cube is composed of the attribute values corresponding to that time, and the values can be differentiated by setting different colors. Because the space-time cube model can ensure the continuity of spatio-temporal data, the space-time cube can show the spatio-temporal characteristics of the whole data when comparing with traditional spatio-temporal analysis [28], instead of the traditional spatio-temporal analysis, which can only visualize a single year, which destroys the continuity of time and ignores the possible interactions between spatio-temporal data [29]. As a temporal variable pattern, spatio-temporal analysis or model persistence metrics are considered worth exploring [30]. The model uses the geometric properties of the time dimension. Spatial entity is a concept of space-time body, and the description of geographic change is simple and straightforward [14]. Three-dimensional visualization of the space-time cube makes it easy to explain trends and patterns of big data over long time scales [31]. The spatio-temporal distribution characteristics, spatio-temporal evolution process, time series clustering analysis and emerging hot spot analysis analysis of drought in the Yangtze River Basin were explored by combining the thinking mode of spatio-temporal analysis, which can provide a scientific basis for the research on spatio-temporal changes of drought for relevant departments [28].

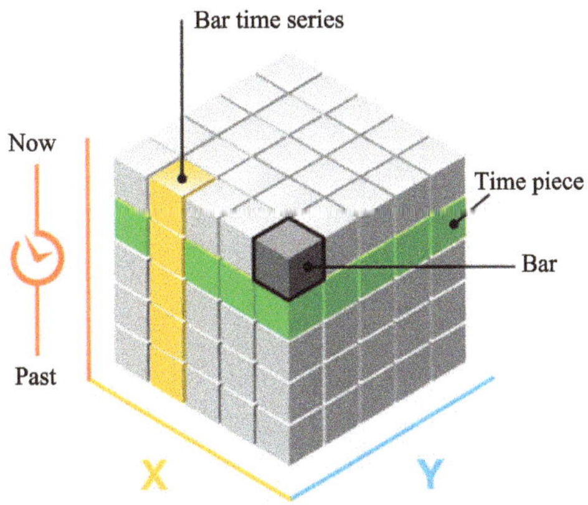

Figure 2. Schematic diagram of the space-time cube.

2.3.3. Time Series Clustering Analysis

Time series clustering groups regions with similar trends and patterns into a common cluster. These clusters are unlabeled and simply indicate the similarity of trends and patterns between different regions [32,33]. It is very difficult to analyze and mine the large amount of data and high-dimensional time series, which will affect the overall analysis results [34]. Due to the various applications of time series cluster analysis, there are many different TSC methods [35]. Based on the similarity of time series features, the time series set stored in the space-time cube is divided. It can aggregate time series based on three conditions: having similar time values, tending to increase and decrease at the same time and having similar repeating patterns [23] (Figure 3). In this paper, the SPEI data in the Yangtze River Basin from 1980 to 2019 was combined with the space-time cube model for time series clustering.

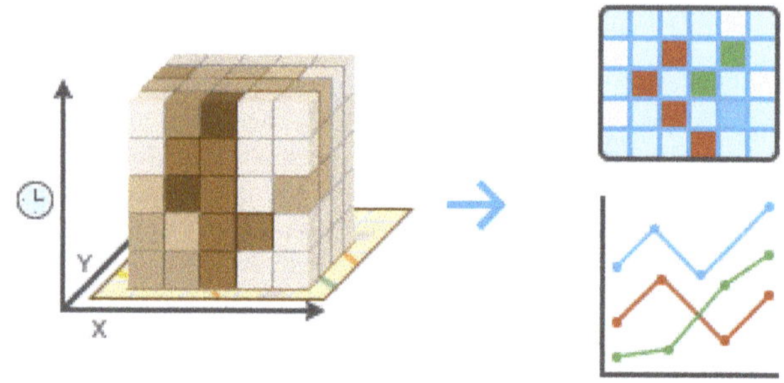

Figure 3. Schematic diagram of time series clustering.

2.3.4. Emerging Hot Spot Analysis

Emerging hot spot analysis can identify the spatio-temporal trend and patterns of change in data [36], and analyze the hot or cold spots of a certain feature at the spatio-temporal scale. The Getis-Ord Gi^* statistic is calculated for each cube bar by specific neighborhood distance and prodomain time step parameters [37]. Gi^* statistic is the z-score. The obtained z-score allows to know where the clustering of high- and low-valued elements occurs in space. Mann–Kendall trend test method is used to test the trend of hot spot analysis results [38]. The results are divided into seven categories: new hot spot, sporadic hot spot, oscillating hot spot, new cold spot, sporadic cold spot, oscillating cold spot and no pattern detected [26]. Finally, according to the spatial pattern characteristics of the time series of each research unit, statistical analysis and the results of the Mann-Kendall trend test, the research results are classified into different types of spatio-temporal patterns for comprehensive expression according to certain classification principles [39]. In recent years, emerging hot spot analysis has been applied to different scientific fields [40,41].

The formula for Gi^* is as follows.

$$Gi^* = \frac{\sum_{j=1}^{n} w_{ij}x_j - \bar{X}\sum_{j=1}^{n} w_{ij}}{S\sqrt{\frac{n\sum_{j=1}^{n} w_{ij}^2 - (\sum_{j=1}^{n} w_{ij})^2}{n-1}}} \tag{6}$$

where x_j is the attribute value of element j, w_{ij} is the spatial weight between elements i and j, n is the total number of elements, and

$$\bar{X} = \frac{\sum_{j=1}^{n} x_j}{n} \tag{7}$$

$$S = \sqrt{\frac{\sum_{j=1}^{n} x_j^2}{n} - (\bar{X})^2} \tag{8}$$

2.3.5. Geodetector

Geodetector is a new statistical method for detecting spatial stratified heterogeneity and revealing the driving factors behind it [42]. The core idea is based on the assumption that if an independent variable has a significant effect on a dependent variable, then the spatial distribution of the independent and dependent variables should have similarity [43,44]. Geodetector is good at analyzing type quantities, while sequential, ratio or interval quantities can be analyzed with appropriate discretization [45]. Geodetector can also be used for statistical analysis [42]. The core of the theory is to detect the consistency of spatial distribution pattern between the dependent variable and the independent variable through spatial stratified heterogeneity, and to measure the explanatory power of the independent variable on the dependent variable accordingly. Geodetector includes 4 detectors: factor detector, ecological detector, interaction detector and risk detector. These detectors are mutually perfect and supportive relationships in measuring the explanatory power of the independent variables on the spatial distribution of the dependent variable [46].

(a) Factor detector

Detecting the spatial stratified heterogeneity of the dependent variable Y and detecting the extent to which a factor X explains the spatial stratified heterogeneity of Y (Figure 4). The influence of each detection factor on the drought in the Yangtze River Basin can be calculated through factor detector, namely q. A larger q value means that the influence of a detection factor X on the drought in the Yangtze River Basin is greater. The expression is:

$$q = 1 - \frac{\sum_{h=1}^{L} N_h \sigma_{h^2}}{N_{\sigma^2}} \quad (9)$$

where h is the stratification of variable Y or factor X, $h = 1, 2, 3, \cdots, L$. N_h and N are the number of units in layer h and the whole area, respectively. σ_{h^2} and σ^2 are the variances of the Y values for layer h and the whole region, respectively.

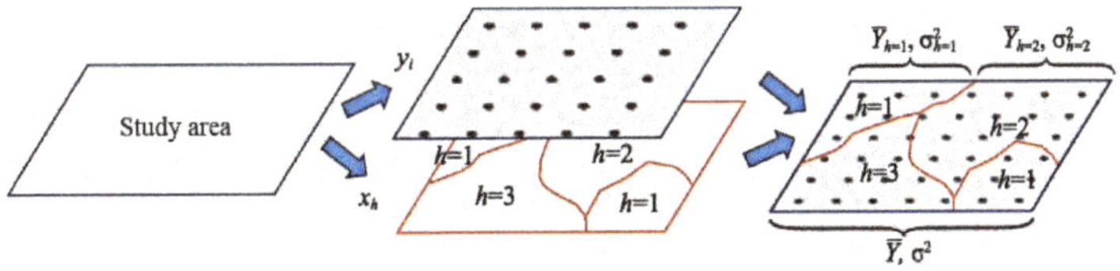

Figure 4. Principle of geodetector.

(b) Interaction detector

It is used to analyze the interaction between the factors [42], that is, to assess whether the factors X1 and X2 together increase or decrease the explanatory power of the dependent variable Y, or whether the effects of these factors on Y are independent of each other. The method of evaluation is to first calculate the q-values of the two factors $X1$ and $X2$ on Y: $q(X1)$ and $q(X2)$, and calculate the value of q when they interact (Figure 5): $q(X1 \cap X2)$. Compare $q(X1)$, $q(X2)$ and $q(X1 \cap X2)$.

Drought formation is the result of a combination of drivers [47]. Referring to existing studies [48–51], 4 major aspects were selected from natural factors (topographic and meteorological factors) and human factors (socio-economic factors and traffic factors), and a total of 12 detection factors X were selected (Table 2). Based on the study of drought differences in the Yangtze River Basin using the factor detector method, the strength of the two-factor effect on drought differences was studied using the interaction detector analysis.

Table 2. Drought classification based on SPEI.

Level	Type	SPEI
1	No drought	$SPEI \geq -0.5$
2	Mild drought	$-1.0 \leq SPEI < -0.5$
3	Moderate drought	$-1.5 \leq SPEI < -1.0$
4	Severe drought	$-2.0 \leq SPEI < -1.5$
5	Extreme drought	$SPEI \leq -2.0$

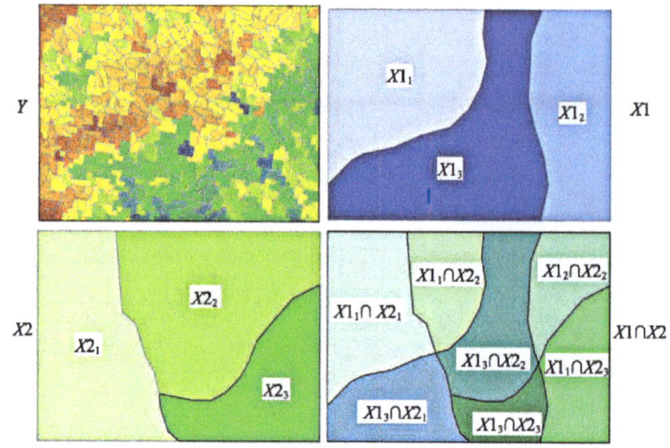

Figure 5. Interaction detector. Y is the dependent variable. $X1$ and $X2$ are evaluation factors. Overlay the two layers $X1$ and $X2$ to get the new layer $X1 \cap X2$.

3. Results and Discussion

3.1. Temporal Variation Characteristics of Drought

As shown in Figure 6, from 1980 to 2019, the annual SPEI of the Yangtze River Basin showed a obvious form of positive and negative oscillations in the short term. The trend line shows that the overall rate of increase is 0.01/10a, indicating a wet trend in the climate of the Yangtze River Basin, which is consistent with the findings of Zhang et al. [52]. In the past 40 years, the drought periods in the Yangtze River Basin were mainly concentrated in 1986–1988 and 2006–2013, among which the drought intensity was higher in 1986, 1988 and 2006, and with SPEI values of −0.75, −0.73 and −0.62, respectively, indicating Mild drought. The wet periods were mainly concentrated in 1980–1983 and 1989–2005, among which 1983 and 1998 were relatively wet, with SPEI values reaching 0.59 and 0.68, respectively.

This paper counted the area of drought areas in each year, as shown in Figure 7. It can be found that the percentage of drought areas in 1986, 1988 and 2006 were 0.68, 0.70 and 0.67, respectively, which indicates that the majority of areas in that time node were in drought.

The seasons are defined according to the meteorological division method. The division rules of different seasons and months are in the order of March to May (spring), June to August (summer), September to November (autumn), and December to February of the next year (winter).

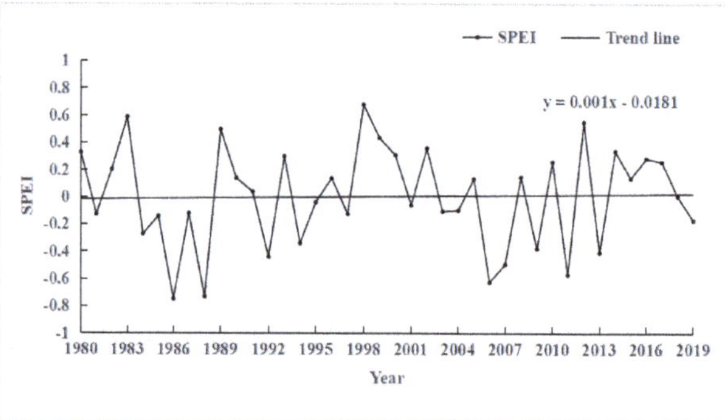

Figure 6. Changes in annual mean SPEI in the Yangtze River Basin from 1980 to 2019.

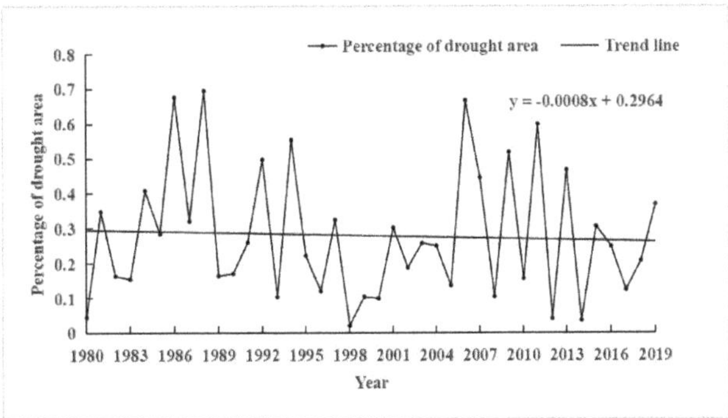

Figure 7. Percentage of drought area.

As show in Figure 8, the seasonal time scale results showed that the overall SPEI of each seasonal scale in the Yangtze River Basin from 1980–2019 showed significant positive and negative fluctuations compared to the annual scale. The following conclusions can be drawn from the trend line: in spring and summer, the SPEI values showed no significant increasing trend, with an increasing rate of 0.061/10a and 0.003/10a, respectively; in autumn and winter, the SPEI values showed no significant decreasing trend, with decreasing rates of 0.006/10a and 0.077/10a, respectively. Compared with other seasons, the frequency of winter drought in the Yangtze River Basin from 1980 to 2019 was higher, and the variation trend of summer SPEI was more similar to that of the annual scale.

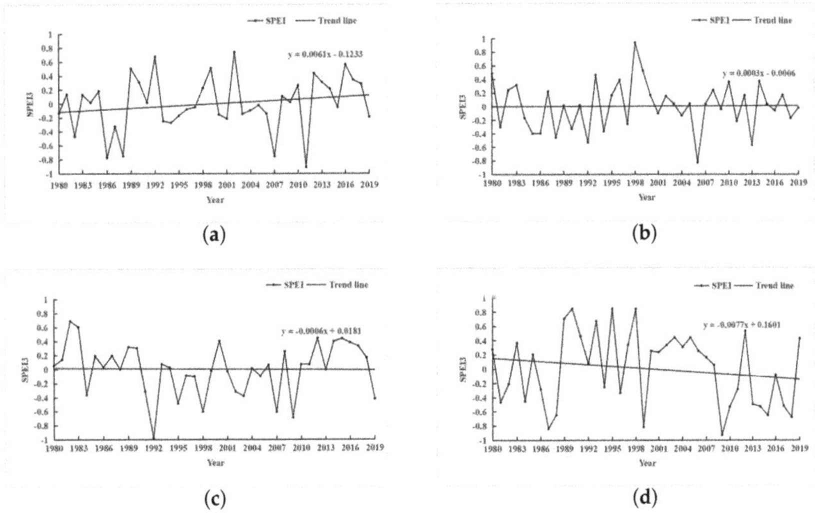

Figure 8. Variation of seasonal SPEI values in the Yangtze River Basin from 1980 to 2019. (**a**) Spring; (**b**) summer; (**c**) autumn; and (**d**) winter.

3.2. *Spatial Variation Characteristics of Drought*

3.2.1. Space-Time Cube for Multi-Scale SPEI

This paper combined with space-time cube model to demonstrate the spatio-temporal distribution of multi-scale SPEI of 114 meteorological stations in the Yangtze River Basin. Figure 9 shows the spatio-temporal monitoring of drought in the Yangtze River Basin at the seasonal scales of spring, summer, autumn, and winter, respectively. As a whole, most areas of the Yangtze River Basin and most of the time show light drought or no drought. From the perspective of time and space, there are obvious seasonal differences in drought conditions in the Yangtze River Basin. There is no perennial drought in the same season in different years, but the drought conditions has gradually improved in recent years, and summer is wetter than other seasons. From the annual scale (Figure 9e), the areas with severe drought were mainly distributed in the southwest and east of the Yangtze River Basin. On the whole, drought occurred at each meteorological station, and there were three main conditions: early drought conditions were more severe and gradually improved, early drought conditions were good but gradually deteriorated, and always in no drought or mild drought state.

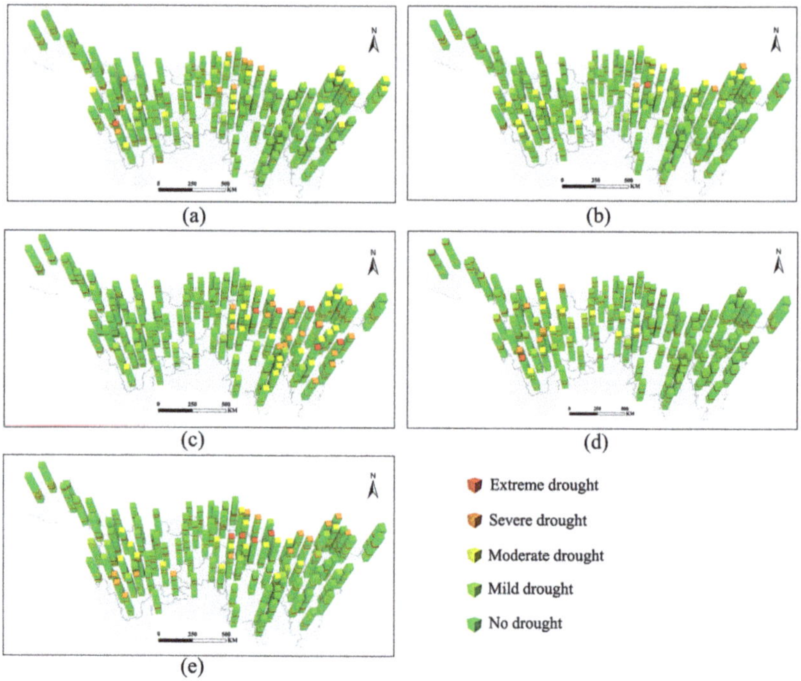

Figure 9. Space-time cube results of multi-scale SPEI in Yangtze River Basin. (**a**) Spring; (**b**) summer; (**c**) autumn; and (**d**) winter; (**e**) Year.

3.2.2. Result of Time Series Clustering Analysis

In this paper, the spatio-temporal distribution of the drought in the Yangtze River Basin in the past 40 years was clustered by the space-time cube results, and the results are shown in Figure 10. The time series clustering results with SPEI seasonal scale and SPEI annual scale are in Figure 10. The number of clusters in Figure 10 refers to the same color region as one class, for example, Figure 10a is two colors, so the number of clusters is 2. As shown in Figure 10, the number of SPEI seasonal-scale and annual-scale clusters is small, indicating that drought conditions do not significantly differ among regions in the Yangtze River Basin. Compared with spring, summer and winter, the number of clusters in autumn is higher and mainly concentrated in the western part of the Yangtze River Basin, because the Yangtze River Basin spans the eastern, central and western parts of China. There are significant differences in precipitation and temperature in autumn in the west compared with other regions, and therefore the differences in drought conditions become larger.

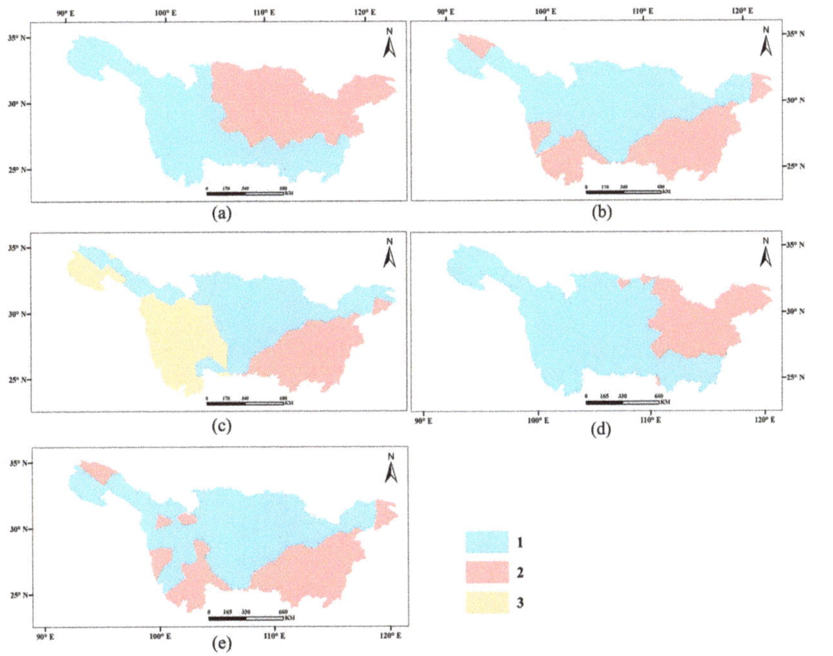

Figure 10. Result of time series clustering analysis. (**a**) Spring; (**b**) summer; (**c**) autumn; and (**d**) winter; (**e**) Year.

3.2.3. Result of Emerging Hot Spot Analysis

Combined with the space-time cube results, the emerging hot spots of multi-scale drought in the Yangtze River Basin in recent 40 years were analyzed (Figure 11). From the annual scale of SPEI (Figure 11e), there is an oscillating hot spot trend in the northwest and northeast of the Yangtze River Basin, indicating that severe drought years in these regions occur irregularly in historical years.

From the SPEI seasonal scale (Figure 11a–d), in spring, there is oscillating hot spot trend in the west of the Yangtze River Basin, indicating that the severe drought years in these areas occurred irregularly in historical years; in the southeast of the Yangtze River Basin, there is oscillating cold spot trend, indicating that the drought in these areas is not severe, but has historically occurred irregularly. In summer, there is an oscillating hot spot trend in the west, east and southeast of the Yangtze River Basin, indicating that the severe drought years in these areas occurred irregularly in historical years; in the central and east of the Yangtze River Basin, there is oscillating cold spot trend, indicating that the drought in these areas is not severe, but has historically occurred irregularly. In autumn, there is new hot spot trend in the northwest of Yangtze River Basin, the drought was not serious in the region previously, but in recent years, the drought is serious; there is oscillating hot spot trend in the east and northwest of Yangtze River Basin, indicating that the severe drought years in these areas occurred irregularly in historical years; in the west of Yangtze River Basin, there is oscillating cold spot trend, indicating that the drought in these areas is not severe, but has historically occurred irregularly. In winter, there is oscillating hot spot trend in the east of Yangtze River Basin, indicating that the severe drought years in these areas occurred irregularly in historical years; in the west of Yangtze River Basin, there is oscillating cold spot trend, indicating that the drought in these areas is not severe, but has historically occurred irregularly.

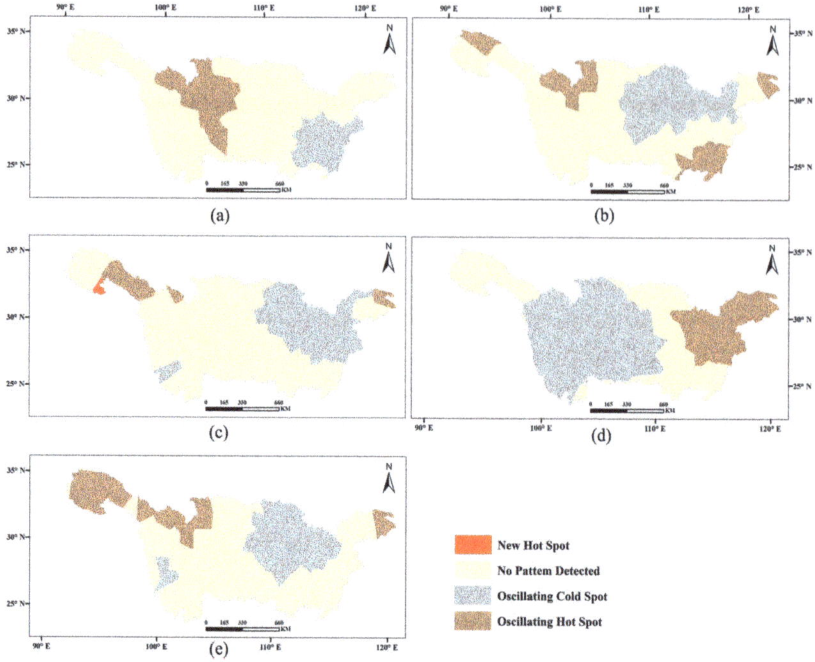

Figure 11. Result of emerging hot spot analysis. (**a**) Spring; (**b**) summer; (**c**) autumn; and (**d**) winter; (**e**) Year.

3.3. Analysis of Drought Drivers in the Yangtze River Basin

3.3.1. Factor Detector

In this paper, time cross-sectional data of 2000, 2005, 2010, and 2015 were selected for factor detection of each impact factor (the coding meanings are shown in Table 1), and the results are shown in Table 3. This paper selected 12 indicators that may affect drought differences in four dimensions. The average q values of each dimension of the indicators in each period were summed to obtain the effect intensity of different dimensions on drought differences in the Yangtze River Basin, while each effect intensity was divided into two equal effect levels, and it was defined as a strong effect dimension layer when $q > 0.5$, otherwise it was a weak effect intensity layer. The ranking of dimensional effect results is meteorology (0.59) > topography (0.55) > traffic location (0.19) > socio-economic (0.16). Meteorology and topography are the strong dimensional layers, while transportation location and socio-economics are the weak dimensional layers, which is consistent with the actual situation and confirms that meteorology and topography play a significant role in affecting drought differences in the Yangtze River Basin.

From the results of the q values of each influencing factor, the top three influencing factors with q values were considered as the dominant factors. The top 3 in 2000 are elevation (0.18), distance to railroad (0.18) and average annual temperature (0.16). The top 3 in 2005 are average annual temperature (0.32), soil type (0.30) and elevation (0.28). The top 3 in 2010 are average annual precipitation (0.43), elevation (0.19) and soil type (0.10). The top 3 in 2015 are average annual precipitation (0.70), soil type (0.44) and average annual temperature (0.42).

Table 3. Factor detection results of drought differences in the Yangtze River Basin.

Factor	2000	2005	2010	2015	q (Average)	q (Sum)
X_1	0.18	0.28	0.19	0.41	0.27	0.56
X_2	0.06	0.00	0.05	0.07	0.05	
X_3	0.13	0.30	0.10	0.44	0.24	
X_4	0.16	0.32	0.09	0.42	0.25	0.59
X_5	0.01	0.20	0.43	0.70	0.34	
X_6	0.01	0.02	0.00	0.04	0.02	0.16
X_7	0.02	0.02	0.02	0.02	0.02	
X_8	0.07	0.03	0.01	0.09	0.05	
X_9	0.05	0.09	0.00	0.15	0.07	
X_{10}	0.00	0.00	0.00	0.00	0.00	0.19
X_{11}	0.01	0.06	0.00	0.08	0.03	
X_{12}	0.18	0.16	0.09	0.22	0.16	

The results of the average q values of the influencing factors show that the average annual precipitation (0.34), elevation (0.27), average annual temperature (0.25) and soil type (0.24) are dominant, among which the factor with the strongest effect is the average annual precipitation.

From the changes in the q values of the selected influencing factors in each period, the more obvious change is the average annual precipitation, which is gradually dominating over time; the elevation, soil type and average annual temperature show an increasing trend and dominate in each period, which indicates that the influence of human behavioral activities on the environment is gradually increasing.

3.3.2. Interaction Detector

The interaction detector was used to detect the drought differential influencing factors (the coding meanings are shown in Table 2) in the Yangtze River Basin in 2000, 2005, 2010 and 2015, respectively, and the results are shown in Figure 12. According to the results of the interaction detector, the influence of double factor interaction on drought differences in the Yangtze River Basin is greater than that of single-factor interaction, and the types of effects include non-linear enhancement and double factor enhancement, that is, the drought differences in the Yangtze River Basin are the result of the combined effect of multiple factors. In 2000, the best double factor combination is elevation and average annual precipitation (0.32). In 2005, the best double factor combination is soil type and average annual precipitation (0.40). In 2010, the best double factor combination is soil type and average annual precipitation (0.52). In 2015, the six best combinations of double factor combination effects are the combinations of average annual precipitation with elevation, soil type, average annual temperature, population density, night light, and human footprint, respectively, and the detection values were 0.72. It can be found that the combination of annual average precipitation and other factors all dominate the influence of drought variation in the Yangtze River Basin from 2010 onwards. This indicates that the difference of drought in the Yangtze River Basin is not the result of a single factor or dimension, but the comprehensive effect of multiple factors and systems.

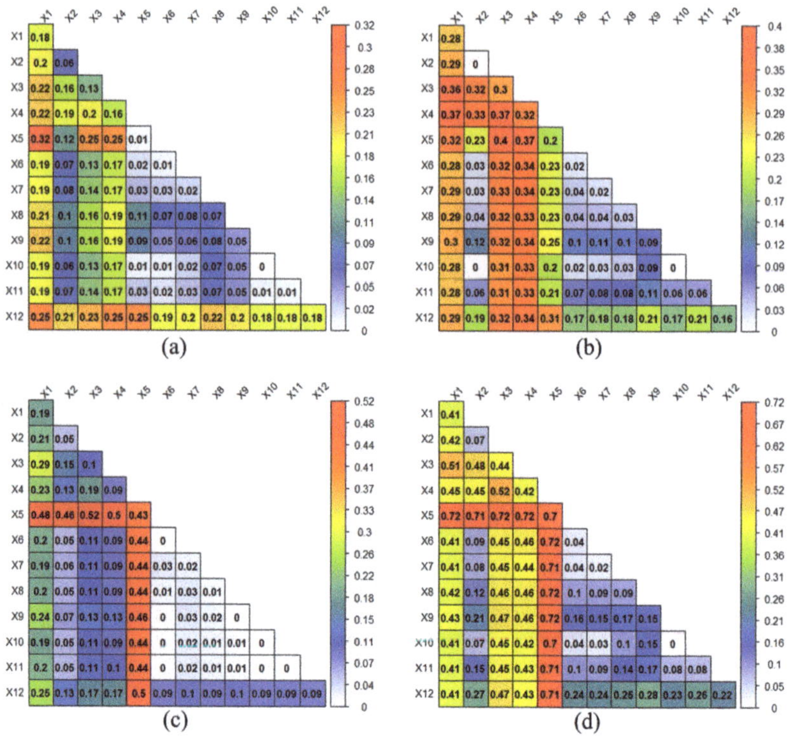

Figure 12. Interaction detector results of drought differences in the Yangtze River Basin. (**a**) 2000; (**b**) 2005; (**c**) 2010; and (**d**) 2015.

4. Discussion

This study was based on SPEI_PM, using space-time cube, time cluster analysis and emerging hot spot methods to analyze the spatial and temporal evolution characteristics of drought in the Yangtze River Basin over the past 40 years and to study the drivers of the Yangtze River Basin using geodetector.

Our analysis demonstrates that the climate of the Yangtze River Basin is an overall wet trend and most areas in the Yangtze River Basin showed mild drought or no drought. Similarly, Huang et al. [24] analyzed the drought characteristics of the Yangtze River Basin based on PDSI and found that there is an overall wet trend in the Yangtze River Basin. In addition, this paper finds the following results. (1) The areas with drought condition are mainly distributed in the southwest and east of the Yangtze River Basin. (2) There are significant seasonal differences in drought conditions in all regions, and the drought condition is more different in autumn compared to spring, summer and winter. (3) The difference of drought condition in the Yangtze River Basin is the result of the combination of multiple factors. Currently, many scholars have been studying the analysis of drought in the Yangtze River Basin. Li et al. [22] calculated SPI and SPEI based on month-by-month meteorological data, and then analyzed the annual variation characteristics of drought in the Yangtze River Basin using SPEI_PM, which did not analyze the drought characteristics of the Yangtze River Basin from multiple time scales and did not further analyze the factors affecting drought in the Yangtze River Basin. Tian et al. [23] studied the historical spatial and temporal evolution of agricultural drought in the Yangtze River Basin based on long time series CCI soil moisture data, and found that the area of agricultural drought in the Yangtze River Basin showed a trend of increasing and then decreasing, with spring and winter droughts dominating in the seasonal scale; however, this study was mainly limited to agricultural drought studies, and did not use SPEI to conduct a comprehensive study of drought in the Yangtze River Basin and did not investigate the mechanism of multiple factors affecting drought in the Yangtze River Basin. Huang et al. [24] explored the spatial and temporal characteristics of drought in the Yangtze River Basin and its evolutionary trends based on PDSI; however, the PDSI used in this study lacked multi-scale characteristics and did not effectively analyze the multi-scale drought characteristics of the Yangtze River Basin. Compared with the current studies on drought in the Yangtze River Basin by other scholars [53], this study not only provided a multi-scale analysis on drought characteristics, but also analyzed the main influencing factors and mechanisms that cause drought changes [54].

Drought is the most severe meteorological disasters to impact human society and occur widely and frequently in China causing considerable damage to the living environment of humans [52]. They have become stronger in frequency [55], severity and duration under the rapid development of the economy and society [56]. To explore the characteristics of drought and its drivers in the Yangtze River Basin, which plays a pivotal role in reducing natural disasters and production and livelihood stress in the study area [57], thus providing a theoretical and decision-making basis for early warning management of meteorological disasters in the Yangtze River Basin.

More detailed studies on drought in the Yangtze River Basin are limited by the difficulty of obtaining more accurate meteorological data. In the subsequent study, we not only want to improve the data accuracy, but also to make a long time series prediction of drought in the study area based on the deep learning model.

5. Conclusions

Based on the SPEI_PM drought index method, this paper analyzed the spatio-temporal drought evolution characteristics and the driving factors of the Yangtze River Basin at multiple time scales from 1980–2019 using space-time cube, time series clustering analysis, emerging hot spot analysis and geodetector. The main conclusions are as follows.

In terms of temporal variation, the annual-scale SPEI values in the Yangtze River Basin from 1980–2019 show obvious forms of positive and negative oscillations in the short term, with an overall upward trend and an increase rate of 0.01/10a, indicating the wet trend of the Yangtze River Basin climate. From 1980–2019, the overall trend of spring and summer SPEI values in the Yangtze River Basin show a non-significant upward trend; the overall trend of autumn and winter SPEI values show a non-significant downward trend. Compared with other seasons, droughts occurred more frequently in the Yangtze River Basin in winter from 1980–2019, and the trend of SPEI values in summer is more similar to the trend of annual scale changes.

In terms of spatial variation, according to the results of space-time cube, it can be seen that most areas of the Yangtze River Basin and most of the time show mild drought or no drought, and the areas with severe annual drought are mainly distributed in the southwest and east of the Yangtze River Basin. At the seasonal scale, summer is wetter than other seasons, and there are obvious seasonal differences in drought conditions among regions in the Yangtze River Basin. The time series clustering analysis results show that the number of SPEI seasonal-scale and annual-scale clusters is small, indicating that drought conditions do not significantly differ among regions in the Yangtze River Basin, and the drought situation in autumn is relatively different from that in spring, summer and winter. The method can cluster areas with similar drought conditions into one category, and the higher the density of stations, the better the results. The results of emerging hot spot analysis can visualize the overall spatial and temporal trends of the drought in the past 40 years, and the trend of drought increase and decrease in the Yangtze River Basin area can be obtained, providing a theoretical basis for drought prevention and relief in the Yangtze River Basin.

By analyzing the drivers of drought variation in the Yangtze River Basin, it can be obtained that topography and meteorology have the greatest influence on drought, among which the average annual precipitation and elevation factors are dominant. In the interaction detection, the influence of double factor interaction on drought change in the Yangtze River Basin is greater than that of single factor, which indicates that the differences of drought conditions in the Yangtze River Basin are the result of the combination of multiple factors.

Author Contributions: Conceptualization, J.W. and Z.W.; methodology, J.W., Z.W. and J.S.; software, J.W. and Z.W.; validation, J.W., Z.W. and L.H.; formal analysis, J.W.; investigation, J.W.; resources, Z.W. and B.Z.; data curation, J.W. and Z.W.; writing—original draft preparation, J.W. and Z.W.; writing—review and editing, J.W., Z.W., J.S. and L.H.; visualization, J.W., Z.W. and B.Z.; supervision, J.W.and Z.W.; project administration, J.W., J.S. and L.H.; funding acquisition, J.S. and L.H. All authors have read and agreed to the published version of the manuscript.

Funding: This research was funded by Major Science and Technology Project of Henan Province, China, grant number "221100210600", "201400211300" and "201400210600".

Institutional Review Board Statement: Not applicable.

Informed Consent Statement: Not applicable.

Data Availability Statement: The datasets from 1980 to 2019 were obtained from the National Meteorological Data Center (http://data.cma.cn/ accessed on 2 March 2020). The driving factors data provincial boundaries in the Yangtze River Basin and the boundary of the Yangtze River Basin were obtained from the Resource and Environmental Science and Data Center of Chinese Academy of Sciences (https://www.resdc.cn/ accessed on 26 March 2022) and Geospatial Data Cloud (https://www.gscloud.cn/ accessed on 26 March 2022).

Conflicts of Interest: The authors declare no conflict of interest.

References

1. Haghighi, A.T.; Zaki, N.A.; Rossi, P.M.; Noori, R.; Hekmatzadeh, A.A.; Saremi, H.; Klove, B. Unsustainability Syndrome-From Meteorological to Agricultural Drought in Arid and Semi-Arid Regions. *Water* **2020**, *12*, 838. [CrossRef]
2. Tng, D.; Apgaua, D.; Paz, C.P.; Dempsey, R.W.; Cernusak, L.A.; Liddell, M.J.; Laurance, S.G. Drought reduces the growth and health of tropical rainforest understory plants. *For. Ecol. Manag.* **2022**, *511*, 120128. [CrossRef]
3. Atherwood, S. Does a prolonged hardship reduce life span? Examining the longevity of young men who lived through the 1930s Great Plains drought. *Popul. Environ.* **2022**, *43*, 530–552. [CrossRef]
4. Wu, J.; Yao, H.; Yuan, X.; Lin B. Dissolved organic carbon response to hydrological drought characteristics: Based on long-term measurements of headwater streams. *Water Res.* **2022**, *215*, 118252. [CrossRef] [PubMed]
5. Wilhite, D.A. Drought and water crises: Science, technology, and management issues. *Bull. Am. Meteorol. Soc.* **2007**, *88*, 1444–1445.
6. Li, Z.; Li, Y.; Li, H.; Liu, Y.; Wang, C. Analysis of Drought Change and Its Impact in Central Asia. *Adv. Earth Sci.* **2022**, *37*, 37–50.
7. Chiang, F.; Mazdiyasni, O.; Aghakouchak, A. Evidence of anthropogenic impacts on global drought frequency, duration, and intensity. *Nat. Commun.* **2021**, *12*, 2754. [CrossRef] [PubMed]
8. Alahacoon, N.; Edirisinghe, M. A comprehensive assessment of remote sensing and traditional based drought monitoring indices at global and regional scale. *Geomat. Nat. Hazards Risk* **2022**, *13*, 762–799. [CrossRef]
9. Stagge, J.H.; Tallaksen, M.; Xu, C.Y.; Van Lanen, H.A. Standardized Precipitation-evapotranspiration Index (SPEI): Sensitivity to Potential Evapotranspiration Model and Parameters. In *Hydrology in a Changing World—Copernicus GmbH (IAHS-AISH Proceedings and Reports)*; International Association of Hydrological Sciences: Wallingford, UK, 2014.
10. Saeed, S.; Mohammadi, G.M.; Saviz, S. Spatial and temporal analysis of drought in various climates across Iran using the Standardized Precipitation Index (SPI). *Arab. J. Geosci.* **2022**, *15*, 1279.
11. Sergio, M.V.; Santiago, B.; Ji, L.-M. A Multiscalar Drought Index Sensitive to Global Warming: The Standardized Precipitation Evapotranspiration Index. *J. Clim.* **2010**, *23*, 1696–1718.
12. Palmer, W.C. *Meteorological Drought*; U.S. Department of Commerce Weather Bureau Research Paper: San Diego, CA, USA, 1965.
13. Narasimhan, B.; Srinivasan, R.Development and evaluation of Soil Moisture Deficit Index (SMDI) and Evapotranspiration Deficit Index (ETDI) for agricultural drought monitoring. *Agric. For. Meteorol.* **2005**, *133*, 69–88. [CrossRef]
14. Chen, H.; Sun, J. Changes in Drought Characteristics over China Using the Standardized Precipitation Evapotranspiration Index. *J. Clim.* **2015**, *28*, 5430–5447. [CrossRef]
15. Ma, X.; Zhu, X.; Zhao, J.; Zhao, N.; Shi, Y. Analysis of Drought Characteristics and Driving Forces in the Urban Belt Along the Yellow River in Ningxia Based on SPEI. *Res. Soil Water Conserv.* **2022**, *29*, 1–10.
16. Yang, R.; Geng, G.; Zhou, H.; Wang, T. Spatial-temporal Evolution of Meteorological Drought in the Wei River Basin Based on SPEI_PM. *Chin. J. Agrometeorol.* **2021**, *42*, 962–974.
17. Mobilia, M.; Longobardi, A. Prediction of Potential and Actual Evapotranspiration Fluxes Using Six Meteorological Data-Based Approaches for a Range of Climate and Land Cover Types. *Isprs Int. J. -Geo-Inf.* **2021**, *10*, 192. [CrossRef]

18. Liu, K.; Jang, D. Analysis of Dryness/Wetness over China Using Standardized Precipitation Evapotranspiration Index Based on Two Evapotranspiration Algorithms. *Chin. J. Atmos. Sci.* **2015**, *39*, 23–36.
19. Ling, M.; Guo, X.; Shi, X.; Han, H. Temporal and spatial evolution of drought in Haihe River Basin from 1960 to 2020. *Ecol. Indic.* **2022**, *138*, 108809. [CrossRef]
20. Men, B.; Cai, B.; Tian, W.SPEI-based Analysis of Temporal and Spatial Characteristics of Meteorological Drought in the Chaobai River Basin. *J. North China Univ. Water Resour. Electr. Power (Nat. Sci. Ed.)* **2022**, *43*, 10–20.
21. Wang, S.; Li, R.; Wu, Y.; Zhao, S. Effects of multi-temporal scale drought on vegetation dynamics in Inner Mongolia from 1982 to 2015, China. *Ecol. Indic.* **2022**, *136*, 108666. [CrossRef]
22. Li, X.; Sha, J.; Wang, Z.L. Comparison of drought indices in the analysis of spatial and temporal changes of climatic drought events in a basin. *Environ. Sci. Pollut. Res.* **2019**, *26*, 10695–10707. [CrossRef]
23. Tian, Q.; Lu, J.; Chen, X. Spatio-temporal Evolution of Agricultural Drought in the Yangtze River Basin Based on Long-term CCI Soil Moisture Data. *Resour. Environ. Yangtze Basin* **2022**, *31*, 472–481.
24. Huang, T.; Xu, L.; Fan, H.; Meng, Y. Temporal and Spatial Variation Characteristics and the Evolution Trends of Droughts in the Yangtze River Basin. *Res. Environ. Sci.* **2018**, *31*, 1677–1684.
25. Jiang, W.; Niu, Z.; Wang, L.; Yao, R.; Gui, X.; Xiang, F.; Ji, Y. Impacts of Drought and Climatic Factors on Vegetation Dynamics in the Yellow River Basin and Yangtze River Basin, China. *Remote Sens.* **2022**, *14*, 930. [CrossRef]
26. Huang, H.; Zhang, B.; Ma, S.; Ma, B.; Cu, Y.Q.; Wang, X.D.; Ma, C.R.; Chen, K.Q.; Zhang, T. Temporal and Spatial Variations of Meteorological Drought and Drought Risk Analysis in Hedong Area of Gansu Province. *Chin. J. Agrometeorol.* **2020**, *41*, 459–469.
27. Santiago, B.; Sergio, M.V.; Fergus, R.; Borja, L. Standardized precipitation evapotranspiration index (SPEI) revisited: Parameter fitting, evapotranspiration models, tools, datasets and drought monitoring. *Int. J. Climatol.* **2014**, *34*, 3001–3023.
28. Mitchell, A. *The ESRI Guide to GIS Analysis*; ESRI Press: Redlands, CA, USA, 2012; Volume 3.
29. Chen, Y. An Analysis of Spatial-temporal Distribution Characteristics of Housing Prices in Nanchang Based on Space-time Cube. *Jiangxi Sci.* **2019**, *37*, 371–377.
30. Zhang, Q.; Xu, D.; Ding, Y. Spatio-temporal pattern mining of the last 40 years of drought in China based on SPEI index and spatio-temporal cube. *Agric. Res. Arid. Areas* **2021**, *39*, 194–201.
31. Mo, C.B.; Tan, D.C.; Mai, T.Y.; Bei, C.; Qin, J.; Pang, W.; Zhang, Z. An analysis of spatiotemporal pattern for COVID-19 in China based on space-time cube. *J. Med. Virol.* **2020**, *92*, 1587–1595. [CrossRef] [PubMed]
32. Mcleod, G. Exploring spatial patterns of Virginia tornadoes using kernel density and space-time cube analysis (1960–2019). *ISPRS Int. J. -Geo-Inf.* **2021**, *10*, 310.
33. Ahmadi, H.; Argany, M.; Ghanbari, A.; Ahmadi, M. Visualized spatiotemporal data mining in investigation of Urmia Lake drought effects on increasing of PM10 in Tabriz using space-time cube (2004–2019). *Sustain. Cities Soc.* **2022**, *76*, 103399. [CrossRef]
34. Chen, T.; Shi, X.; Wong, Y.D.A lane-changing risk profile analysis method based on time-series clustering. *Phys. A Stat. Mech. Its Appl.* **2021**, *565*, 125567. [CrossRef]
35. Duan, L.Z.; Yu, F.S.; Pedrycz, W.; Wang, X.; Yang, X. Time-series clustering based on linear fuzzy information granules. *Appl. Soft Comput.* **2018**, *73*, 1053–1067. [CrossRef]
36. Lahreche, A.; Boucheham, B. A fast and accurate similarity measure for long time series classification based on local extrema and dynamic time warping. *Expert Syst. Appl.* **2021**, *168*, 114374. [CrossRef]
37. Delforge, D.; Watlet, A.; Kaufmann, O.; Van Camp M. Time-series clustering approaches for subsurface zonation and hydro facies detection using a real time-lapse electrical resistivity dataset. *J. Appl. Geophys.* **2021**, *184*, 104203. [CrossRef]
38. Xu, B.; Qi, B.; Ji, K.; Liu, Z.; Deng, L.; Jiang, L. Emerging hot spot analysis and the spatial–temporal trends of NDVI in the Jing River basin of China. *Environ. Earth Sci.* **2022**, *81*, 1–15. [CrossRef]
39. Getis, A.; Ord, J.K. The Analysis of Spatial Association by Use of Distance Statistics. *Geogr. Anal.* **1992**, *24*, 189–206. [CrossRef]
40. Purwanto, P.; Utaya, S.; Handoyo, B. Spatiotemporal analysis of Covid-19 spread with emerging hotspot analysis and space-time cube models in east java, Indonesia. *Int. J. -Geo-Inf.* **2021**, *10*, 133. [CrossRef]
41. Shan, B.; Zhang, Z.; Chen, Y. Analysis Methods of Spatio-temporal Patterns and Its Empirical Applications-A Case Study of Manufacturing Industry of Shandong Province. *J. Geomat. Sci. Technol.* **2021**, *38*, 624–630+638.
42. Betty, E.L.; Bollard, B.; Murphy, S.; Ogle, M.; Hendriks, H.; Orams, M.B.; Stockin, K.A. Using emerging hot spot analysis of stranding records to inform conservation management of a data-poor cetacean species. *Biodivers. Conserv.* **2019**, *29*, 643–665. [CrossRef]
43. Chambers, S.N. The spatiotemporal forming of a state of exception: Repurposing hot-spot analysis to map bare-life in southern Arizona's borderlands. *Geo J.* **2020**, *85*, 1373–1384. [CrossRef]
44. Wang, J.; Xu, C. Geodetector: Principle and prospective. *Acta Geogr. Sin.* **2017**, *72*, 116–134.
45. Wang, J.F.; Li, X.H.; Christakos, G.; Liao, Y.L.; Zhang, T.; Gu, X.; Zheng, X.Y. Geographical detectors-based health risk assessment and its application in the neural tube defects study of the heshun region, china. *Int. J. Geogr. Inf. Sci.* **2010**, *24*, 107–127. [CrossRef]
46. Wang, J.F.; Hu, Y. Environmental health risk detection with Geogdetector. *Environ. Model. Softw.* **2012**, *33*, 114–115. [CrossRef]
47. Cao, F.; Ge, Y.; Wang, J.F. Optimal discretization for geographical detectors-based risk assessment. *Mapp. Sci. Remote Sens.* **2013**, *50*, 78–92. [CrossRef]
48. Li, C.; Wu, Y.; Gao, B.; Wu, Y.; Zheng, K.; Li, C. Spatial Differentiation and Driving Factors of Rural Settlement in Plateau Lake: A Case Study of the Area Around the Erhai. *Econ. Geogr.* **2022**, *42*, 220–229.

49. Chen, W.; Li, J.; Zeng, J.; Ran, D.; Yang, B. Spatial heterogeneity and formation mechanism of eco-environmental effect of land use change in China. *Geogr. Res.* **2019**, *38*, 2173–2187.
50. Tian, F.; Yang, J.; Liu, L.; Wu J. Progress of research on the conception, characteristic, and influencing factors of drought propagation from the perspective of geographic sciences. *Prog. Geogr.* **2022**, *548*, 202–205. [CrossRef]
51. Yang, Q.; Luo, G.; Gao, C. Research Progress of Agricultural Drought from the Perspective of Geography. *J. North China Univ. Water Resour. Electr. Power (Nat. Sci. Ed.)* **2020**, *168*, 27–34+64.
52. Zhang, H.; Ding, J.; Wang, Y.; Zhou, D.; Zhu, Q. Investigation about the correlation and propagation among meteorological, agricultural and groundwater droughts over humid and arid/semi-arid basins in China. *J. Hydrol.* **2021**, *603*, 127007. [CrossRef]
53. Schwalm, C.R.; Anderegg, W.R.; Michalak, A.M.; Fisher, J.B.; Biondi, F.; Koch, G.; Litvak, M.; Ogle, K.; Shaw, J.D.; Wolf, A.; et al. Global patterns of drought recovery. *Nature* **2017**, *548*, 202–205. [CrossRef]
54. Xia, J.; Li, Z.; Zeng, S.; Zou, L.; She, D.; Cheng, D. Perspectives on eco-water security and sustainable development in the Yangtze River Basin. *Geosci. Lett.* **2021**, *8*, 1–9. [CrossRef]
55. Ren, Y.; Liu, J.; Liu, S.; Wang, Z.; Liu, T.; Shalamzari, M.J. Effects of Climate Change on Vegetation Growth in the Yellow River Basin from 2000 to 2019. *Remote Sens.* **2022**, *14*, 687. [CrossRef]
56. Ren, Y.; Liu, J.; Shalamzari, M.J.; Arshad, A.; Liu, S.; Liu, T.; Tao, H. Monitoring Recent Changes in Drought and Wetness in the Source Region of the Yellow River Basin, China. *Water* **2022**, *14*, 861. [CrossRef]
57. Liu, J.; Ren, Y.; Tao, H.; Shalamzari, M.J. Spatial and Temporal Variation Characteristics of Heatwaves in Recent Decades over China. *Remote Sens.* **2021**, *13*, 3824. [CrossRef]

atmosphere

MDPI

Article

Remote-Sensing Drought Monitoring in Sichuan Province from 2001 to 2020 Based on MODIS Data

Yuxin Chen [1,2], Jiajia Yang [1,2], Yuanyuan Xu [1,2], Weilai Zhang [1,2], Yongxiang Wang [1,2], Jiaxuan Wei [1,2] and Wuxue Cheng [1,2,*]

1 The Faculty Geography Resources Sciences, Sichuan Normal University, Chengdu 610101, China
2 Key Laboratory of Land Resources Evaluation and Monitoring in Southwest China, Sichuan Normal University, Chengdu 610066, China
* Correspondence: cwx714826@sicnu.edu.cn

Abstract: In this study, four drought monitoring indices were selected to simulate drought monitoring in the study area and a correlation analysis was conducted using the self-calibrated Palmer Drought Index (sc-PDSI) to screen for the most suitable drought monitoring index for the study area. Then, the spatio-temporal variation characteristics of drought in the study area were discussed and analyzed. The results showed that the Crop Water Stress Index (CWSI) was most suitable for drought monitoring in the Sichuan Province. CWSI had the best monitoring in grasslands (r = 0.48), the worst monitoring in woodlands (r = 0.43) and the highest fitting degree of overall correlation (r = 0.47). The variation of drought time in the Sichuan Province showed an overall trend of wetting and the drought situation was greatly alleviated. In the past 20 years, the dry years in the Sichuan Province were from 2001 to 2007, in which the driest years were 2006 and 2007; 2012–2013 was the transition interval between drought and wet; any year from 2013 to 2020 was a wet year, showing a transition trend of "drought first and then wet". The spatial distribution of drought was greater in the south than in the north and greater in the west than in the east. Panzhihua City and the southern part of the Liangshan Prefecture were the most arid areas, while the non-arid areas were the border zone between the western Sichuan Plateau and the Sichuan Basin. Looking at the spatial distribution of drought, "mild drought" accounted for the largest percentage of the total area (60%), mainly concentrated in the western Sichuan plateau. The second largest was "drought free" (33%), mostly concentrated in the transition area between the western Sichuan Plateau and the Sichuan Basin (western Aba Prefecture, Ya'an City, Leshan City and northern Liangshan Prefecture). The area of "moderate drought" accounted for a relatively small proportion (6%), mainly concentrated in Panzhihua City, the surrounding areas of Chengdu City and the southern area of the Liangshan Prefecture. The area of severe drought accounted for the least (1%), mostly distributed in Panzhihua City and a small part in the southern Liangshan Prefecture. The drought center ranged from 101.8° E to 103.6° E and 28.8° N to 29.8° N, with the movement trend of the drought center moving from the northeast to the southwest to the northeast.

Keywords: Sichuan Province; drought index; MODIS data; drought monitoring

Citation: Chen, Y.; Yang, J.; Xu, Y.; Zhang, W.; Wang, Y.; Wei, J.; Cheng, W. Remote-Sensing Drought Monitoring in Sichuan Province from 2001 to 2020 Based on MODIS Data. *Atmosphere* **2022**, *13*, 1970. https://doi.org/10.3390/atmos13121970

Academic Editors: Jinping Liu, Quoc Bao Pham, Arfan Arshad and Masoud Jafari Shalamzari

Received: 11 October 2022
Accepted: 18 November 2022
Published: 25 November 2022

1. Introduction

The Sichuan Province lies in the transition zone between the Qinghai–Tibet Plateau and the middle and lower reaches of the Yangtze River, with the characteristics of being high in the west and low in the east. The western part of Sichuan Province is characterized by a fragile plateau climate and ecological environment and the vertical zonal difference is great, while the eastern part is characterized by an abundant monsoon climate and precipitation. The climate and lower pad surface properties of the study areas are very different, which is of great reference value for research. In recent years, frequent droughts occurred in the Yangtze River Basin (especially in Sichuan and Chongqing in 2006), which posed a serious threat to agricultural and forestry production in this region. In order to better cope with and study the impact of drought on the study area, it is necessary to adopt effective monitoring methods to accurately analyze the drought situation in the study area.

The application of remote-sensing technology in drought monitoring has become mainstream [1]. Remote-sensing technology can make up for the shortage of meteorological station data and obtain meteorological data over a long time and a wide range [2]. At present, the meteorological drought index is based on the data of meteorological stations at different time scales. It uses mathematical and physical methods to calculate the drought index, so as to monitor the drought caused by climate anomalies in specific regions and specific periods [3]. The earliest drought indices used to characterize drought conditions were the Vegetation Condition Index (VCI) and the Temperature Condition Index (TCI) developed by Kogan [4,5]. Based on the Normalized Difference Vegetation Index (NDVI) and Land Surface Temperature (LST) changes in different time series. The Palmer Drought Severity Index (PDSI) developed by meteorologists Wayne Palmer et al. [6] put forward a drought index based on water supply and demand. The PDSI is widely used in drought assessment because it considers the temperature factor, can effectively reflect the impact of climate change on droughts and, at the same time, can consider the water supply and their relationship for regional drought assessment. However, there are differences in drought analysis in different spaces, so it is not always suitable to assess drought in different regions [7]. Compared with PDSI, the self-calibrating Palmer Drought Severity Index (sc-PDSI) is a great improvement and the calculation of evapotranspiration using the FAOPM formula has higher accuracy [7]. At the same time, sc-PDSI uses the meteorological data of the respective stations for the calculations, giving fewer regional constraints and high spatial comparability [8]. The application of sc-PDSI in regional drought analysis is relatively mature [9,10]. The Standard Precipitation Evapotranspiration Index (SPEI) is the degree of deviation between precipitation and evapotranspiration by Vicente Serrano et al. [11] to characterize the drought of a given area. In recent years, the application of SPEI to analyze regional drought has been increasing [12]. The Drought Severity Index (DSI) was proposed by Mu et al. [13]. Further, Jakson et al. [14] proposed considering energy and water exchange between the vegetation, soil and atmosphere and the related Crop Water Stress Index (CWSI). This is a standardized index according to the variation of the degree of water deficit in different time series compared with the standard state. Considering the comprehensive impact of NDVI and LST on drought, Sand Holt et al. [15] proposed the Temperature Vegetation Dryness Index (TVDI), Carlson et al. [16] proposed the Vegetation Supply Water Index (VSWI), Wang et al. [17] proposed the Vegetation Temperature Condition Index (VTCI). Thereafter, the development of a drought index combined with remote sensing technology involved various meteorological and hydrological elements such as soil water content, elevation, LST and NDVI. Liu et al. [18] used BP neural network to propose Integrated Agricultural Drought Index (IDI).

VCI and TCI are easy to calculate and mature in application but data from long time series are easily affected by non-drought stress factors [19]. DSI has unique advantages for global drought monitoring but there are great differences in its regional application. Although it is simple to calculate, it is still affected by non-drought stress factors based on historical data [20]. Therefore, the above three indices do not have universality in drought monitoring in agriculture and forestry. IDI has significant advantages for regional drought simulation but cannot be widely used due to the complexity of its calculation. Most of the previous studies considered the applicability of the index but not under the influence of different lower pad surfaces. Therefore, the Sichuan Province, where the lower pad surface is relatively complex, is selected as the study area. In order to better simulate drought in the study area, CWSI, VSWI, VTCI and TVDI are selected as the fitting models for the study area by comprehensively considering multiple drought stress factors. (The above indices have been widely used due to their simple calculation, easy access to data, and not being easily affected by non-drought stress factors [21,22]). The correlational analysis is conducted using the self-calibrated Palmer Drought Severity Index (sc-PDSI) to comprehensively select an appropriate drought index for the study area; this will provide a scientific basis for drought monitoring and management in the study area.

2. Data and Methods

2.1. Overview of the Study Area

The Sichuan Province is selected as the study area in this paper. The Sichuan Province is located in southwest China and consists of two major areas: the Sichuan Basin and the Western Sichuan Plateau. The terrain of the Sichuan Province is in the transition zone of the first and second steps in China and the lower pad surface has various properties. It is a key development province in southwest China and an economic and cultural center in the region. It is also an important grain-producing area, meaning Sichuan's ecological environment is vulnerable to the impact of human activities. In the past 20 years, the frequent droughts in the Sichuan Province (represented by high temperatures and drought in 2001, 2006 and 2022) have caused great losses to agricultural production and people's lives in the Sichuan Province.The geographical location of Sichuan Province is shown in Figure 1.

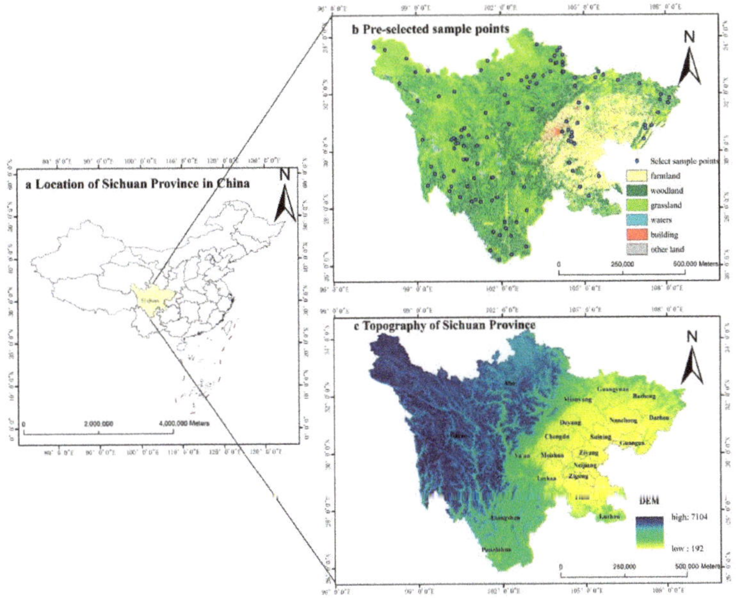

Figure 1. Study area land type zoning and selection sample points.

2.2. Data Sources and Research Methods

Using MODIS data and sc-PDSI, the drought model was constructed after preprocessing the data and its accuracy was verified to select a model with good fitting to analyze the spatial and temporal pattern of drought in the study area. The framework of the study was divided into three main parts: data preprocessing, model construction and applicability evaluation, and drought pattern analysis, as shown in Figure 2.

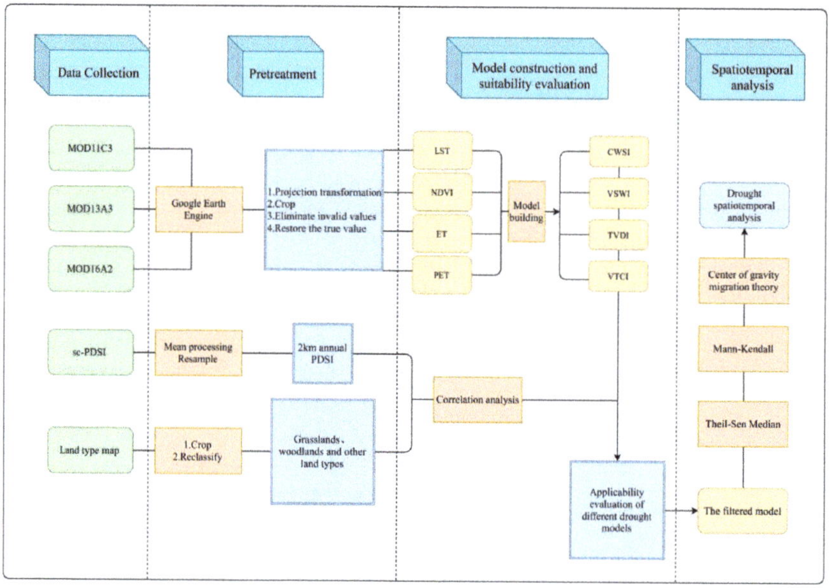

Figure 2. Flowchart of drought estimation based on MODIS remote sensing data.

2.2.1. Data Sources

ET, PET, NDVI, and LST data from 2001 to 2020 were obtained from MODIS image data obtained from the National Aeronautics and Space Administration (NASA) (https://ladsweb.modaps.eosdis.nasa.gov/, accessed on 15 April to 20 May 2022). Among them, the MOD11C3 product contains the synthesized LST data on a monthly basis, with a resolution of 0.05° × 0.05°. MOD13A3 had monthly synthesized NDVI with a resolution of 1 km. MOD16A2 synthesized Evapotranspiration (ET) and Potential Evapotranspiration (PET) in 8 days with a resolution of 0.5 km. Google Earth Engine (GEE) was used for data preprocessing and clipping to output Geo-Tiff format with a resolution of 0.5 km. From 2001 to 2020 the sc-PDSI data are from the Climatic Research Unite (https://crudata.uea.ac.uk/, accessed on 1 May 2022) with a spatial resolution of 0.5° × 0.5°. The average calculated from 2001 to 2020 is the average PDSI and sampling to 2 km. The annual mean sc-PDSI from 2001 to 2020 was calculated and the tool "Create fishing nets" in ArcGIS10.8 was used to create 0.5 km×0.5 km fishing nets and their annotation points in the study area. The annotated points were used to obtain the attributes of the source data at each annotated point through the 'value extraction to point' tool, and the fishing nets were resampled to 0.5 km by attributing the attributes. The Land cover types data (with a resolution of 1 km) for the Sichuan Province were obtained from the Resource and Environmental Science and Data Center of the Chinese Academy of Sciences (http://www.resdc.cn, accessed on 13 May 2022.). The range of forest, grassland, cultivated land and other land types could be obtained by reclassification.

2.2.2. Research Methods

Four remote sensing-based indices, i.e., the *CWSI*, *VSWI*, *TVDI*, and *VTCI*, were selected to detect the drought in the study area. According to the principle of water balance, *CWSI* determines the drought degree of the region according to the soil evapotranspiration deficit. It also involves a variety of agronomic and meteorological factors with clear physical meaning and high reliability. *CWSI* is defined as:

$$CWSI = 1 - ET/PET \quad (1)$$

where *ET* is the actual evapotranspiration and *PET* is the potential evapotranspiration. *CWSI* returns a value between 0 and 1, the larger the value is, the more arid and water scarce the region is, and vice versa.

The physical meaning of *VSWI* is that vegetation index and canopy temperature remain within a certain range when plants' water supply is normal, while an insufficient water supply affects plant growth. In order to reduce water loss, foliar stomata will partially close, resulting in a canopy temperature rise.

VSWI is defined as:

$$VSWI = NDVI/T_c \tag{2}$$

where, *NDVI* is the normalized vegetation index, and T_c is the canopy temperature of vegetation. Since it is difficult to obtain the canopy temperature, *LST* is used to replace it. VSWI values are between 0 and 1; the smaller the value is, the more arid and water-scarce the area is, and vice versa.

TVDI (Sandholt et al. [15]) is used in the study of soil moisture. It was found that there were many contour lines in the feature space of TS-NDVI, based upon which the concept of *TVDI* was proposed. Later, Carlson [16] found that when the vegetation coverage of the study area is large, the scatter plot is obtained by using the *LST* and *NDVI*, obtained from remote-sensing data, as the horizontal and vertical coordinates are triangular. The value of *TVDI* was calculated from the vegetation index and the land surface temperature. Meanwhile, Wang et al. [17] proposed *VTCI* based on *NDVI* and *LST* feature space. The two are defined as:

$$VTCI = (LST_{NDVI,max} - LST)/(LST_{NDVI,max} - LST_{NDVI,min}) \tag{3}$$

$$VTCI = (LST - LST_{NDVI,min})/(LST_{NDVI,max} - LST_{NDVI,min}) \tag{4}$$

$$LST_{NDVI,max} = a_1 + b_1 \times NDVI \tag{5}$$

$$LST_{NDVI,min} = a_2 + b_2 \times NDVI \tag{6}$$

where, *LST* is the surface temperature, and $LST_{NDVI,min}$ and $LST_{NDVI,max}$ represent the corresponding minimum and maximum. They correspond to "dry edge" and "wet edge" and a_1, a_2 and b_1, b_2, are the fitting coefficients of dry and wet edges, respectively. *TVDI* and *VTCI* are both between 0 and 1. The smaller the *VTCI* value is, the more arid and water scarce the region is, and vice versa. The smaller the *TVDI* value, the wetter the region is, and vice versa.

The Theil-sen Median method, also known as Sen slope estimation, is a robust non-parametric statistical trend calculation method. This method has high computational efficiency and is insensitive to measurement errors and outliers so it is often used in trend analysis of long-time series data [23]. The Mann–Kendall (MK) test is a non-parametric trend test method for time series, proposed by Mann in 1945 and further improved by Kendall and Sneyers. It does not require the measurement values to follow a normal distribution and is not affected by missing values and outliers so it is suitable for trend significance tests for long time series data. The Sen slope estimation is used to calculate the trend value, which is usually used in combination with the MK nonparametric test; that is, the Sen trend value is calculated first and then the trend significance is determined using the MK method.

According to the center of gravity transfer theory. the geographical center of gravity can reflect the spatial and temporal distribution characteristics of an element. The relative transfer distance and direction to its center of gravity can reflect the variation amplitude and spatial difference of the geographical element in this period. It is often used to reflect the transformation of the economy and population [22]. The application of this technique in drought monitoring can effectively describe the spatial location of the center of gravity shift in arid areas and provide a basis for monitoring research in arid areas.

3. Suitability Assessment and Drought Classification

3.1. Correlation between Remote-Sensing Drought Index and Sc-PDSI

In order to verify the accuracy of spatial and temporal monitoring of the four remote-sensing drought indices, a Pearson coefficient correlation analysis was performed between the four indices and sc-PDSI data (Figure 3). According to the statistical analysis, the correlation coefficients of CWSI, VSWI, TVDI and VTCI are −0.44, 0.32, −0.28 and 0.28, respectively. On the whole, CWSI and TVDI are negatively correlated with sc-PDSI while the other indices are positively correlated with sc-PDSI.

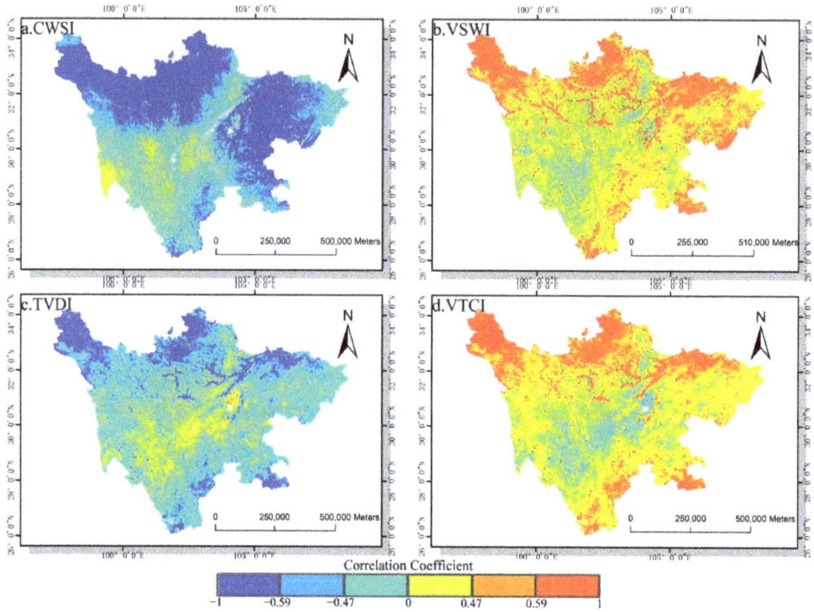

Figure 3. Correlation coefficient between remote-sensing drought Indices and sc-PDSI in 2001–2020.

Among them, only the correlation between CWSI and sc-PDSI passes the significance test ($p < 0.05$) in most regions, which indicates that CWSI has a better fit to the inter-annual variation in soil drought in the study area.

For the correlation coefficients of different land types (Figure 4), it can be seen that CWSI has a higher fitting degree in the steppe. The mean of the correlation coefficient of the steppe is 0.48, which passes the significance test ($p < 0.05$). Although the mean of woodland is 0.43, it is still much higher than the other three indices. The comprehensive analysis shows that CWSI has great advantages in drought monitoring and simulation in the Sichuan Province.

3.2. Drought Classification

The above correlational analysis shows that the CWSI index has better applicability than other indices in the study area; the CWSI was selected to analyze the spatial and temporal characteristics of drought in the study area. First of all, the drought grade criteria should be divided. In this study, sc-PDSI data were used to classify drought grades (Table 1). Most sc-PDSI data from 2001 to 2020 are between −3 and 2. There are values −3 to −4 in sc-PDSI data of some years, but they are few and they do not exist after the 20-year mean treatment. Therefore, according to the sc-PDSI criteria for drought classification, the drought grade is divided into four classes. In addition, 120 sample points are selected according to the area proportion of different land types in Sichuan Province (Figure 1b). The specific number of sample points is 48 (45.56%) forestland, 20 (13.85%) arable land,

38 (25.12%) grassland, 4 (3.23%) building land and 10 (7.86%) other land. Considering the existence of unsuitable land types such as water area and traffic land and the highest fitting degree of grassland, the number of selected grassland sites is increased and the above 120 sample sites are obtained through data screening of alternative sample sites. We then perform a unary linear regression (Figure 5) to obtain the partition thresholds of CWSI corresponding to different grades (Table 1).

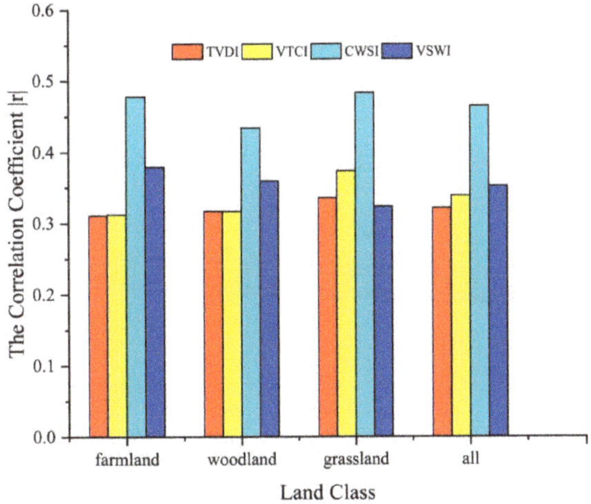

Figure 4. Correlation coefficient between remote-sensing drought index and sc-PDSI in different vegetation zones in 2001–2020.

Table 1. Drought categories.

Drought Rating	sc-PDSI	CWSI
No drought	>0	0~0.59
Mild drought	−1~0	0.59~0.72
Moderate drought	−2~−1	0.72~0.85
Severe drought	−2~−3	0.85~0.92

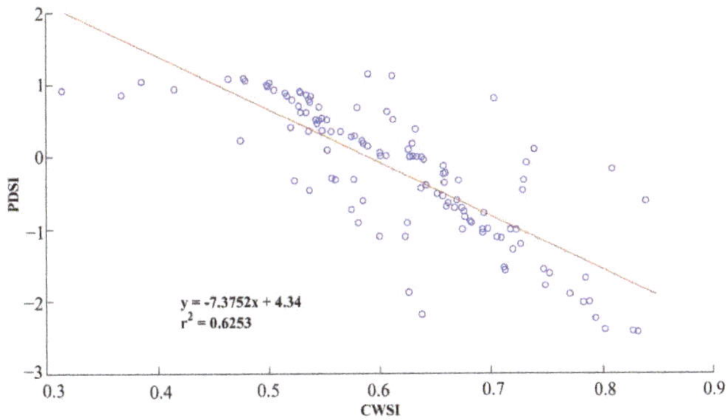

Figure 5. Linear regression analysis of PDSI and CWSI in the Sichuan Province from 2001 to 2020.

4. Spatial-Temporal Pattern Analysis of Drought in the Sichuan Province

4.1. Variation Characteristics of Drought Time

According to the statistics of CWSI and sc-PDSI index data from each year, Figure 6 shows that the fluctuation range of CWSI is between 0.53 and 0.62 and the fluctuation range of sc-PDSI is between −1.48 and 1.21. CWSI shows an obvious downward trend ($p < 0.05$), while sc-PDSI shows an obvious upward trend ($p < 0.05$), indicating that CWSI is consistent with sc-PDSI in terms of the time development trend of drought; that is, the trend of drought in the study area slow down. Based on previous studies, the cumulative anomaly value is determined to be stable; that is, the changing trend does not pass the significance test $p < 0.05$, which is regarded as the turning interval. According to Figure 6, the drought trend of CWSI and sc-PDSI tend to be consistent, with CWSI on the whole in a downward trend and sc-PDSI in an upward trend (the smaller the CWSI value is, the wetter it is, while the larger sc-PDSI value is, the wetter it is), so the drought situation has been greatly improved. At the same time, the drought and wetness transition intervals of the two indices are both in 2012–2013 and there are significant abrupt changes in 2006 and 2007 (drought caused by high temperatures in the study area in 2006 and 2007). Therefore, the reliability of CWSI for drought monitoring and simulation in the study area is strong.

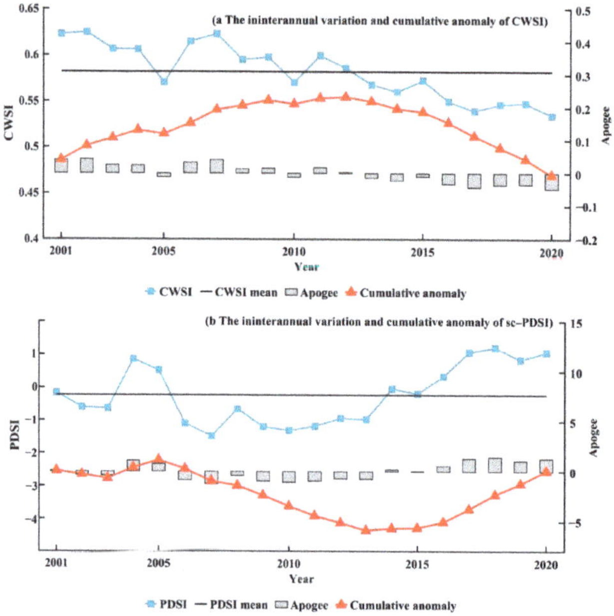

Figure 6. The interannual variation and cumulative anomaly of CWSI and sc-PDSI.

4.2. Spatial Variation Characteristics of Drought

Figure 7 shows the spatial distribution of the multi-annual mean of CWSI and the spatial distribution of the drought grade from 2001 to 2020. The available CWSI values ranged from 0.06 to 0.91. Most of the low values of CWSI are concentrated in the central belt of Sichuan Province; namely, the junction of the plateau and basin. The high-value areas are concentrated in Panzhihua, Xichang, Chengdu and other cities and their surrounding areas, as well as the hinterland of the western Sichuan Plateau. The spatial pattern of high values in the plateau basin junction zone, low values in the two sides of the plateau basin, and low values in the southern Sichuan plateau is generally formed. In terms of the spatial distribution of drought classes, Panzhihua City and the Liangshan Prefecture are the most severe drought areas in the study area, followed by Chengdu City and its surrounding areas and the central region of the Garze Prefecture. The border areas of the plateau basin and Luzhou, Yibin, Dazhou, Bazhong, Guang'an and part of the western Garze Prefecture were drought-free areas. From Figure 7, it can be found that the proportion of mild drought in the total area of the CWSI drought spatial distribution is the largest (60%), which is mainly concentrated in the western Sichuan plateau, followed by the central area of the

Sichuan Basin. Secondly, the percentage that is drought-free of the total area is relatively large (33%), and mostly concentrated in the border zone between the western Sichuan Plateau and the Sichuan Basin (western Aba Prefecture, Ya'an, Leshan City, and northern Liangshan Prefecture). The area of moderate drought is relatively small (6%), mainly concentrated in Panzhihua City, the surrounding areas of Chengdu City and the southern areas of the Liangshan Prefecture. The area of severe drought is the smallest (1%) and is mostly distributed in Panzhihua City and a small part of the southern Liangshan Prefecture. Based on the above analysis, the percentages of drought grades in the study area are, in descending order, mild drought > no drought > moderate drought > severe drought.

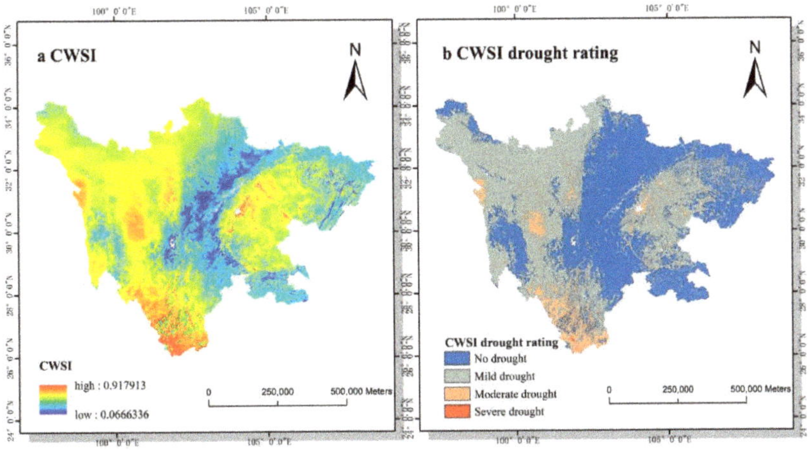

Figure 7. Trend of drought based on CWSI in Sichuan Province from 2001 to 2020.

4.3. Spatial-Temporal Evolution of Drought in the Sichuan Province

The Sen trend and Mann–Kendall method is used to obtain the variation trend of CWSI and the spatial distribution of its significance Figure 8a,b). According to Table 2, the spatial distribution of the trend and significance of CWSI are superimposed and analyzed to obtain the spatial distribution of the detailed changes of drought (Figure 8).

Table 2. Category of significant variation of drought trend.

CWSI Slope	Z	Trend Type	Trend Features
Slope > 0	2.58 < |Z|	3	Significantly dried
	1.96 < |Z| 2.58 or less	2	Dry
	1.65 < |Z| 1.96 or less	1	Slightly dried
Slope = 0	Z	0	Stable and unchanged
Slope < 0	1.65 < |Z| 1.96 or less	−1	Slightly wet
	1.96 < |Z| 2.58 or less	−2	Wet
	2.58 < |Z|	−3	Significantly wet

According to Figure 8, it can be seen that the change rate of CWSI from 2001 to 2020 ranged from −0.0366 to 0.0208 and the overall spatial distribution show that the drought mitigation degree in the western region was greater than in the eastern region (Figure 8a). The areas with significant drought changes in the last 20 years are the eastern part of the Sichuan Basin and the northern region of the Garze Prefecture and Aba Prefecture (Figure 8b). Most of the areas with significant changes are "significantly wetter" and the proportion of the area is the largest (46.5%) compared with other areas, followed by "stable and unchanged" (25%), and "wet", "slightly wet", "slightly dried", "dry" and "significantly dried". Among them, the wetting trend accounts for 79.5% of the total area while the drying trend accounts for only 1.5%. The comprehensive analysis shows that the trend of drought change in the study area is overall wetting, indicating that the drought situation in the study area is getting better overall from 2001 to 2020, and the drought level is easing in most areas (Figure 8c).

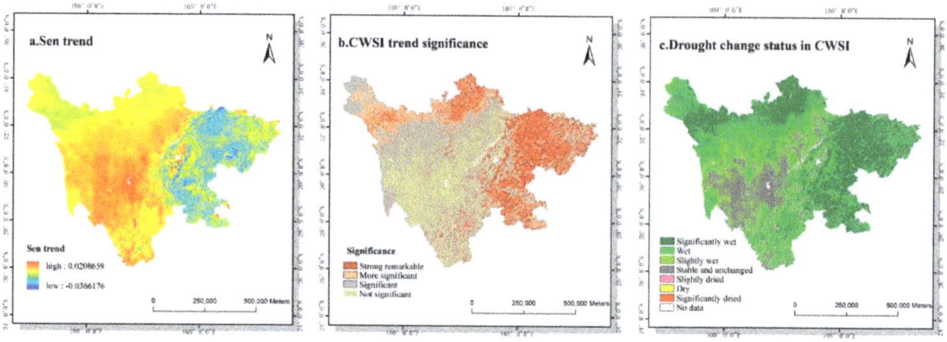

Figure 8. The trend of drought based on CWSI in 2001–2020.

4.4. *Analysis of the Change of the Drought Center of Gravity in the Sichuan Province*

The larger the CWSI value, the greater the degree of drought in the region, so this paper selects the CWSI of drought-prone areas (CWSI > 0.72) as the weight and calculates the distribution of the center of gravity of drought-prone areas every 4 years.

The Gration trajectory of the center of gravity in the drought-prone areas is shown in Figure 9. From the figure, it can be seen that the center of gravity in the drought-prone region of CWSI is concentrated between 101.8° E to 103.6° E and 28.8° N to 29.8° N. On the whole, the center of gravity shifts southward in latitude and westward in longitude. In terms of spatial distribution, the center of gravity shifts from the area around Chengdu in 2001 to Leshan in 2020. Although the drought-prone areas are not distributed in the driest regions, the trajectory of their center of gravity shift can reveal the pattern of drought migration. The overall migration trend of drought-prone areas is from the northeast to the southwest, and then to the northeast.

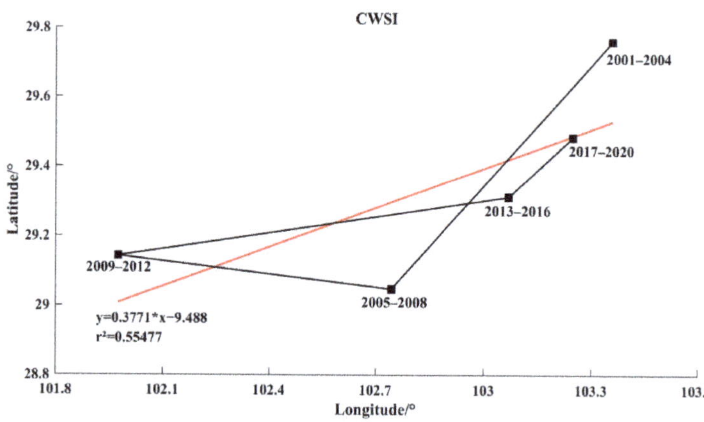

Figure 9. Comparison of barycenter trajectories.

5. Discussion

The correlation analysis between the four indices and sc-PDSI showed that CWSI is more suitable for drought monitoring in the study area, followed by VSWI. However, since the VSWI index is concentrated in the range 0~0.2, they lack discriminative power for drought class classification, and therefore, cannot meet the adaptation requirements of the study area. CWSI has a greater advantage in regional drought monitoring because ET/PET reflects the energy and water exchange between vegetation, soil, and atmosphere and can better describe soil moisture information [24,25]. NDVI/LST reflects the stress effect of drought on vegetation and can better reflect the effect of soil water deficit on vegetation growth [26,27]. However, it has also been suggested that the dry and wet side fitting equations of LST-NDVI and VTCI of LST and NDVI in TVDI are more influenced by regional differences and vegetation types, and the dry and wet side fitting is not good [28]. Therefore, the drought monitoring of TVDI and VTCI in the study area is not effective and widely affected by regional differences.

The lowest values of CWSI are concentrated in the northwestern part of Sichuan. This is mainly due to the low local temperature, sufficient precipitation, abundant groundwater sources, and a high proportion of irrigated farmland [8]. Similarly, in other areas of the study area with irrigated farmland networks, there are lower values of CWSI. The high values of CWSI are concentrated in Panzhihua city because the subtropical climate of Panzhihua city is controlled by subtropical high pressure, so the climate is dry and rainy, with high variability in the subsurface layer and increased evapotranspiration, resulting in high values of regional CWSI. Since CWSI is based on the vegetation evapotranspiration theory, the external water supply can increase the actual evapotranspiration when the actual evapotranspiration does not reach PET; this may lead to lower CWSI values. CWSI is calculated based on canopy temperature. Canopy temperature is inversely proportional to leaf stomatal closure and evapotranspiration. Stomatal closure is a result of crop water stress, which in turn reduces the transpiration rate of the plant. A low transpiration rate reduces plant cooling; therefore, an increase in canopy temperature is seen as an indicator of water stress. If a meteorological drought occurs due to insufficient precipitation, climate change-induced temperature increases will exacerbate the drought by increasing evapotranspiration. The areas with significant changes in drought trends in Sichuan Province (the eastern part of Sichuan Province, i.e., the Sichuan Basin area) are mostly monsoonal in climate, and their high precipitation and evapotranspiration are highly adaptable to CWSI, forming the advantage of CWSI in drought monitoring simulations in Sichuan Province.

The drought in the study area is mainly concentrated in the southern part of Panzhihua city and its surrounding areas. As a traditional industrial city, the mining of mineral resources in the Panzhihua area has to some extent destroyed the nature of the substratum and weakened the soil water exchange between ET/PET, thus exacerbating the drought in the area. As a result, the drought in the region has been aggravated by human actions. It is closely followed by the central areas of Chengdu and Garze. The drought in Chengdu and its surrounding areas is caused by the change in the nature of the substratum due to the combined effects of urban expansion and the urban heat island effect. However, the drought in Garze is caused by natural factors such as climate and topography; the sparse precipitation and long sunshine hours in the plateau region result in reduced soil moisture. Therefore, the drought monitoring of CWSI shows the drought status. The drought-free areas in the study area are mainly concentrated at the junction of the western Sichuan plateau and the Sichuan basin (western Aba, Ya'an, Leshan, and northern Liangshan). The area is dominated by forest cover with high vegetation coverage and significant elevation differences; high precipitation results in good soil moisture retention. In general, the CWSI results of drought monitoring in Sichuan Province show a good trend, thanks to the emphasis on environmental protection in recent years; for example, the implementation of the policy of "returning farmland to forest and grass" and "Sichuan Ecological Protection Red Line". As a result, the ecological environment in Sichuan province has been improving and the drought has been alleviated. The center of gravity of drought areas shifted from northeast to southwest to northeast. The specific shift is divided into two phases, 2001–2009 and 2010–2020, reflecting the trend of the center of gravity shifting from the area around Chengdu to southwest Sichuan and then back to Chengdu. This effect is mainly due to the over-exploitation of mineral resources and environmental changes caused by human activities in Panzhihua City before 2009, which formed the drought center and easily shifted to the southwest. From 2010 to 2020, the South Asian high-pressure and subtropical high-pressure systems were active, resulting in high temperatures in the interior. In addition, the eastern plain of Sichuan Province is densely populated, with over-exploited resources and a lack of water conservation projects. Due to natural and man-made causes, drought has shifted eastward.

Different drought monitoring indices have different adaptation statuses in different study areas. In this paper, four drought monitoring indices are constructed, their correlation with sc-PDSI is verified, and the more appropriate CWSI index is selected as the drought monitoring index for the study area. Although index screening is conducted based on the lower pad surface of the study area, it is limited to four factors, ET, PET, NDVI and LST, to analyze the degree of drought. The effects of other factors such as precipitation [29], extreme hazards [30] and topography [31] on drought conditions are not considered.

6. Conclusions

In most areas of the Sichuan Province, woodlands and grasslands are more sensitive to water exchange between vegetation, soil and atmosphere, which means that ET and PET can better reflect the physical processes and thus CWSI has a greater advantage for drought monitoring simulations. Although VSWI has a high correlation in correlation analysis, it is difficult to classify drought classes due to its over-concentration (mostly between 0 and 0.2). The fitted equations of dry and wet edges of TVDI and VTCI are influenced by regional differences and vegetation types, and the dry and wet edges are not well fitted on a large scale. Therefore, CWSI is selected as the best-fitting drought monitoring index in the study area.

The drought monitoring results show that the drought conditions in the study area are gradually improving. From 2001 to 2011 it was relatively dry, with the most severe years ranging from 2006 to 2007 (influenced by the high-temperature drought in Sichuan in 2006); 2012 to 2013 was the transitional interval between dry and wet in the study area. From 2014 to 2020, the study area showed a stable wet trend, especially in the eastern part of the basin. This is mainly due to the improving environmental and drought conditions in Sichuan Province as a result of China's ecological protection policies. In the eastern plateau, drought occurs only in the central part of the Garze Prefecture, and the drought-intensive areas of the Sichuan Basin are concentrated in Chengdu City and its surrounding areas. In general, Panzhihua City and the southern part of Liangshan Prefecture are the most severely drought-stricken areas. The ratio of drought-rated areas is as follows: mild drought > no drought > moderate drought > severe drought.

The overall drought trend in the study area is improving and the drought mitigation trend is effective, especially in the Sichuan basin. Drought-prone areas are concentrated in the range of 101.8° E–103.6° E and 28.8° N–29.8° N. The center of gravity of drought in drought-prone areas tends to move to the southwest, showing a shift from northeast to southwest to northeast.

Author Contributions: Conceptualization, W.C.; methodology, Y.C.; software, Y.C.; validation, W.C., J.Y. and Y.X.; formal analysis, Y.C. and J.Y.; investigation, W.Z., Y.X. and J.W.; resources, W.C. and Y.W.; data curation, Y.C.; writing—original draft preparation, W.C., Y.C. and J.Y.; writing—review and editing, J.Y., Y.C. and W.Z.; visualization, W.C. and Y.W.; supervision, J.W.; project administration, W.C.; funding acquisition, W.C. All authors have read and agreed to the published version of the manuscript.

Funding: This research was funded by the Humanities and Social Sciences, Ministry of Education of The People's Republic of China (grant number: 18YJC850004) and the National Natural Science Foundation of China (grant number: 32060370).

Institutional Review Board Statement: Not applicable.

Informed Consent Statement: Not applicable.

Data Availability Statement: The MODIS image data obtained from the National Aeronautics and Space Administration (NASA) (https://ladsweb.modaps.eosdis.nasa.gov/, accessed on 15 April to 20 May 2022). sc-PDSI data comes from the Climatic Research Unit (https://crudata.uea.ac.uk/, accessed on 1 May 2022). Land use data of Sichuan Province are obtained from the Resource and Environmental Science and Data Center of the Chinese Academy of Sciences (http://www.resdc.cn, accessed on 13 May 2022).

Conflicts of Interest: The authors declare no conflict of interest, and the funders had no role in the design of the study; in the collection, analyses, or interpretation of data; in the writing of the manuscript; or in the decision to publish the results.

References

1. Cheng, Q.P.; Gao, L.; Zhong, F.L.; Zuo, X.A.; Ma, M.M. Spatiotemporal variations of drought in the Yunnan-Guizhou Plateau, southwest China, during 1960–2013 and their association with large-scale circulations and historical records. *Ecol. Indic.* **2020**, *112*, 106041. [CrossRef]
2. Gu, Z.J.; Duan, X.W.; Liu, B.; Hu, J.M.; He, J.N. The spatial distribution and temporal variation of rainfall erosivity in the Yunnan Plateau, Southwest China: 1960–2012. *Catena* **2016**, *145*, 291–300. [CrossRef]
3. Um, M.J.; Kim, Y.; Park, D. Evaluation and modification of the Drought Severity Index (DSI) in East Asia. *Remote Sens. Environ.* **2018**, *209*, 66–76. [CrossRef]
4. Kogan, F.N. Droughts of the late 1980s in the United States as derived from NOAA polar-orbiting satellite data. *Bull. Am. Meteorol. Soc.* **1995**, *76*, 655–668. [CrossRef]
5. Kogan, F.N. Application of vegetation index and brightness temperature for drought detection. *Adv. Space Res.* **1995**, *15*, 91–100. [CrossRef]
6. Palmer, W.C. *Meteorological Drought*; US Department of Commerce, Weather Bureau: Washington, DC, USA, 1965; Volume 30.
7. van der Schrier, G.; Barichivich, J.; Briffa, K.; Jones, P. A scPDSI-based global data set of dry and wet spells for 1901–2009. *J. Geophys. Res. Atmos.* **2013**, *118*, 4025–4048. [CrossRef]
8. Li, C.B.; Adu, B.; Li, H.H.; Yang, D.H. Spatial and temporal variations of drought in Sichuan Province from 2001 to 2020 based on modified temperature vegetation dryness index (TVDI). *Ecol. Indic.* **2022**, *141*, 109106. [CrossRef]
9. Lewińska, K.E.; Ivits, E.; Schardt, M.; Zebisch, M. Alpine forest drought monitoring in South Tyrol: PCA based synergy between scPDSI data and MODIS derived NDVI and NDII7 time series. *Remote Sens.* **2016**, *8*, 639. [CrossRef]
10. Gaire, N.P.; Dhakal, Y.R.; Shah, S.K.; Fan, Z.-X.; Bräuning, A.; Thapa, U.K.; Bhandari, S.; Aryal, S.; Bhuju, D.R. Drought (scPDSI) reconstruction of trans-Himalayan region of central Himalaya using Pinus wallichiana tree-rings. *Palaeogeogr. Palaeoclimatol. Palaeoecol.* **2019**, *514*, 251–264. [CrossRef]
11. Vicente-Serrano, S.M.; Beguería, S.; López-Moreno, J.I. A multiscalar drought index sensitive to global warming: The standardized precipitation evapotranspiration index. *J. Clim.* **2010**, *23*, 1696–1718. [CrossRef]
12. Sein, Z.M.M.; Zhi, X.; Ogou, F.K.; Nooni, I.K.; Lim Kam Sian, K.T.; Gnitou, G.T. Spatio-temporal analysis of drought variability in myanmar based on the standardized precipitation evapotranspiration index (SPEI) and its impact on crop production. *Agronomy* **2021**, *11*, 1691. [CrossRef]
13. Mu, Q.; Zhao, M.; Kimball, J.S.; McDowell, N.G.; Running, S.W. A remotely sensed global terrestrial drought severity index. *Bull. Am. Meteorol. Soc.* **2013**, *94*, 83–98. [CrossRef]
14. Jackson, R.D.; Idso, S.; Reginato, R.; Pinter, P., Jr. Canopy temperature as a crop water stress indicator. *Water Resour. Res.* **1981**, *17*, 1133–1138. [CrossRef]
15. Sandholt, I.; Rasmussen, K.; Andersen, J. A simple interpretation of the surface temperature/vegetation index space for assessment of surface moisture status. *Remote Sens. Environ.* **2002**, *79*, 213–224. [CrossRef]
16. Carlson, T.N.; Gillies, R.R.; Perry, E.M. A method to make use of thermal infrared temperature and NDVI measurements to infer surface soil water content and fractional vegetation cover. *Remote Sens. Rev.* **1994**, *9*, 161–173. [CrossRef]
17. Wang, P.-x.; Li, X.-w.; Gong, J.-y.; Song, C. Vegetation temperature condition index and its application for drought monitoring. In Proceedings of the IGARSS 2001. Scanning the Present and Resolving the Future. Proceedings. IEEE 2001 International Geoscience and Remote Sensing Symposium (Cat. No.01CH37217), Sydney, NSW, Australia, 9–13 July 2001; pp. 141–143.
18. Liu, X.; Zhu, X.; Zhang, Q.; Yang, T.; Pan, Y.; Sun, P. A remote sensing and artificial neural network-based integrated agricultural drought index: Index development and applications. *Catena* **2020**, *186*, 104394. [CrossRef]
19. Bokusheva, R.; Kogan, F.; Vitkovskaya, I.; Conradt, S.; Batyrbayeva, M. Satellite-based vegetation health indices as a criteria for insuring against drought-related yield losses. *Agric. For. Meteorol.* **2016**, *220*, 200–206. [CrossRef]
20. Li, Z.; Han, Y.; Hao, T. Assessing the consistency of remotely sensed multiple drought indices for monitoring drought phenomena in continental China. *IEEE Trans. Geosci. Remote Sens.* **2020**, *58*, 5490–5502. [CrossRef]
21. Balsa-Barreiro, J.; Li, Y.; Morales, A. Globalization and the shifting centers of gravity of world's human dynamics: Implications for sustainability. *J. Clean. Prod.* **2019**, *239*, 117923. [CrossRef]
22. Dhorde, A.; Patel, N. Spatio-temporal variation in terminal drought over western India using dryness index derived from long-term MODIS data. *Ecol. Inform.* **2016**, *32*, 28–38. [CrossRef]

23. McLeod, A.I. Kendall Rank Correlatoin and Mann-Kendall Trend Test. Available online: https://cran.r-project.org/web/packages/Kendall/index.html (accessed on 5 July 2022).
24. Allen, R.; Irmak, A.; Trezza, R.; Hendrickx, J.M.; Bastiaanssen, W.; Kjaersgaard, J. Satellite-based ET estimation in agriculture using SEBAL and METRIC. *Hydrol. Process.* **2011**, *25*, 4011–4027. [CrossRef]
25. Cooke, W.H.; Mostovoy, G.V.; Anantharaj, V.G.; Jolly, W.M. Wildfire potential mapping over the state of Mississippi: A land surface modeling approach. *GIScience Remote Sens.* **2012**, *49*, 492–509. [CrossRef]
26. Park, S.; Im, J.; Jang, E.; Rhee, J. Drought assessment and monitoring through blending of multi-sensor indices using machine learning approaches for different climate regions. *Agric. For. Meteorol.* **2016**, *216*, 157–169. [CrossRef]
27. Ren, Y.; Liu, J.; Liu, S.; Wang, Z.; Liu, T.; Shalamzari, M.J. Effects of Climate Change on Vegetation Growth in the Yellow River Basin from 2000 to 2019. *Remote Sens.* **2022**, *14*, 687. [CrossRef]
28. Huang, J.; Zhuo, W.; Li, Y.; Huang, R.; Sedano, F.; Su, W.; Dong, J.; Tian, L.; Huang, Y.; Zhu, D. Comparison of three remotely sensed drought indices for assessing the impact of drought on winter wheat yield. *Int. J. Digit. Earth* **2020**, *13*, 504–526. [CrossRef]
29. Tirivarombo, S.; Osupile, D.; Eliasson, P. Drought monitoring and analysis: Standardised precipitation evapotranspiration index (SPEI) and standardised precipitation index (SPI). *Phys. Chem. Earth Parts A/B/C* **2018**, *106*, 1–10. [CrossRef]
30. Liu, J.; Ren, Y.; Tao, H.; Shalamzari, M.J. Spatial and Temporal Variation Characteristics of Heatwaves in Recent Decades over China. *Remote Sens.* **2021**, *13*, 3824. [CrossRef]
31. Awange, J. Drought Monitoring: Topography and Gauge Influence. In *Food Insecurity & Hydroclimate in Greater Horn of Africa*; Springer: Cham, Switzerland, 2022; pp. 387–420.

atmosphere

MDPI

Article

Impact of Climate Change and Drought Attributes in Nigeria

Akinwale T. Ogunrinde [1,2], Philip G. Oguntunde [3], Akinola S. Akinwumiju [4], Johnson T. Fasinmirin [3], David A. Olasehinde [1], Quoc Bao Pham [5], Nguyen Thi Thuy Linh [5] and Duong Tran Anh [6,7,*]

1 Department of Agricultural and Biosystems Engineering, Landmark University, Omu Aran P.M.B. 1001, Kwara State, Nigeria
2 Department of Agricultural Engineering, University of Benin, Benin-City P.M.B. 1154, Edo State, Nigeria
3 Department of Agricultural and Environmental Engineering, Federal University of Technology, Akure P.M.B. 704, Ondo State, Nigeria
4 Department of Remote Sensing and GIS, Federal University of Technology, Akure P.M.B. 704, Ondo State, Nigeria
5 Institute of Applied Technology, Thu Dau Mot University, Thu Dau Mot 75000, Vietnam
6 Laboratory of Environmental Sciences and Climate Change, Institute for Computational Science and Artificial Intelligence, Van Lang University, Ho Chi Minh City 700000, Vietnam
7 Faculty of Environment, Van Lang University, Ho Chi Minh City 700000, Vietnam
* Correspondence: duong.trananh@vlu.edu.vn

Abstract: Data from historical observatories and future simulations were analyzed using the representative concentration pathway (RCP) 8.5 scenario, which covered the period from 1951 to 2100. In order to characterize the drought, three widely used drought indicators were used: the standardized precipitation index (SPI), the reconnaissance drought index (RDI), and the standardized precipitation and evapotranspiration index (SPEI). The ensemble of the seven (7) GCMs that used RCA-4 was able to capture several useful characteristics of Nigeria's historical climatology. Future climates were forecasted to be wetter than previous periods during the study period based on the output of drought characteristics as determined by SPI. SPEI and RDI predicted drier weather, in contrast. SPEI and RDI's predictions must have been based on the effect of rising temperatures brought on by global warming as depicted by RCP 8.5, which would then have an impact on the rate of evapotranspiration. According to drought studies using the RCP 8.5 scenario, rising temperatures will probably cause more severe/extreme droughts to occur more frequently. SPEI drought frequency changes in Nigeria often range from 0.75 (2031–2060) to 1.80 (2071–2100) month/year, whereas RDI changes typically range from 0.30 (2031–2060) to 0.60 (2071–2100) month/year. The frequency of drought incidence has recently increased and is now harder to forecast. Since the Sendai Framework for Disaster Risk Reduction 2015–2030 (SFDRR) and the Sustainable Development Goals (SDGs) have few more years left to be completed, drastic efforts must be made to create climate-resilient systems that can tackle the effects that climate change may have on the water resources and agricultural sectors.

Keywords: climate change; mitigation; drought characterization; evapotranspiration; Nigeria

check for updates

Citation: Ogunrinde, A.T.; Oguntunde, P.G.; Akinwumiju, A.S.; Fasinmirin, J.T.; Olasehinde, D.A.; Pham, Q.B.; Linh, N.T.T.; Anh, D.T. Impact of Climate Change and Drought Attributes in Nigeria. *Atmosphere* **2022**, *13*, 1874. https://doi.org/10.3390/atmos13111874

Academic Editor: Tianbao Zhao

Received: 15 October 2022
Accepted: 3 November 2022
Published: 10 November 2022

1. Introduction

Nigeria has been identified as being particularly vulnerable to the effects of climate change, which primarily appear as natural disasters such as flood or drought. Because hydroelectric power and food production are heavily dependent on precipitation in this area, any considerable decrease in precipitation will have a long-term dramatic impact on people's socioeconomic activities. Over the past 50 years, there have been significant changes in the country's land and water resource management policies as a result of the precipitation declines in the 1970s and 1980s [1–4]). According to the geographical zone, the future climate outlook is expected to have a variety of extreme characteristics in different parts of the world [5]. Nigeria is the key stakeholder, and West Africa has been regarded as one of the most susceptible regions [6,7]. Since the majority of Nigerians work in industries that are susceptible to climate change, any tilt toward dryness in the future climate may enhance this area's susceptibility to severe drought. Therefore, it is important to assess and have knowledge about the expected characteristics of the region's climate in the future in order to improve the impact and vulnerability evaluations.

Drought has been directly or indirectly linked to water scarcity as well as many other environmental problems. *The Intergovernmental Panel on Climate Change Assessment Report of 2021* has indicated various impact of changes in climate on global and regional water resources [8]. Many studies have also indicated an increasing hydrological cycle as a result of variability in the frequency and magnitude of precipitation on a regional scale [9–12]. However, it is important to recognize that this type of alteration will have a regional impact on the frequency of drought or flood conditions. An evaluation of regional climate can either be conducted through statistical analysis of historical data or mathematical modelling of the physical hydrological procedures for future forecasting purposes [13].

Modeling for West African countries, or basins using global circulation models, (GCMs) has highlighted concerns about the potential connection between precipitation and temperature change [14,15]. When it comes to the timing of changes in rainfall and evapotranspiration, predictions made using the Coupled Model Inter-comparison Project Phases 5 and 6 (CMIP5 and CMIP6) are out of sync [16,17], thereby crippling investment strategies [14,18]. Additionally, a review by Druyan [19] of studies based on eight GCMs and two RCMs on Sahel precipitation during the past century indicated a lack of consensus, underscoring the necessity for additional studies that can provide hydrological pointers for climate change in the region. Shiru et al. [20]; Ogunrinde et al. [21]; Ajayi and Ilori [18]; Adeyeri et al. [22]; Kumi and Abiodun [23]; Oguntunde et al. [14] and Sylla et al. [24] emphasized both historical and future prediction changes based on GCMs in extreme climate variability and trends over either some sub-regions in West Africa or the whole West Africa.

One of the studies [18] was based on a single global climate model (GCM), without taking into account the biases in the projection. The indices computed in some of the studies include the length of warm spell, string of rainy and dry days, the start and end of precipitation, and the length of the growing season; however, it does not provide enough information on the local drought conditions. Additionally, there are no studies specifically on the projection of drought features based on climate change scenarios on a national scale. Furthermore, the specific atmospheric circulation and climate of Nigeria prevent the straightforward transfer or application of projections from other nations to the development of mitigation mechanisms.

This study focuses on the investigation of important drought characteristics on the long seasonal (hydrological) drought type over Nigeria. The investigations that completely understudy drought occurrence in the country using both observatory and simulated dataset-based climate change scenario are not readily available. Thus, studies such as this one can support adequately organizing agricultural and water resources under global warming conditions. Consequently, the specific aims of this study are: (i) to define drought in Nigeria using three drought indices; and (ii) to analyze the influences of climate change on the probable increases in drought frequency and intensity over Nigeria under the RCP 8.5 scenario.

2. Materials and Methods

2.1. Study Area

Nigeria has a total land mass area of about 930,000 km^2 and is situated between latitudes 4°15′ and 13°55′ N and longitudes 2°40′ and 14°45′ E (Figure 1). Nigeria has only two distinct seasons (dry and wet) because it is a tropical country. The dry season is marked by low relative humidity and higher temperatures than during the wet season due to strong winds from the Sahara Desert. The mean annual temperature in the country is usually around 25 °C. It is usually very hot in the northern region between March and June and between February and April in the south, as the temperature may rise above 30 °C during the day and drop a little during the night period. Rainfall magnitude and frequency in Nigeria relies primarily on the latitude. As the latitude increases, so the rainfall strength also increases. The southern region generally sees convectional rains because of the region's closeness to the equatorial belt.

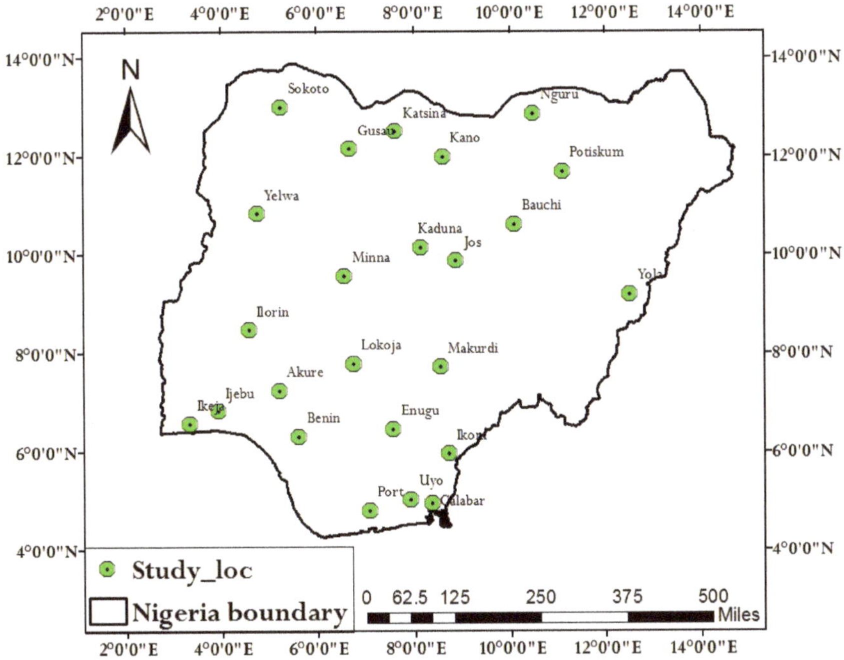

Figure 1. The study area indicating the meteorological stations used.

2.2. Datasets and Analytical Procedure

Both observatory (NIMET) and numerical models are used in this study's analysis. The two datasets include minimum and maximum temperatures as well as rainfall. The NIMET datasets were gathered at the Nigeria Meteorological Agency's Abuja, Nigeria, headquarters. Twenty-five stations were covered using historical NIMET datasets with a daily temporal resolution between 1981 and 2015. The retrieved NIMET datasets underwent a homogeneity and quality control test. The datasets for the models were developed using a dynamic downscaling strategy. This is one of the techniques for producing a higher resolution climatic reality from global circulation models with a lower resolution (GCMs). In this study, the Rossby Center Regional Climate Model (RCA-4) as reported by Nikulin et al. [25] was used to downscale seven GCMs spanning the CORDEX-African region. The historical period, which runs from 1951 to 2005, and the future period, which runs from 2006 to 2100, are both included in the downscaling's output.

The RCA-4 historical simulations between 1981 and 2005 were compared with observatory datasets in order to gauge the effectiveness of the model in reproducing the historical climate of Nigeria within the study area (NIMET). The high emission scenario included in the simulations, known as RCP 8.5, covers the period from 1951 to 2100. According to the RCP8.5 scenario, CO^2 levels will be around 950 ppm by 2100 [14].

The standardized precipitation index (SPI), standardized precipitation evapotranspiration index (SPEI), and reconnaissance drought index (RDI) are the three drought indices that have been studied. SPI is a probability-based indicator that assesses how much a given period's rainfall deviates from the long-term average value, which is ideally not less than 25 to 30 years [26]. The rainfall data series was fitted using the gamma distribution.

The greatest likelihood technique was used to determine the distribution factors and the probability density function (PDF) of the gamma distribution can be written as follows:

$$f(x;\alpha;\beta) = \frac{1}{\beta^{\alpha}\Gamma\ (\alpha)}\ x^{\alpha-1}e^{-x/\beta}\ for\ x,\alpha,\beta > 0 \tag{1}$$

where α is the shape factor, β is the scale factor and x is the rainfall value and $\Gamma\ (\alpha)$ is the gamma function.

In terms of computation, the SPI and SPEI are similar. The main modification is the inclusion of PET data, which is a measure of evaporation demand. When PET is subtracted from rainfall values, the climatic water balance during the time period being evaluated is calculated [27]. Next, a three-parameter log-logistic distribution is fitted to the difference. In this work, the Hargreaves empirical model was used to calculate the PET [28]. The SPI and SPEI within the study region were calculated using the R programming language over a 12-month timeline, which is crucial for hydrological drought evaluation.

The basic form of RDI, like the SPEI, is the ratio of total rainfall to PET for a particular reference time. Because it includes the primary input and outflow of a natural water system, the RDI may be a suitable index for assessing water availability. The DrinC (Drought Indices Calculator) program was utilized to streamline the RDI computing process. The NIMET and RCA-4 model datasets were used to generate the indices individually. Using monthly, seasonal, and yearly timesteps, the initial value (a_k) is displayed in aggregated form. The a_k is calculated as follows for the year i and a time basis of k (months):

$$a_k^{(i)} = \frac{\sum_{j=1}^{k} P_{ij}}{\sum_{j=1}^{k} ET_{ij}},\ i = 1\ to\ N\ and\ j = 1\ to\ 12 \tag{2}$$

The preparation of RDIst used the assumption that a_k values follow a lognormal distribution, and RDIst is computed as follows:

$$RDI_{st}^{(i)} = \frac{y^{(i)} - \overline{y}}{\sigma_y} \tag{3}$$

where $y^{(i)}$ is the ln $a_k^{(i)}$, y is mean and σ_y is its standard deviation.

3. Results and Discussion

3.1. Assessment of Climate Models over Nigeria

This section briefly discusses the capacity of the climate models to accurately represent the climate of the research area. Prior to the model's assessment, Figure 2a shows the long-term historical average monthly distribution of *T*min and *T*max, while Figure 2b displays that of the precipitation (PRCP) and evapotranspiration (ET) of the study area. The performance of each model dataset in relation to the observed datasets is shown in Figure 3. Since the correlation coefficient (r) value is more than 0.6, the performance of each of the seven models is adequate. The observed data in Figure 4 provide as an example of how well the models captured the important aspects of the annual seasonal cycle of the climate parameters and the size of the cycles. For example, the highest temperatures are recorded in both observed and model data during the months of March and November, which also happen to be transition months between the Dry–Wet and Wet–Dry seasons, respectively (Figure 4b). The models also correctly predicted the minimum local temperature in August, which also happens to be the month with the most clouds (Figure 4c). The measured air temperature does, however, exhibit some bias according to the models. For *T*min from January to December and for *T*max from April to November, the models showed a cold bias. Except for August, the observed *T*max curve coincided with the ensemble members of the model datasets, and values have generally been near to the ensemble median throughout the year. With few exceptions, the models also accurately depicted the observed annual seasonal rainfall cycle over Nigeria (Figure 4a). For instance, the observed rainfall pattern peaked in July, whereas the simulated rainfall pattern shows two modal peaks in April and August. The observed peak can be connected to the Intertropical Convergence Zone (ITCZ) movement because Nigeria is known for having numerous climate belts, which can produce volatility over the area between April and August, which may not have been considered in the development of the climate models [21]. The observed rainfall values only deviated from the ensemble members of the models in the months of May and June. According to [14], the pattern of rainfall distribution in the West African region might either have a single modal characteristic or a double modal characteristic. This demonstrates that the single modal peak reported by the NIMET observation dataset and the twin modal peaks suggested by the ensembles of RCA-4 are nonetheless consistent with the research area's climatology.

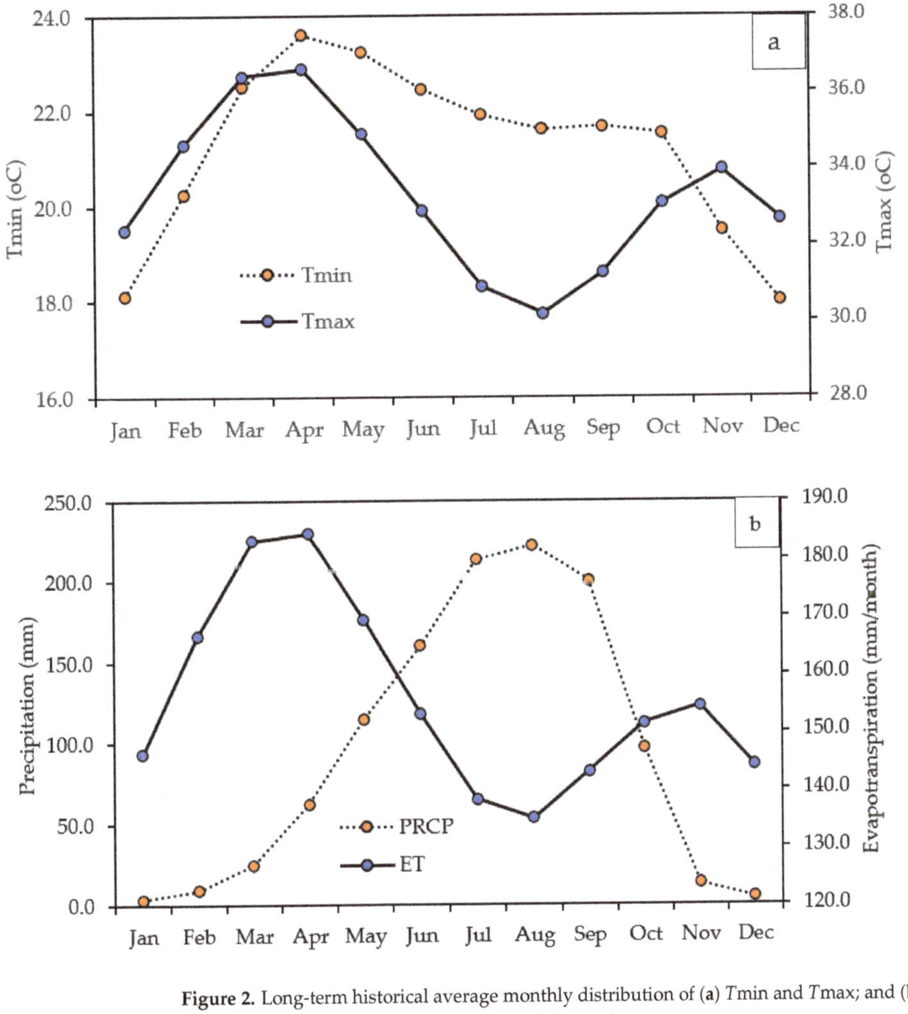

Figure 2. Long-term historical average monthly distribution of (**a**) *T*min and *T*max; and (**b**) precipitation (PRCP) and evapotranspiration (ET) of the study area.

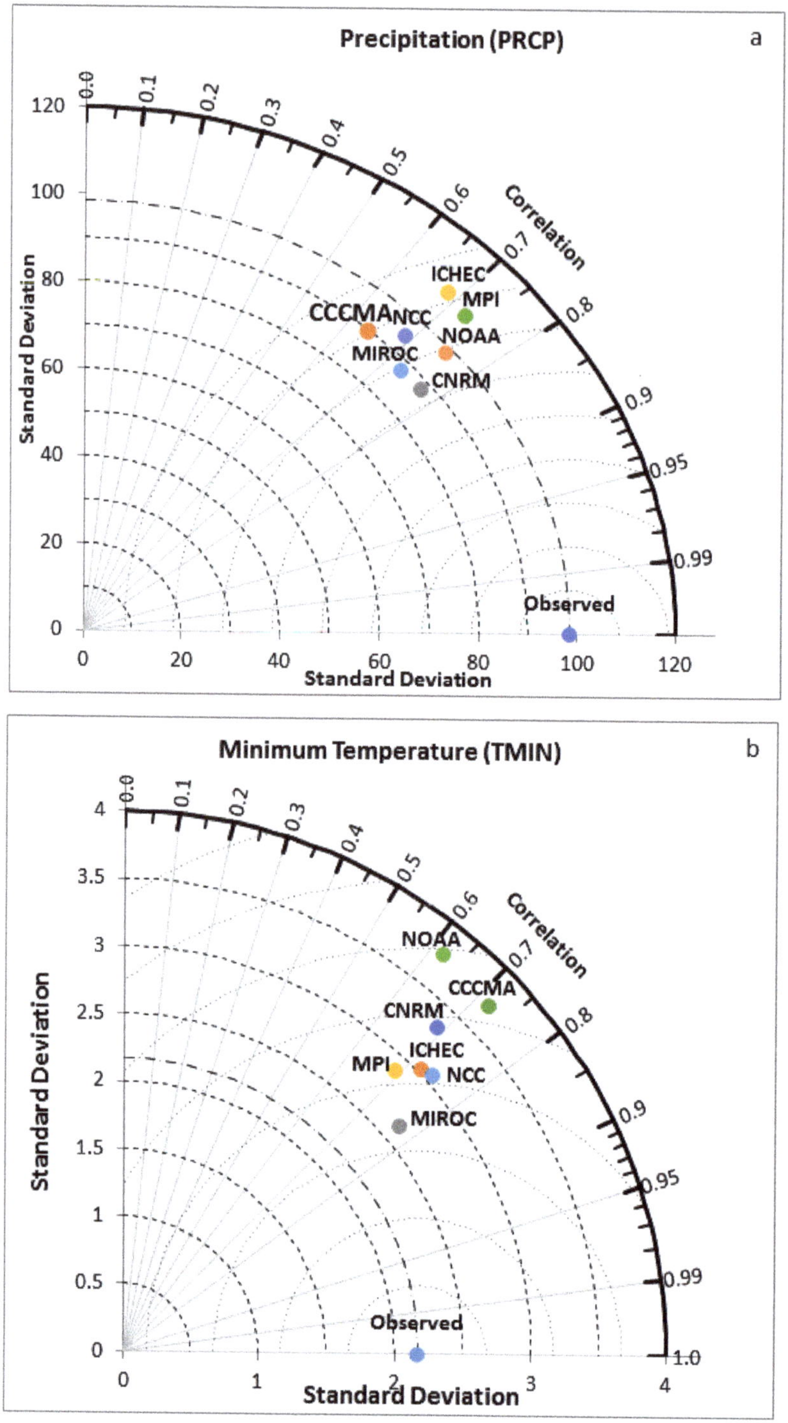

Figure 3. *Cont.*

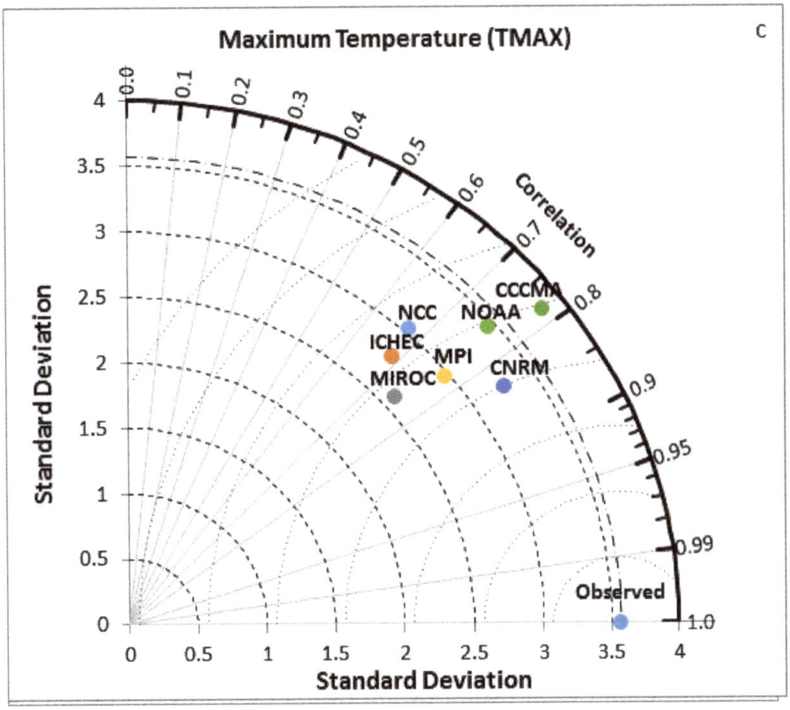

Figure 3. Taylor Diagram to compare the seven models and their combinations with respect to the observed data in terms of correlation coefficient and standard deviation. (**a**) for precipitation, (**b**) for minimum temperature, and (**c**) for maximum temperature.

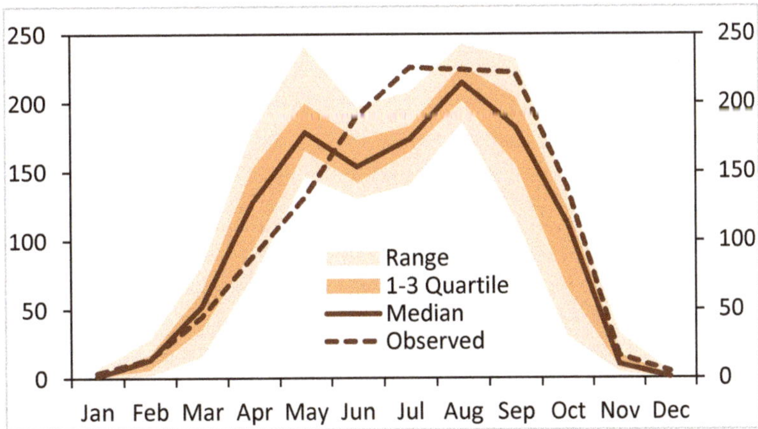

(**a**) PRCP

Figure 4. *Cont.*

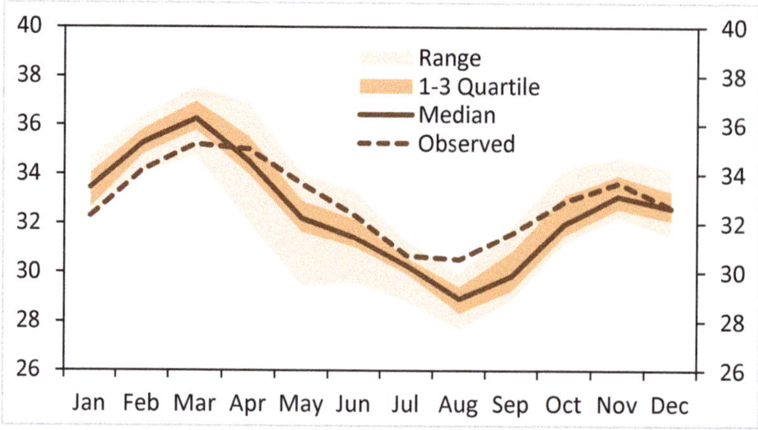

(b) *T*max

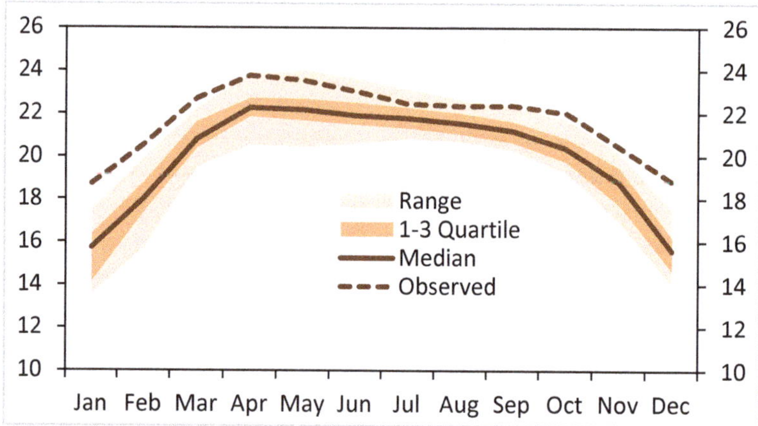

(c) *T*min

Figure 4. The annual seasonal cycle of observed and modelled of (**a**) Precipitation (PRCP) in mm/month; (**b**) Maximum Temperature (*T*max) in °C; and (**c**) Minimum Temperature (*T*min) in °C over Nigeria between 1981 and 2005. **Note:** The light pink color indicates the range between the maximum and minimum values; the dark pink shows the range between the 25th and 75th percentiles, while the thick line indicates the models' median and the dashed line shows the NIMET observation.

3.2. Future Prediction of Drought Characteristics

Figure 5a–c display predicted variations in drought frequency and intensity over Nigeria using SPI, SPEI, and RDI. The RCA-4 ensemble median over the research area suggests that the SPEI will see greater negative oscillations under the RCP 8.5 scenario, especially toward the end of the twenty-first century. According to this pattern, from 2035 to 2100, Nigeria would experience an increase in the frequency and severity of droughts. A comparable circumstance is also noticed with RDI (Figure 5c). On the other hand, as the twenty-first century goes on, the RCA-4 ensembles median shows a relatively stable to minimal changes in SPI for RCP8.5 scenario (Figure 5a). According to some earlier studies [27,29], drought indices that included PET (SPEI and RDI) are preferable to rainfall-based (SPI) indices in conditions of high temperatures. The current study's future drought projections are in good agreement with the findings of Oguntunde et al. [14] which state that, under historical climatic conditions, there is a strong correlation and similar trend pattern between the rainfall-based index (SPI) and the PET drought indices (SPEI and RDI). However, there are significant deviations in both the frequency and intensity of drought under RCP 8.5, particularly from the middle to the end of the twenty-first century.

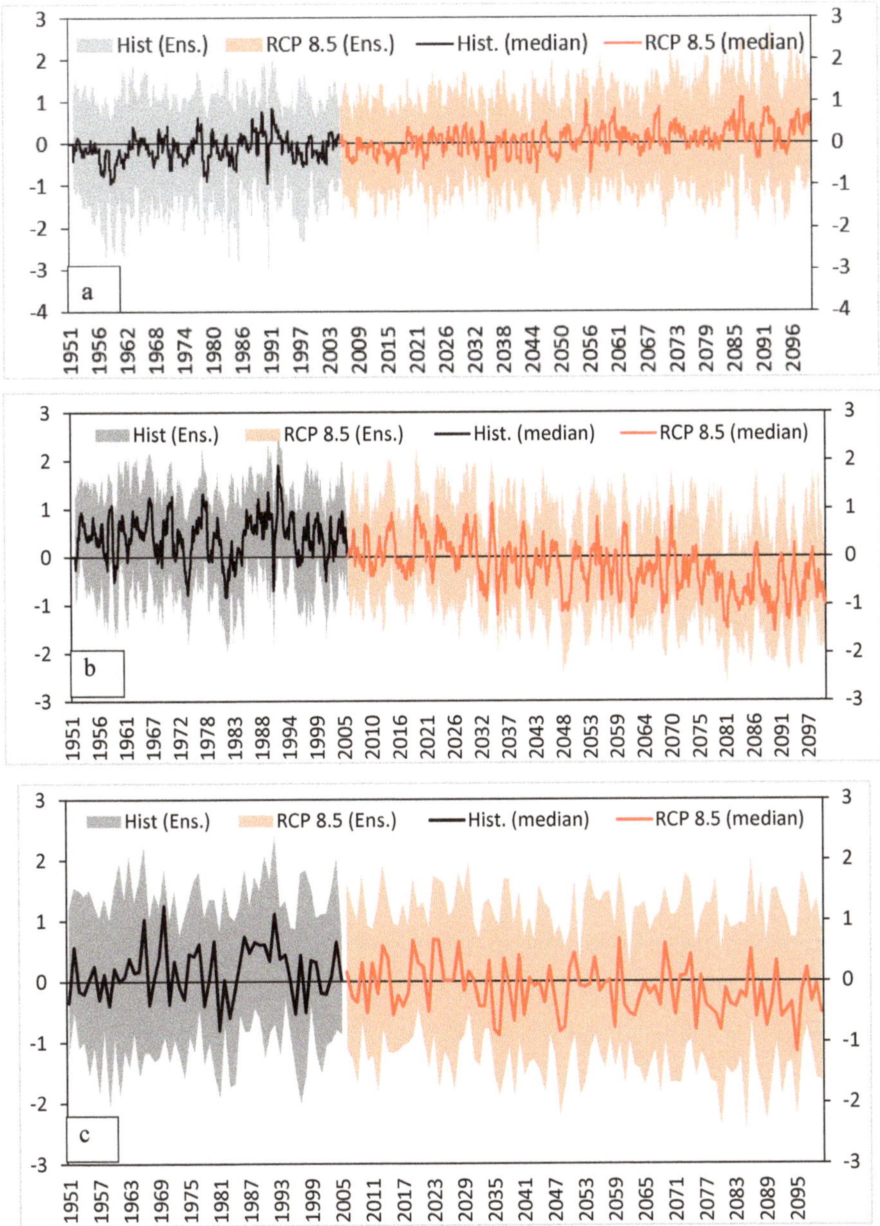

Figure 5. Historical and projected future changes in 12-month drought intensity over Nigeria for RCP 8.5 climate scenario. The droughts are characterized using (**a**) SPI; (**b**) SPEI; and (**c**) RDI.

In Figure 6, the 12-month RDI, SPEI, and SPI drought indices for the near and far eras are used to indicate the change in the frequency of drought. The SPEI indicates 0.75 month/year during 2031–2060 (relative to the reference era, 1976–2005) and 1.80 month/year in 2071–2100 (relative to the far era), whereas the RDI indicates 0.3 month/year in 2031–2060 and 0.6 month/year in 2071–2100, although the variations from the SPI are negligible. Projections of the change in drought frequency based on the RDI or SPEI are anticipated to be higher than estimates based on the SPI, indicating a higher atmospheric water demand as a result of global warming (RCP 8.5). Under the RCP 8.5, changes in drought frequency based on SPI are forecast to decrease during the near and far eras whereas drought occurrence based on RDI and SPEI is projected to increase. The SPI's reduction in the frequency of droughts indicates that precipitation is expected to rise, especially as the twenty-first century comes to a close.

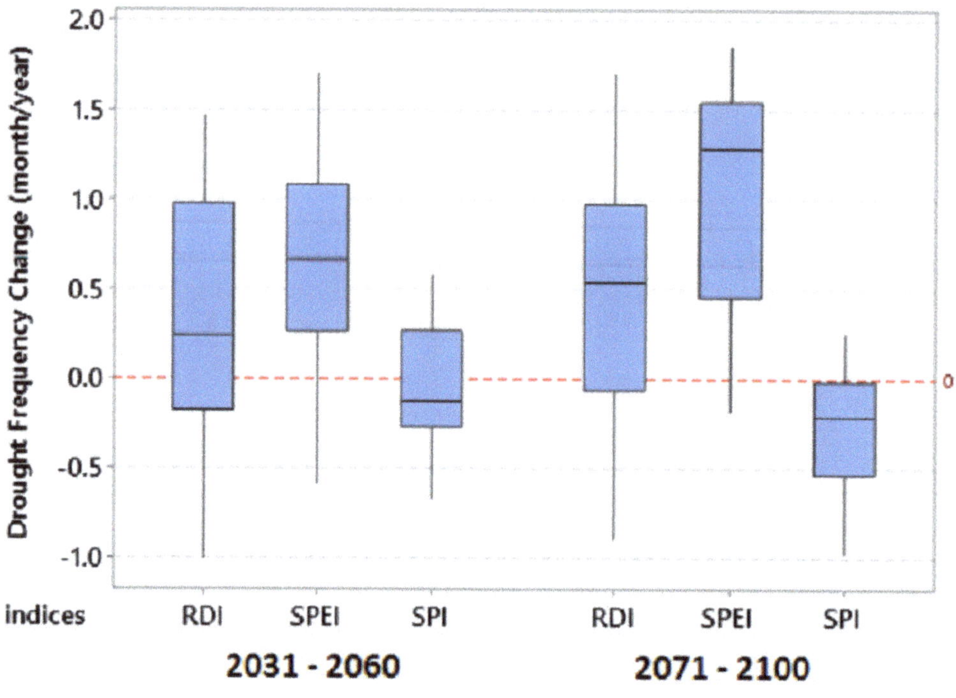

Figure 6. Projected future changes in 12-month drought frequency over Nigeria.

This shows that a good mitigation approach must be carefully considered when assessing the danger of future droughts. With the exception of flash floods caused by an abrupt rise in sea level, an increase in temperature of between 2 and 4 °C has the power to cancel out the effects of any increase in the intensity and duration of precipitation within a short period. In order to better understand how precipitation, evaporative demand, and streamflow might affect drought conditions, drought indices can be used [11].

Global natural disasters have become more frequent since the turn of the century, according to data from CRED [30], and they have in fact become the new normal. The frequency of natural disasters has been steadily rising over the past 20 years at a significant rate. The rate of occurrence was 360 incidents per year on average in the 2010s, compared to 100 and 90 events on average in the 1980s and 1970s, respectively. Natural catastrophes are occurring more frequently than ever, and this has a tremendous impact on both the economies and people's way of life. Globally, the cost of major natural disasters (related to geophysics, biology, hydrology, and climate) has been projected to be above USD 170 billion annually for the past twelve years. Since the global average surface temperature has been steadily rising at an alarming rate over the past five decades compared to the rate of growth between 1961 and 1980, there has been a clear correlation between climate-related disasters and climate change. We are starting to notice how climate change is causing an increase in the frequency and intensity of extreme weather events (for example, severe drought episodes), even as many studies on the impact of climate change are still in the early stages of development.

Particularly in Nigeria, the overpopulation, changing climate, and lack of effective river development strategies are projected to produce issues in the agricultural and water resources sectors in the future [14]. According to earlier climate simulation studies over West Africa, there is a high likelihood that as the current century goes on, temperatures will rise and there will be an increase in the number of extreme climate events (such as droughts, floods, and protracted dry spells during the rainy season) [15,31]. In a few decades, a fierce competition for water resources is anticipated to lead to disputes between upstream and downstream parties, underlining the necessity of sound water policy and effective drought mitigation measures in Nigeria.

More regional climate models (RCMs) could improve the assessment of prediction uncertainty in the future. Additionally, CORDEX's higher resolution models will make it possible to spatially analyze future projections, providing more helpful information on high-risk areas within Nigeria. Other climatic scenarios, such as RCP 4.5 and 6.0, should be taken into account.

4. Conclusions

Natural disasters have always posed a danger to society, the economy, and the environment. The frequency of incidence has recently increased and is now harder to forecast. Since the Sendai Framework for Disaster Risk Reduction 2015–2030 (SFDRR) and the Sustainable Development Goals (SDGs) have less than ten years left to be completed, serious efforts must be made to create climate-resilient systems that will battle the effects that climate change may have on the water resources and agricultural sectors. The study's key findings suggest that the ensemble of RCA-4 offers an accurate representation of Nigeria's seasonal changes in its yearly cycle climatology. The variation of drought using NIMET data is comparable to the deviation in the drought indices derived using models from RCA-4 ensemble members. The frequency and severity of droughts will also rise as the twenty-first century goes on due to the effects of climate change, according to RDI and SPEI under the RCP 8.5 climatic scenario. The current analysis will help governments and stakeholders at all levels design tactical adaptation and mitigation systems, according to its forecasts.

Author Contributions: A.T.O., P.G.O., Q.B.P., N.T.T.L. and D.T.A.: Original project, Conceptualization, Methodology, Software, Writing—Original version, Visualization, Reviewing and Editing. A.S.A., J.T.F., D.A.O. and Q.B.P.: Conceptualization, Methodology, and Software. D.T.A.: Formal analysis, Original version, Conceptualization, Writing, Supervision, and Revision. J.T.F., D.A.O., Q.B.P. and N.T.T.L.: Formal analysis, Revision, and Writing. All authors have read and agreed to the published version of the manuscript.

Funding: This research received no external funding.

Institutional Review Board Statement: Not applicable.

Informed Consent Statement: Not applicable.

Data Availability Statement: The data that support the findings of this study are available upon reasonable request.

Conflicts of Interest: The authors declare that they have no known competing financial interests or personal relationships that could have appeared to influence the work reported in this paper. This manuscript has not been published or presented elsewhere in part or entirety and is not under consideration by another journal. There are no conflicts of interest to declare.

References

1. Ogunrinde, A.T.; Oguntunde, P.G.; Akinwumiju, A.S.; Fasinmirin, J.T. Analysis of Recent Changes in Rainfall and Drought Indices in Nigeria, 1981–2015. *Hydrol. Sci. J.* **2019**, *64*, 1755–1768. [CrossRef]
2. Oguntunde, P.G.; Lischeid, G.; Abiodun, B.J. Impacts of climate variability and change on drought characteristics in the Niger River basin, West Africa. *Stoch. Environ. Res. Risk Assess.* **2018**, *32*, 1017–1034. [CrossRef]
3. van de Giesen, N.; Liebe, J.; Jung, G. Adapting to climate change in the Volta Basin. *West Afr. Curr. Sci.* **2010**, *98*, 1033–1038.
4. Ogilvie, A.; Mahé, G.; Ward, J.; Serpantié, G.; Lemoalle, J.; Morand, P.; Barbier, B.; Diop, A.T.; Caron, A.; Namarra, R.; et al. Water, agriculture and poverty in The Niger River basin. *Water Int.* **2010**, *5*, 594–622. [CrossRef]
5. Alfieri, L.; Bisselink, B.; Dottori, F.; Naumann, G.; de Roo, A.; Salamon, P.; Wyser, K.; Feyen, L. Global projections of river flood risk in a warmer world. *Earth's Futur.* **2017**, *5*, 171–182. [CrossRef]
6. IPCC. *Climate Change 2007: The Physical Science Basis: Working Group I Contribution to the Fourth Assessment Report of the IPCC*; Solomon, S., Qin, D., Manning, M., Chen, Z., Marquis, M., Averyt, K.B., Tignor, M., Miller, H.L., Eds.; Cambridge University Press: Cambridge, UK, 2007.
7. IPCC. *Impacts, Adaptation, and Vulnerability, Part A: Global and Sectoral Aspects: Working Group II to the Fifth Assessment Report of the Intergovernmental Panel on Climate Change*; Field, C.B., Barros, V.R., Dokken, D.J., Mach, K.J., Mastrandrea, M.D., Bilir, T.E., Chatterjee, M., Ebi, K.L., Estrada, Y.O., Genova, R.C., et al., Eds.; Cambridge University Press: Cambridge, UK; New York, NY, USA, 2014; p. 1132.
8. Intergovernmental Panel on Climate Change. Summary for Policymakers. In *Climate Change 2021: The Physical Science Basis. Contribution of Working Group I to the Sixth Assessment Report of the Intergovernmental Panel on Climate Change*; Cambridge University Press: Cambridge, UK; New York, NY, USA,, 2021; pp. 3 32.
9. Yao, N.; Zhao, H.; Li, Y.; Biswas, A.; Feng, H.; Liu, F.; Pulatov, B. National-scale variation and propagation characteristics of meteorological, agricultural, and hydrological droughts in China. *Remote Sens.* **2020**, *12*, 3407. [CrossRef]
10. Rudd, A.C.; Kay, A.L.; Bell, V.A. National-scale analysis of future river flow and soil moisture droughts: Potential changes in drought characteristics. *Clim. Chang.* **2019**, *156*, 323–340. [CrossRef]
11. Meresa, H.K.; Osuch, M.; Romanowicz, R. Hydro-Meteorological Drought Projections into the 21-st Century for Selected Polish Catchments. *Water* **2016**, *8*, 206. [CrossRef]
12. Quan, Q.; Liang, W.; Yan, D.; Lei, J. Influences of joint action of natural and social factors on atmospheric process of hydrological cycle in Inner Mongolia, China. *Urban Clim.* **2022**, *41*, 101043. [CrossRef]
13. Commonwealth Scientific and Industrial Research Organization. Hydrological consequences of climate change-summary of outcomes for a scientific meeting. In Proceedings of the CSIRO Discovery Centre, Canberra, Australia, 15–16 November 2007; Commonwealth Scientific and Industrial Research Organization: Canberra, Australia, 2011; pp. 1–6.
14. Oguntunde, P.G.; Abiodun, B.J.; Lischeid, G. Impacts of climate change on hydro-meteorological drought over the Volta Basin. *West Afr. Glob. Planet. Change* **2017**, *155*, 121–132. [CrossRef]
15. Patricola, C.M.; Cook, K.H. Sub-Saharan Northern African climate at the end of the twenty-first century: Forcing factors and climate change processes. *Clim. Dyn.* **2010**, *37*, 1165–1188. [CrossRef]
16. Taylor, K.E.; Stouffer, R.J.; Meehl, G.A. An Overview of CMIP5 and the Experiment Design. *Bull. Am. Meteorol. Soc.* **2012**, *93*, 485–498. [CrossRef]
17. Eyring, V.; Bony, S.; Meehl, G.A.; Senior, C.A.; Stevens, B.; Stouffer, R.J.; Taylor, K.E. Overview of the Coupled Model Intercomparison Project Phase 6 (CMIP6) experimental design and organization. *Geosci. Model Dev.* **2016**, *9*, 1937–1958. [CrossRef]
18. Ajayi, V.O.; Ilori, O.W. Projected Drought Events over West Africa Using RCA4 Regional Climate Model. *Earth Syst. Environ.* **2020**, *4*, 329–348. [CrossRef]
19. Druyan, L.M. Studies of twenty-first-century precipitation trends over West Africa. *Int. J. Climatol.* **2010**. [CrossRef]
20. Shiru, M.S.; Chung, E.-S.; Shahid, S.; Alias, N. GCM selection and temperature projection of Nigeria under different RCPs of the CMIP5 GCMS. *Theor. Appl. Climatol.* **2020**, *141*, 1611–1627. [CrossRef]
21. Ogunrinde, A.T.; Oguntunde, P.G.; Akinwumiju, A.S.; Fasinmirin, J.T. Evaluation of the impact of climate change on the characteristics of drought in Sahel Region of Nigeria: 1971–2060. *Afr. Geogr. Rev.* **2020**, *40*, 192–210. [CrossRef]
22. Adeyeri, O.; Lawin, A.; Laux, P.; Ishola, K.; Ige, S. Analysis of climate extreme indices over the Komadugu-Yobe basin, Lake Chad region: Past and future occurrences. *Weather Clim. Extrem.* **2019**, *23*, 100194. [CrossRef]
23. Kumi, N.; Abiodun, B.J. Potential impacts of 1.5C and 2C global warming on rainfall onset, cessation and length of rainy season in West Africa. *Environ. Res. Lett.* **2018**, *13*, 055009. [CrossRef]
24. Sylla, M.B.; Nikiema, P.M.; Gibba, P.; Kebe, I.; Klutse, N.A.B. Climate Change over West Africa: Recent Trends and Future Projections. In *Adaptation to Climate Change and Variability in Rural West Africa*; Springer: Berlin/Heidelberg, Germany, 2016; pp. 25–40. [CrossRef]

25. Nikulin, G.; Jones, C.; Giorgi, F.; Asrar, G.; Büchner, M.; Cerezo-Mota, R.; Christensen, O.B.; Déqué, M.; Fernandez, J.; Hänsler, A.; et al. Precipitation climatology in an ensemble of CORDEX-Africa regional climate simulations. *J. Clim.* **2012**, *25*, 6057–6078. [CrossRef]
26. McKee, T.B.; Doesken, N.J.; Kleist, J. The relationship of drought frequency and duration to time scales. In Proceedings of the 8th Conference on Applied Climatology, Anaheim, CA, USA, 17–22 January 1993; American Meteorological Society: Boston, MA, USA, 1993; pp. 179–184.
27. Vicente-Serrano, S.M.; Beguería, S.; López-Moreno, J.I. A multi-scalar drought index sensitive to global warming: The standardized precipitation evapotranspiration index. *J. Clim.* **2010**, *23*, 1696–1718. [CrossRef]
28. Hargreaves, G.H.; Samani, Z.A. Reference Crop Evapotranspiration from Temperature. *Appl. Eng. Agric.* **1985**, *1*, 96–99. [CrossRef]
29. Beguería, S.; Vicente-Serrano, S.M.; Reig, F.; Latorre, B. Standardized precipitation evapotranspiration index (SPEI) revisited: Parameter fitting, evapotranspiration models, tools, datasets and drought monitoring. *Int. J. Climatol.* **2014**, *34*, 3001–3023. [CrossRef]
30. Centre for Research on the Epidemiology of Disasters. International Disaster Database-EM-DAT CRED. EM-DAT [Online]. Brussels. Updated as at 18 January 2021. Available online: http://www.emdat.be/database (accessed on 30 August 2022).
31. Oguntunde, P.G.; Abiodun, B.J. The impact of climate change on the Niger River Basin hydroclimatology, West Africa. *Clim. Dyn.* **2013**, *40*, 81–94. [CrossRef]

atmosphere

MDPI

Article

Estimating FAO Blaney-Criddle b-Factor Using Soft Computing Models

Suthira Thongkao [1,2], Pakorn Ditthakit [1,3,*], Sirimon Pinthong [1,3], Nureehan Salaeh [1,3], Ismail Elkhrachy [4], Nguyen Thi Thuy Linh [5] and Quoc Bao Pham [5,*]

1 Center of Excellence in Sustainable Disaster Management (CESDM), Walailak University, 222, Thaiburi, Thasala, Nakhon Si Thammarat 80160, Thailand
2 School of Languages and General Education, Walailak University, 222, Thaiburi, Thasala, Nakhon Si Thammarat 80160, Thailand
3 School of Engineering and Technology, Walailak University, 222, Thaiburi, Thasala, Nakhon Si Thammarat 80160, Thailand
4 Civil Engineering Department, College of Engineering, Najran University, King Abdulaziz Road, P.O. Box 1988, Najran 66291, Saudi Arabia
5 Institute of Applied Technology, Thu Dau Mot University, Thu Dau Mot 75000, Binh Duong, Vietnam
* Correspondence: dpakorn@mail.wu.ac.th (P.D.); phambaoquoc@tdmu.edu.vn (Q.B.P.)

Abstract: FAO Blaney-Criddle has been generally an accepted method for estimating reference crop evapotranspiration. In this regard, it is inevitable to estimate the b-factor provided by the Food and Agriculture Organization (FAO) of the United Nations Irrigation and Drainage Paper number 24. In this study, five soft computing methods, namely random forest (RF), M5 model tree (M5), support vector regression with the polynomial function (SVR-poly), support vector regression with radial basis function kernel (SVR-rbf), and random tree (RT), were adapted to estimate the b-factor. And Their performances were also compared. The suitable hyper-parameters for each soft computing method were investigated. Five statistical indices were deployed to evaluate their performance, i.e., the coefficient of determination (r^2), the mean absolute relative error (MARE), the maximum absolute relative error (MXARE), the standard deviation of the absolute relative error (DEV), and the number of samples with an error greater than 2% (NE > 2%). Findings reveal that SVR-rbf gave the highest performance among five soft computing models, followed by the M5, RF, SVR-poly, and RT. The M5 also derived a new explicit equation for b estimation. SVR-rbf provided a bit lower efficacy than the radial basis function network but outperformed the regression equations. Models' Applicability for estimating monthly reference evapotranspiration *(ETo)* was demonstrated.

Keywords: Blaney-Criddle b-Factor; machine learning; M5 model tree; random forest; random tree; reference crop evapotranspiration; support vector regression

check for updates

Citation: Thongkao, S.; Ditthakit, P.; Pinthong, S.; Salaeh, N.; Elkhrachy, I.; Linh, N.T.T.; Pham, Q.B. Estimating FAO Blaney-Criddle b-Factor Using Soft Computing Models. *Atmosphere* **2022**, *13*, 1536. https://doi.org/10.3390/atmos13101536

Academic Editor: Gianni Bellocchi

Received: 31 July 2022
Accepted: 15 September 2022
Published: 20 September 2022

1. Introduction

Reference evapotranspiration *(ETo)* estimation is imperative information to serve water resources planning, management, and operation [1]. The Blaney-Criddle method, as proposed by the Food and Agricultural Organization (FAO) of the United Nations, is a well-known temperature-based reference crop evapotranspiration. This method gives more advantage when having limitations of the measured data than the Penman-Monteith method, which requires many meteorological data [2–4]. Many attempts [5–9] have been made to evaluate the efficiency of the FAO Blaney-Criddle method in estimating reference crop evapotranspiration for many regions. The study results by Jhajharia, Ali, DebBarma, Durbude, and Kumar [6] revealed that in humid locations, the Blaney-Criddle method was superior to other temperature-based methods, such as Hargreaves and Thornth–Waite. This is because it offered an approximate solution of the reference crop evapotranspiration closest to the FAO Penman-Monteith. However, the calibration of FAO-Blaney-Criddle parameters by the use of meteorological data in the corresponding region is important as they differ from location to location [10]. In addition, the Blaney-Criddle approach has been updated to suit the environment [11,12].

For many years, soft computing methods have been applied to manage water resource problems and related hydrology issues, especially for predicting evapotranspiration [13–15]. Tzimopoulos, Mpallas, and Papaevangelou [14] applied fuzzy logic to establish a temperature-based approach for estimating possible evapotranspiration and compared it to the Blaney-Criddle method. Ramanathan, Saravanan, Adityakrishna, Srinivas, and Selokar [13] found that artificial neural networks (ANN), wavelet neural networks (WNN), and fuzzy logic (FL) yielded better results in estimating ET_0 compared to traditional approaches, such as the Penman-Monteith method, the Blaney-Criddle method, and the Hargreaves method. Ferreira et al. [16] indicated that clustering weather stations with analogous hydrological characteristics and lagged time data improved the performance of ANN and support vector machine (SVM) for estimating ET_0. Yu et al. [17] pointed out the importance of selecting input patterns by studying its sensitivity analysis and concluded two crucial weather variables for modeling ET_0, i.e., maximum and minimum temperature. Shabani et al. [18] indicated that Gaussian Process Regression (GPR) outperformed K-Nearest Neighbors (KNN), Random Forest (RF), and Support Vector Regression (SVR) in predicting pan evaporation. They also emphasized the necessity of suitably choosing weather variables depending on the unique weather station features. Mohammadi and Mehdizadeh [19] revealed hybrid support vector regression with a whale optimization algorithm outperformed a sole support vector regression in modeling reference evapotranspiration, which is a function of air temperatures, relative humidity, solar radiation, sunshine duration, and wind speed. Granata and Di Nunno [20] applied recurrent neural networks to forecast actual evapotranspiration in short term ahead. Their study revealed that in subtropical climatic conditions of South Florida, long short-term memory (LSTM) gave better efficiency than a nonlinear autoregressive network with exogenous inputs (NARX), and there was no significant effect of sensible heat flux and relative humidity on actual evapotranspiration forecasting. On the other hand, in the semi-arid climate of Central Nevada, NARX outperformed LSTM, and there were slight effects due to relative humidity, sensible heat flux, and forecast horizon. Our best literature reviews pointed out the research gap in estimating the b factor of the FAO Blaney-Criddle formula using the soft computing method since only one soft computing method, the Radial Base Function (RBF) network, was researched.

This research article intends to investigate the applicability of soft computing methods in estimating the b factor of the FAO Blaney-Criddle formula, which is advantageous for hydrology and agriculture-related issues. The novelty of this research is the first attempt to use random forest (RF), M5 model tree (M5), support vector regression (SVR) with two kernel functions (i.e., polynomial and radial), and random tree (RT) for estimating the b factor. Their performance was compared with the previous studies. Each model's weaknesses and advantages were discussed. The rest of this article is organized as follows: the next section explains the method and data used, including FAO Blaney-Criddle b factor, soft computing models, Weka machine learning tool, tuning hyper-parameters, data used, and statistical model performance indices. Section 3 provides the significant finding results of the suitable hyper-parameters for each soft computing method and their comparative performance among five soft computing methods as conducted in the present study and the previous studies. Our main study's finding is concluded and recommended in the final section.

2. Materials and Methods

2.1. FAO Blaney-Criddle B Factor

The original Blaney-Criddle equation requires information on the average daily percentage of total daily hours and mean daily air temperature for predicting reference crop evapotranspiration. Its formula is expressed as follows, by [21].

$$ET_0 = a + b[p(0.46T + 8.13)] \quad (1)$$

where ET_0 is reference crop evapotranspiration (mm/d); a and b are calibrated constants; p is the average daily percentage of total annual daytime hours; and T is the average daily air temperature (°C). The a factor can be derived by:

$$a = 0.0043RH_{min} - (n/N) - 1.41 \quad (2)$$

where RH_{min} is the lowest daily relative humidity (%); and n/N is the average ratio of actual to possible sunshine hours. The p and N values can be received from tables when specifying latitudes and months [21,22]. They can be obtained using formulas as proposed by [23,24].

For determination of the value of b factor, Doorenbos, Pruitt, and Agl [21] proposed it in tabular form. It depends on the lowest daily relative humidity (RH_{min}), daytime wind speed (U_d), and the average ratio of actual to possible sunshine hours (n/N) (see detailed information in Table 1). The authors can simply utilize the technique of table interpolation to obtain the b value. However, it needs seven interpolation times for getting that value, leading to lead to considerable error [25]. To defeat such drawback, Frevert et al. [26] first proposed a regression equation (see Equation (3)) and, later, it was improved by Allen and Pruitt [27] (see Equation (4)). Nevertheless, it was still an error in estimating the b value of approximately 10% compared to the tabular values. Ambas and Evangglos [28] used weighted least squares to estimate b factor of the FAO24 Blaney-Criddle method as shown in Equation (5). It gave close results as compared to the previous studies. Equations (3)–(5) still have an error in estimating the b value as compared to the tabular values. Hence, it requires other techniques to decrease the error.

$$\begin{aligned} b &= 0.81917 - 0.0040922(RH_{min}) + 1.0705\left(\tfrac{n}{N}\right) + 0.065649(U_d) - 0.0059684(RH_{min})(n/N) \\ &\quad -0.0005967(RH_{min})(U_d) \end{aligned} \quad (3)$$

$$\begin{aligned} b &= 0.908 - 0.00483(RH_{min}) + 0.7949\left(\tfrac{n}{N}\right) + 0.0768[\ln(U_d + 1)]^2 - 0.0038(RH_{min})\left(\tfrac{n}{N}\right) \\ &\quad -0.000433(RH_{min})Ud + 0.281 ln(Ud + 1) In\left(\tfrac{n}{N} + 1\right) - 0.00975\, ln(Ud + 1) \\ &\quad [ln(RH_{min} + 1)^2 In\ (n/N + 1)] \end{aligned} \quad (4)$$

$$\begin{aligned} b &= 0.88165 + 0.857596\left(\tfrac{n}{N}\right) - 0.00454(RH_{min}) + 0.093803(U_d) - 0.00405(RH_{min})\left(\tfrac{n}{N}\right) \\ &\quad -0.00087(RH_{min})(U_d) \end{aligned} \quad (5)$$

2.2. Soft Computing Models

Soft computing models refer to a data analysis of a complex system in order to discover the relationship between system state variables, i.e., independent and dependent variables, without explicit knowledge of the physical nature of the system [29]. In this section, four data-driven models, e.g., random forest (RF), M5 model tree (M5), support vector regression (SVR), and random tree (RT) are briefly explained, as follows.

2.2.1. Random Forest (RF)

The Random Forest (RF) was first introduced by Breiman [30] and has been a common modification of decision trees, which is one of the collections of techniques for data classification and regression [31]. There are two major phases of model construction. In the first step, RF generates a number of individual trees based on the decision tree process. Each tree is created by randomly selecting different sampled training data sets from the entire training data set (also known as the bagging method or bootstrap aggregation) and sub-attributes (or features) from all attributes in the training data set. Second, the voting method is applied, that is, the model prediction is finally achieved by voting for the classification problem or by using the mean value for the regression problem from the predictive performance of each tree generated. In comparison to the M5, full-grown RF trees are not pruned back. This is one of the key benefits of the regression of RF over the M5. As the number of trees increases, the error of speculation still converges even without pruning the tree, and over-fitting is not a matter of concern in light of the Strong Law of Large Numbers [32]. The RF model was adapted based on regression models in this study.

2.2.2. M5 Model Tree (M5)

The M5 Model Tree (M5) model tree was first implemented by Quinlan [33]. It applies a divide-and-conquer method to the creation of a relationship between independent and dependent variables and can be applied to both qualitative (categorical) and quantitative variables. Building M5 involves three stages. The first stage involves the development of a decision tree by dividing the data set into subsets (or leaves). Second, the overgrown tree is reduced, and linear regression functions substitute the plucked sub-trees to avoid overfitting the structure or a weak generalizer. The merging of certain lower sub-trees into one node is processed as part of the pruning approach. The smoothing procedure is finally employed to reduce the serious discontinuities between the linear models in the leaves of the trimmed trees, especially for models created from a small number of training samples.

2.2.3. Support Vector Regression (SVR)

Support Vector Regression (SVR) was developed by Vapnik [34] and his colleagues. This is the adaptation of the support vector machine (SVM) for regression. The basic idea of SVR learning is to solve the separation hyperplane that can correctly divide the training data set and has the largest geometric interval [35,36]. Using the automated conversion of nominal values to numerical values, SVM may be both numerical and nominal. Normalization or standardization shall be processed for all input data prior to the corresponding step. Unlike Support Vector Machine (SVM) for a classifier, which finds a line that best divides training data into classes, SVR processes the best line that separates the training data set by having a minimal error in the cost function. For this reason, an optimization algorithm is used to consider those data instances in the training dataset that are nearest to the minimum cost line. These instances are then referred to as support vectors, which is the name of this technique. In the event that a line that matches the data cannot be identified, a margin is inserted along the line to loosen the constraint. This margin helps the overall outcome to be better, but it does offer some poor predictions to be tolerated. Adequate determination of the complexity parameter C is important. Giving a low C value gives a broad minimum margin, otherwise, it gives a smaller minimum margin. In several real-world problems, it has been found that the use of a straight line is not sufficient for separating data sets. It is also more fitting to use curves or even polygonal regions. By converting data into higher dimensional areas, the kernel functions have been meant to draw lines and predict.

2.2.4. Random Tree (RT)

RT is a fundamental decision tree algorithm collaborating with Quinlan C4.5 or Classification and Regression Trees (CART). It chooses a random subset of attributes for each split from the available attributes before it is implemented with a subset size determined by the part ratio parameter. This method constructs a decision tree and chooses the feature to maximize the information gained using a portion of the data as training data. It is strong and straightforward to use, producing extremely accurate forecasts [29,30]. For a regression tree, a dataset is divided into sub-spaces, and fitting a constant is proceeded for each sub-space [32]. Consequently, A single-tree model exhibits a low level of prediction accuracy and a propensity to be very unstable. However, it can produce extremely accurate results via bagging RT as a decision tree method. It is highly flexible and has quick learning.

2.3. Weka Machine Learning Tool

WEKA (Waikato Environment for Knowledge Analysis) is a Java-based open-source machine learning platform released under the GPL (GNU). It was subsequently established by the University of Waikato in New Zealand. WEKA can impose pre-processing, classification, clustering, association rules, and selection of attributes for data. It also has a graphical representation visualization tool. WEKA has four main applications: Explorer, Experimenter, Knowledgeflow, and SimpleCLI. We can use the Explorer environment to explore the data. If we want to conduct experiments and conduct statistical tests between learning methods, the authors can use an experimenter. Knowledgeflow essentially supports the same features as an explorer, but it is a drag-and-drop interface that supports progressive learning. The authors can work on the WEKA command-line interface in a simpleCLI environment.

2.4. Tuning Hyper-Parameters

Developing a soft computing model or machine learning model considers two parameters, i.e., model parameters and hyper-parameters. Unlike model parameters obtained during the training process, hyper-parameters are the pre-setting parameters by the user to determine model structure before training the models. The control of a machine learning model's behavior requires hyperparameter adjustment. Therefore, our predicted model parameters will yield less performance if our hyper-parameters aren't properly tuned to minimize the loss function. In general, the process for tuning hyper-parameters includes defining a model, defining the range of possible values for all hyper-parameters, defining a method for sampling hyper-parameter values, defining evaluative criteria to judge the model, and defining a cross-validation method. In this experiment, a WEKA experimenter was utilized to do a systematic trial and error, that is, varying one interesting parameter and fixing the remained parameters, and repeating this step until covering all parameters. The Root Relative Squared Error (RRSE) with ten-fold cross-validation, given in WEKA, was used as a criterion for selecting the best parameter value for all 216 data sets.

2.5. Data Used

In this study, 216 data sets taken from the b factor tabular of FAO Blaney-Criddle [21] were utilized, coincident with the study purpose. For evaluating the models' performance with the previous studies, the training and testing process data sets were the same as those used in Trajkovic, Stankovic, and Todorovic [25]. They randomly selected 186 of 216 data sets for training models and used all 216 data sets for testing models. Table 1 summarizes the statistical analysis of relevant parameters of FAO Blaney-Criddle b for the training and testing processes. In overall statistic values, they were very similar for both training and testing data sets. However, when considering the Kurtosis value, it indicated all parameters (U_d, n/N, RH_{min}, and b) for both training and testing datasets had platykurtic distributions. Also, the skewness value showed that U_d, n/N, and RH_{min} for both training and testing datasets were approximately symmetric ("-" sign means skewed left and "+" sign means skewed right), while b for both training and testing datasets were moderately skewed right. The correlation analysis was conducted to individually evaluate the strength of the relationship between each input parameter (U_d, n/N, and RH_{min}) and an output parameter (b). A low degree correlation was found for U_d (r = 0.27, and 0.26 for training and testing stages, respectively), and a high degree correlation was obtained for the rest parameters. The n/N gave the correlation coefficient (r) of 0.57 and 0.58 for the training and testing stages, respectively. Additionally, RH_{min} provided a strong negative relationship by giving the correlation coefficient (r) of -0.74 for both the training and testing stages, respectively.

Table 1. Statistical evaluation of FAO Blaney-Criddle b parameters for training and testing data sets.

Statistical Values	Training				Testing			
	U_d	n/N	RH_{min}	b	U_d	n/N	RH_{min}	b
Maximum	10.00	1.00	100.00	2.63	10.00	1.00	100.00	2.63
Minimum	0.00	0.00	0.00	0.38	0.00	0.00	0.00	0.38
Average	5.01	0.51	49.68	1.19	5.01	0.50	49.72	1.18
Standard Deviation	3.42	0.34	34.06	0.47	3.42	0.34	34.07	0.46
Kurtosis	−1.27	−1.28	−1.26	−0.09	−1.27	−1.27	−1.26	0.10
Skewness	−0.01	−0.05	0.01	0.62	−0.01	0.01	0.01	0.67
Correlation Coefficient (r)	0.27	0.57	−0.74	1.00	0.26	0.58	−0.74	1.00
Number of data	186				216			

2.6. Statistical Model Performance Indices

Five statistical indices were deployed to evaluate model performance, i.e., the coefficient of determination (r^2) (Equation (6)), the mean absolute relative error (MARE) (Equation (7)), the maximum absolute relative error (MXARE) (Equation (8)), the standard deviation of the absolute relative error (DEV) (Equation (9)), and the number of samples with an error greater than 2% (NE > 2%). All of these statistical indices were used by Trajkovic, Stankovic, and Todorovic [25] on this particular issue. The r^2 calculates the level of linearity of two variables and its maximum is 1.00. MARE and MXARE determine the difference between the real and the expected b factor and should be as small as possible. The perfect model should have an NE of zero. Finally, a Taylor diagram was proposed to comparatively elaborate and evaluate the efficacy of the developed models. This diagram can simultaneously show three statistic parameters, i.e., correlation, root mean square error, and standard deviation. The equations of statistical indices are given below, where b_{ai} is the actual b-factor, is the estimated b-factor, and n is the number of samples in a data set.

$$r^2 = \left[\frac{\sum_{i=1}^{n} \left(b_{ai} - \overline{b}_{ai}\right)\left(b_{ei} - \overline{b}_{ei}\right)}{\sqrt{\sum_{i=1}^{n} \left(b_{ai} - \overline{b}_{ai}\right)^2} \cdot \sqrt{\sum_{i=1}^{n} \left(b_{ei} - \overline{b}_{ei}\right)^2}} \right]^2 \quad (6)$$

$$MARE = \frac{1}{n} \sum_{i=1}^{n} \left| \frac{b_{ai} - b_{ei}}{b_{ai}} \right| \quad (7)$$

$$MXARE = max\left(\left| \frac{b_{ai} - b_{ei}}{b_{ai}} \right| \right) \text{ for } i = 1, \ldots, n \quad (8)$$

$$DEV = \sqrt{\frac{\sum_{i=1}^{n} \left[\left| \frac{b_{ai} - b_{ei}}{b_{ai}} \right| - MARE \right]^2}{(n-1)}} \quad (9)$$

3. Results and Discussion

3.1. Results of Tuning Hyper-Parameters

Table 2 shows the results of tuning hyper-parameters. Their explanation for each soft computing model is as follows.

Table 2. Summary of the optimal hyper-parameters for soft computing models.

Hyper-Parameter	RF		M5		SVR-poly		SVR-rbf		RT	
	Value	Sensitive	Value	Sensitive	Value	Sensitive	Value	Sensitive	Value	Sensitive
numIteration	300	yes	-	-	-	-	-	-	-	-
batchSize	100	no	100	no	-	-	-	-	100	no
numExecutionSlots	1	no	-	-	-	-	-	-	-	-
minNumInstances	-	-	4	yes	-	-	-	-	-	-
numDecimalPLaces	-	-	4	no	-	-	-	-	2	no
buildRegressionTree	-	-	FALSE	yes	-	-	-	-	-	-
complexity	-	-	-	-	0.8	yes	1.0	yes	-	-
exponent	-	-	-	-	1.0	yes			-	-
gamma	-	-	-	-	-	-	1.0	yes	-	-
minNum	-	-	-	-	-	-	-	-	1.0	yes
numFolds	-	-	-	-	-	-	-	-	0	yes
minVarianceProp	-	-	-	-	-	-	-	-	0.001	yes
RRSE	12.14		11.46		24.21		2.37		24.23	

3.1.1. Random Forest (RF)

In the process of tuning hyper-parameters for RF, some default parameters were selected as default in WEKA software, i.e., (1) infinite maximum tree depth, and (2) int(log 2(#predictors) + 1) function used to set the number of randomly selected attributes. However, three parameters, namely: (1) numIteration, which is the number of trees in the random forest; (2) batchSize, which is the optimum number of instances to be processed when predicting batch; and (3) numExecutionSlots, which is the number of threads available for execution to be used to create the collection, were investigated in our experiment. Findings revealed that numIteration of 300, batchSize of 100 (default value), and numExecutionSlots of 1 (default value) were the suitable hyper-parameters for RF with the testing data set. All cases gave an RRSE value of 12.14. The numIteration was a sensitive parameter, while batchSize and numExecutionSlots were not sensitive.

3.1.2. M5 Model Tree (M5)

The authors experimented tuning hyper-parameters of the M5 model tree using the default parameters in WEKA software of unpruned to be false, and use Unsmoothed to be false. The four parameters, i.e., batchSize, minNumInstances, numDecimalPLaces, and buildRegressionTree were investigated. If the batch prediction is utilized, the bathcSize option specifies the recommended number of instances to process. More or fewer instances are conceivable, but this allows implementations to select the batch size they want. The minimal number of instances to allow at a leaf node is specified by minNumInstances. The number of decimal places to utilize for the model's output is numDecimalPLaces. It can be decided whether to construct a regression tree/rule instead of a model tree/rule using the buildRegressionTree method.

Findings revealed that batchSize of 100 (default value), minNumInstances of 4 (default value), and numDecimalPLaces of 4 (default value), were the suitable hyper-parameters for M5 with the testing data set. All best cases gave an RRSE value of 11.46. By using batchSize of 100, minNumInstances of 4, and numDecimalPLaces of 4, the authors investigated the effect of selecting or not selecting a regression tree/rule. There was no need to generate a regression tree/rule due to giving an RRSE value of 11.46 compared to creating a regression tree/rule, which gave an RRSE value of 56.26. The minNumInstances and buildRegressionTree were sensitive parameters, while batchSize and numDecimalPLaces were not sensitive.

3.1.3. Support Vector Regression (SVR)

For SVR, two kernel functions, namely the polynomial kernel with variable exponent value and the radial basis function kernel with varying gamma value, were investigated. Also, the complexity parameter (C) was varied between 0.0 and 1.0 to determine the optimum value. The gamma parameter represents the influence of a single training reach, with low values indicating 'far' and large values indicating 'close.' The inverse of the impact radius of the samples chosen by the model as support vectors are gamma parameters. The modified sequential minimal optimization (SMO) as an iterative algorithm was used to solve the regression problem for SVR [34]. The authors found the optimal hyper-parameters for SVR with polynomial kernel function were the complexity parameter (C) of 0.8 and the exponent (n) of 1.0. By fixing the complexity parameter (C) value of 0.8 and varying the exponent (n) value from 1.0 to 4.0, it was sensitive to the exponent value for SVR with a polynomial kernel function. The best case gave an RRSE value of 24.21.

Furthermore, the optimal hyper-parameters for the radial basis function kernel were the complexity parameter (C) of 1.0 and the gamma parameter (γ) of 1.0. By fixing the complexity parameter (C) value of 1.0 and varying the gamma parameter (γ) value from 1.0 to 4.0, it was sensitive to the gamma parameter (γ) value for the radial basis function kernel. Additionally, the gamma parameter (γ) value of 1.0 gave the least RRSE. For both cases, the suitable C parameters were equal to or more than 0.8. It indicated that these data sets required a smaller minimum margin to separate the data. From those suitable exponents of the polynomial kernel function and gamma parameter of the radial basis function kernel were equal to 1.0, it also manifested that these data sets are not conglomerate data sets and do not require projecting the data into a higher-dimensional space for data separation. The best case gave an RRSE value of 2.37.

3.1.4. Random Tree (RT)

RT was conducted to determine the suitable hyper-parameters and some default parameters were selected as suggested by WEKA software, i.e., (1) unlimited maximum depth of the tree and (2) int(log 2(#predictors) + 1) function used to set the number of randomly selected attributes. However, five parameters, namely batchSize, numDecimalPlaces, minNum, numFolds, and minVarianceProp, were investigated. The batchSize refers to the preferred number of instances to be processed when batch predictions are made. The numberDecimalPlaces is the number of decimal places to be utilized in model output. The minNum means the minimum total weight of the instances in a leaf. The numFolds is configured to determine the quantity of data used. For backfitting, one fold is utilized, and the other is applied for building the tree. The minVarianceProp represents the smallest variance of all data present at a node in regression trees to be divided.

Findings revealed that batchSize of 100 (default value), numDecimalPlaces of 2 (default value), minNum of 1 (default value), numFolds of 0 (default value), and minVarianceProp of 0.001 (default value), were the suitable hyper-parameters for RT with testing data set. All best cases gave an RRSE value of 24.23. The minNum, numFolds, and minVarianceProp were sensitive parameters, while batchSize and numDecimalPLaces were not sensitive.

3.2. *Model's Performance Comparison*

After getting the most suitable hyper-parameters for each soft computing model, the authors proceeded to assess their performance in estimating the FAO Blaney-Criddle b factor. As explained earlier, to compare the model's performance, five statistical indices were used, i.e., the coefficient of determination (r^2), the mean absolute relative error (MARE), the maximum absolute relative error (MXARE), the standard deviation of the absolute relative error (DEV), and the number of samples with an error greater than 2% (NE > 2%). This evaluation was only conducted for the testing stage following the study by Trajkovic, Stankovic, and Todorovic [25].

Table 3 shows the comparative results of statistical indices getting from the present and previous studies' testing stages. By ranking the model with each statistical index and counting the frequency for five soft computing models, it was found that SVR-rbf outperformed the other methods, followed by RF, RT, M5, and SVR-poly. This is because SVR-rbf has the lowest values of MARE (%), MXARE (%), NE > 2%, and DEV (%) and the highest value of r^2. By doing the same thing, SVR-rbf, RF, and RT gave better results than the regression-based approach proposed by Frevert, Hill, and Braaten [26], Allen and Pruitt [27], and Ambas and Evanggelos [28], while M5 and SVR-poly gave the lower performance. However, SVR-rbf's performance as compared to the RBF network was comparable due to providing a bit lower performance.

Table 3. Statistical indices comparison in a testing stage for the present and previous studies.

Statistical Indices	Present Study					Previous Studies			
	RF	M5	SVR-poly	SVR-rbf	RT	Frevert et al. (1983)	Allen & Pruitt (1991)	Ambas & Evanggelos (2010)	RBF Network
MARE (%)	1.81	2.96	7.52	0.49	1.19	3.07	1.69	5.99	0.34
MXARE (%)	8.1	19.2	58.7	5.0	17.6	14.4	11.8	41.1	1.8
NE > 2%	80	116	171	7	25	126	64	141	0
DEV (%)	1.62	2.97	8.00	0.55	3.16	2.72	1.68	7.22	0.31
r^2	0.997	0.991	0.944	1.000	0.993	0.989	0.998	0.962	1.000

Figure 1 shows the performance of eight models in the testing stage. The left-hand side shows plotting the actual b-factor and estimated b-factor (y-axis) with the data set order (x-axis), and a scatter plot is displayed on the right-hand side. The data set order was received from the b factor tabular of FAO Blaney-Criddle [21] with 216 data sets. The authors could not plot the graph herein for the RBF network due to having no raw predicted data shown in the literature. Figure 2 presents a Taylor diagram to compare the performance of eight models, except for the RBF network, due to the same reason mentioned. Estimating b factor by Frevert, Hill, and Braaten [26], Allen and Pruitt [27], and Ambas and Evanggelos [28] were calculated by Equations (3)–(5), respectively. A Taylor diagram pointed out that SVR-rbf provided the results closest to FAO Blaney-Criddle b parameters obtained from the table as proposed by Doorenbos, Pruitt, and Agl [21], followed by Frevert, Hill, and Braaten [26], RF, RT, M5, Allen and Pruitt [27], Ambas and Evanggelos [28], and SVR-poly. Using the equation proposed by Ambas and Evanggelos [28] and SVR-poly, it gave overestimation and underestimation for the b-factor, respectively, since they have more and less standard deviation (see Figure 2). Consequently, it indicates that these two models gave more uncertainty in estimating the b-factor than other models. Figure 3 shows a set of linear equations obtained from M5. It includes six rule sets.

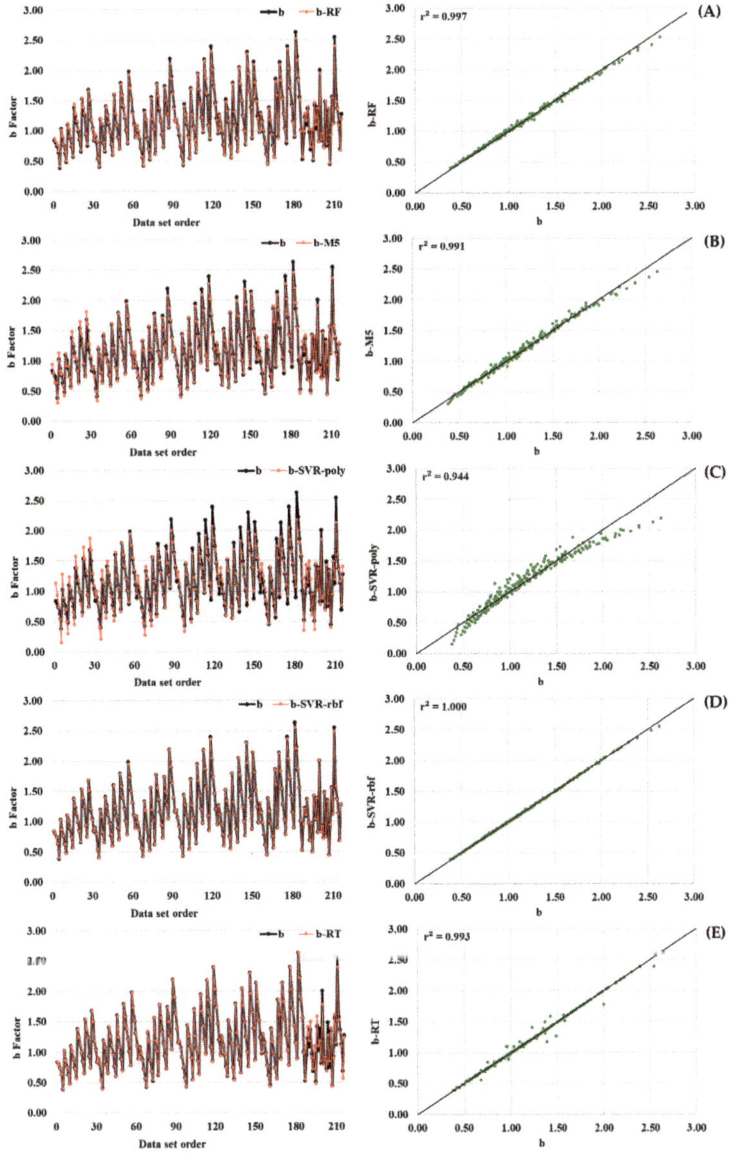

Figure 1. *Cont.*

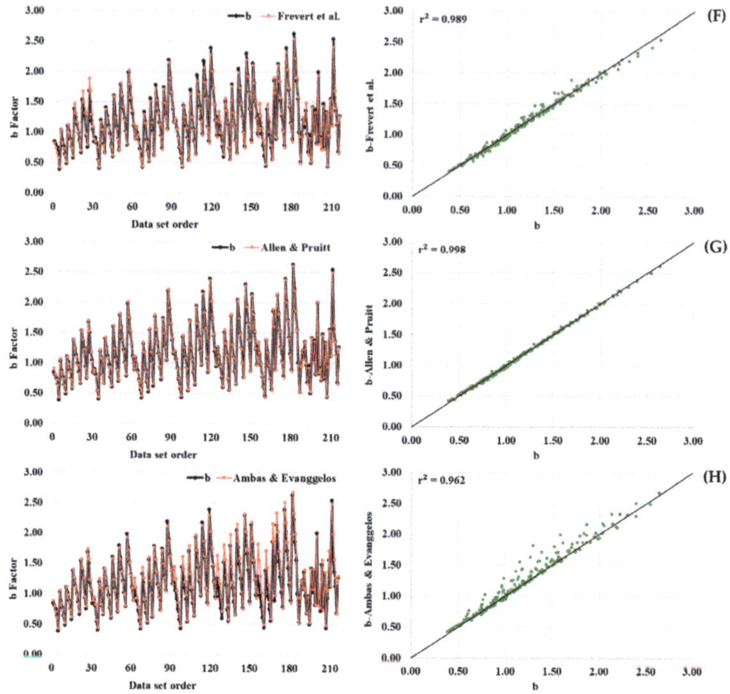

Figure 1. Performance of eight models in testing stage: (**A**) RF, (**B**) M5, (**C**) SVR-poly, (**D**) SVR-rbf, (**E**) RT, (**F**) Frevert et al., (**G**) Allen & Pruitt, and (**H**) Ambas & Evanggelos.

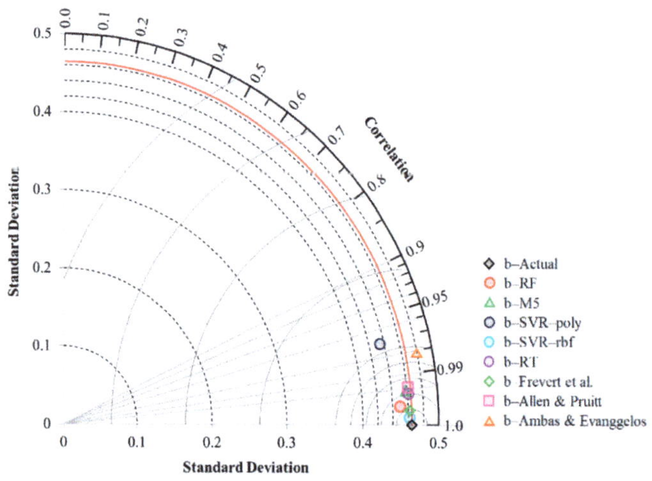

Figure 2. Taylor diagram for seven models in the testing stage.

Rule: 1 If $RH_{min} <= 50$ *and* $n/N > 0.3$ *and* $U_d > 5$ *Then*

$$b = 0.0457 \times U_d + 0.9257 * n/N - 0.0126 \times RH_{min} + 1.0738$$

Rule: 2 If $RH_{min} <= 70$ *and* $n/N <= 0.5$ *and* $RH_{min} > 30$ *Then*

$$b = 0.0282 \times U_d + 0.7857 \times n/N - 0.0081 \times RH_{min} + 1.0545$$

Rule: 3 If $RH_{min} <= 70$ *Then*

$$b = 0.0391 \times U_d + 0.8259 \times n/N - 0.0094 \times RH_{min} + 0.9974$$

Rule: 4 If $n/N > 0.5$ *and* $RH_{min} <= 90$ *Then*

$$b = 0.0169 \times U_d + 0.5448 \times n/N - 0.0069 \times RH_{min} + 1.086$$

Rule: 5 If $n/N > 0.3$ *Then*

$$b = 0.0132 \times U_d + 0.5033 \times n/N - 0.0091 \times RH_{min} + 1.2776$$

Rule: 6 If $RH_{min} <= 90$ Then

$$b = 0.0098 \times U_d + 0.6036 \times n/N - 0.0048 \times RH_{min} + 0.9056$$

Figure 3. The explicit equation rule set obtained from M5.

3.3. Models' Applicability for Estimating Monthly Reference evapotranspiration (ETo)

The monthly climatological variables at Nis, Yugoslavia, given by Trajkovic, Stankovic, and Todorovic [25], were used to demonstrate the model's applicability, as shown in Table 4. Reference evapotranspiration *(ETo)* in January and December is equal to zero. That is why any climatological variables of January and December do not appear in Table 4. Table 5 shows the results of applying the developed soft computing models and compares their performance to a table interpolation method [25] and the regression-based models for estimating the b-factor. Table 6 shows the difference between the b-factor obtained from various methods and a table interpolation method. The positive value means overestimation, and the negative value represents underestimation. The b-factor based on the regression-based models was mainly underestimated by 1.12–6.00% compared to those values obtained by the table interpolation method [25], except for estimating the b-factor in June using the equation developed by Frevert et al. [26]. It gave an overestimation of 1.11%. However, most of the soft computing models overestimated by 0.57–3.92% in the estimation of the b-factor. Some of them, for example, M5 provided underestimated by 0.3% in March, 1.57% in April, 3.02% in October, and 2.21% in November.

Table 4. The monthly climatological variables.

Months	Climatological Variables					
	T (°C)	RH_{min} (%)	U_2 (m/s)	n/N	P	A
Feb.	1.8	65	1.40	0.276	0.240	−1.407
Mar.	8.3	50	1.89	0.366	0.270	−1.561
Apr.	10.5	50	1.65	0.390	0.300	−1.585
May	12.7	61	1.60	0.311	0.330	−1.459
Jun.	20.6	45	0.77	0.636	0.347	−1.853
Jul.	21.4	55	1.17	0.535	0.337	−1.709
Aug.	19.6	56	1.00	0.510	0.310	−1.679
Sep.	17.9	43	1.25	0.626	0.280	−1.851
Oct.	11.6	55	1.44	0.323	0.250	−1.497
Nov.	7.8	63	1.34	0.238	0.220	−1.377

Table 5. The results of the soft computing models, a table interpolation method, and the regression-based models applied for the b-factor estimation.

Months	b									
	Frevert et al. (1983)	Allen & Pruitt (1991)	Ambas & Evanggelos (2010)	Table Interpolation [25]	RBF [25]	RF	M5	SVR-poly	SVR-rbf	RT
Feb.	0.779	0.788	0.803	0.821	0.823	0.846	0.844	0.823	0.823	0.823
Mar.	0.965	0.977	0.989	1.011	1.012	1.002	1.008	1.012	1.012	1.012
Apr.	0.975	0.981	0.993	1.016	1.017	1.000	1.000	1.017	1.022	1.020
May	0.836	0.846	0.860	0.886	0.884	0.888	0.909	0.883	0.884	0.884
Jun.	1.175	1.136	1.149	1.162	1.165	1.174	1.141	1.174	1.165	1.165
Jul.	1.030	1.015	1.025	1.047	1.053	1.047	1.088	1.052	1.054	1.053
Aug.	0.998	0.982	0.994	1.017	1.022	1.035	1.038	1.021	1.022	1.020
Sep.	1.203	1.179	1.185	1.199	1.202	1.202	1.197	1.202	1.202	1.202
Oct.	0.881	0.889	0.903	0.928	0.930	0.940	0.900	0.929	0.930	0.930
Nov.	0.764	0.775	0.791	0.813	0.811	0.831	0.795	0.812	0.818	0.811

Table 6. The difference between the b-factor obtained from various methods and a table interpolation method.

Months	Difference of b-Factor									
	Frevert et al. (1983)	Allen & Pruitt (1991)	Ambas & Evanggelos (2010)	Table Interpolation [25]	RBF [25]	RF	M5	SVR-poly	SVR-rbf	RT
Feb.	−0.042	−0.033	−0.018	0.000	0.002	0.025	0.023	0.002	0.002	0.002
Mar.	−0.046	−0.034	−0.022	0.000	0.001	−0.009	−0.003	0.001	0.001	0.001
Apr.	−0.041	−0.035	−0.023	0.000	0.001	−0.016	−0.016	0.001	0.006	0.004
May	−0.050	−0.040	−0.026	0.000	−0.002	0.002	0.023	−0.003	−0.002	−0.002
Jun.	0.013	−0.026	−0.013	0.000	0.003	0.012	−0.021	0.012	0.003	0.003
Jul.	−0.017	−0.032	−0.022	0.000	0.006	0.000	0.041	0.005	0.007	0.006
Aug.	−0.019	−0.035	−0.023	0.000	0.005	0.018	0.021	0.004	0.005	0.003
Sep.	0.004	−0.020	−0.014	0.000	0.003	0.003	−0.002	0.003	0.003	0.003
Oct.	−0.047	−0.039	−0.025	0.000	0.002	0.012	−0.028	0.001	0.002	0.002
Nov.	−0.049	−0.038	−0.022	0.000	−0.002	0.018	−0.018	−0.001	0.005	−0.002

Table 7 presents the estimated monthly reference evapotranspiration (*ETo*). Using a table interpolation method as a baseline, it is also pointed out that the soft computing models outperformed the regression-based models in *ETo* estimation due to giving a lower percentage of yearly difference. All three regression-based models gave underestimation by 3.2–5.1% in estimating *ETo*, while all six soft computing models provided some overestimation by 0.4–0.9%. Based on the data used in this study, the RBF network and RT models gave the highest performance in estimating *ETo* due to having the lowest percentage of yearly difference.

Table 7. Estimated Monthly Reference evapotranspiration (*ETo*).

Months	*ETo* (mm/month)									
	Frevert et al. (1983)	Allen & Pruitt (1991)	Ambas & Evanggelos (2010)	Table Interpolation [25]	RBF [25]	RF	M5	SVR-poly	SVR-rbf	RT
Feb.	7.5	8.1	8.9	10.0	10.2	11.5	11.4	10.2	10.2	10.2
Mar.	48.1	49.3	50.6	52.7	52.8	51.8	52.4	52.8	52.8	52.8
Apr.	66.1	66.9	68.3	71.0	71.1	69.1	69.1	71.1	71.7	71.4
May	74.3	75.7	77.7	81.4	81.1	81.7	84.7	81.0	81.1	81.1
Jun.	159.8	152.7	155.0	157.4	157.9	159.6	153.5	159.6	157.9	157.9
Jul.	140.4	137.6	139.6	143.6	144.8	143.6	151.3	144.6	145.0	144.8
Aug.	112.3	109.7	111.8	115.5	116.3	118.5	119.0	116.2	116.3	116.0
Sep.	109.8	106.5	107.3	109.3	109.7	109.7	109.0	109.7	109.7	109.7
Oct.	45.6	46.4	47.9	50.5	50.7	51.7	47.5	50.6	50.7	50.7
Nov.	17.8	18.6	19.9	21.6	21.4	23.0	20.2	21.5	22.0	21.4
Yearly	781.7	771.5	786.9	813.0	816.0	820.2	818.2	817.1	817.3	816.0
Yearly Difference (%)	−3.9	−5.1	−3.2	0.0	0.4	0.9	0.6	0.5	0.5	0.4

4. Conclusions

Accuracy of reference evapotranspiration *(ETo)* estimation is of importance for agricultural water management. In this study, five soft computing models, namely RF, M5, SVR-poly, SVR-rbf, and RT, were evaluated and compared their performance for estimating FAO Blaney-Criddle b-factor among themselves and the previous studies conducted by the RBF network and three regression equations (Richard G Allen et al., 1991; Frevert et al., 1983; Ambas & Evanggelos (2010). In addition, tuning hyper-parameters for each soft computing model were experimented with to receive its suitable architecture before applying them. The main findings results revealed the following.

(1) Among five soft computing models, it was found that SVR-rbf gave the highest performance in reference evapotranspiration *(ETo)* estimation, followed by M5, RF, SVR-poly, and RT, respectively.
(2) The new explicit equations for FAO Blaney-Criddle b-factor estimation were proposed herein using the M5 model. It is a rule set, including six linear equations.
(3) Compared to the RBF network [25], SVR-rbf provided a bit lower performance but outperformed three previous regression equations.
(4) The soft computing models outperformed the regression-based models in the b-factor estimation since they gave the lower values of MARE (%), MXARE (%), NE > 2%, and DEV (%) and the higher value of r^2.
(5) Models' Applicability for estimating monthly reference evapotranspiration *(ETo)* revealed that the soft computing models outperformed the regression-based models in *ETo* estimation owing to the lower percentage of yearly difference. All three regression-based models underestimated *ETo*, while all six soft computing models slightly overestimated it.
(6) This work's usefulness is to support a more accurate and convenient evaluation of reference crop evapotranspiration with a temperature-based approach. It leads to agricultural water demand estimation accuracy as necessary data for water resources planning and management.

Author Contributions: Conceptualization, S.T. and P.D.; methodology, S.T. and P.D.; software P.D., S.P., N.S. and I.E.; formal analysis, S.T., S.P., N.S. and I.E.; writing—original draft preparation, S.T., N.T.T.L. and P.D.; writing—review and editing, P.D., N.T.T.L. and Q.B.P.; project administration, P.D. and Q.B.P.; funding acquisition, N.T.T.L. All authors have read and agreed to the published version of the manuscript.

Funding: The authors are thankful to the Deanship of Scientific Research at Najran University for funding this work under the Research Collaboration Funding program grant code (NU/RC/SERC/11/3).

Institutional Review Board Statement: Not applicable.

Informed Consent Statement: Not applicable.

Data Availability Statement: Data can be requested by contacting the corresponding author, upon reasonable request.

Acknowledgments: The authors are thankful to the Deanship of Scientific Research at Najran University for funding this work under the Research Collaboration Funding program grant code (NU/RC/SERC/11/3).

Conflicts of Interest: The author declares that there is no conflict of interests regarding the publication of this manuscript. In addition, the ethical issues, including plagiarism, informed consent, misconduct, data fabrication and/or falsification, double publication and/or submission, and redundancies have been completely observed by the authors.

References

1. Xiong, Y.; Luo, Y.; Wang, Y.; Seydou, T.; Xu, J.; Jiao, X.; Fipps, G. Forecasting daily reference evapotranspiration using the Blaney-Criddle model and temperature forecasts. *Arch. Agron. Soil Sci.* **2015**, *62*, 790–805. [CrossRef]
2. Tabari, H.; Grismer, M.E.; Trajkovic, S. Comparative analysis of 31 reference evapotranspiration methods under humid conditions. *Irrig. Sci.* **2013**, *31*, 107–117. [CrossRef]
3. Mobilia, M.; Longobardi, A. Prediction of potential and actual evapotranspiration fluxes using six meteorological data-based approaches for a range of climate and land cover types. *ISPRS Int. J. Geo Inf.* **2021**, *10*, 192. [CrossRef]
4. Hafeez, M.; Chatha, Z.A.; Khan, A.A.; Bakhsh, A.; Basit, A.; Tahira, F. Estimating reference evapotranspiration by Hargreaves and Blaney-Criddle methods in humid subtropical conditions. *Curr. Res. Agric. Sci.* **2020**, *7*, 15–22. [CrossRef]
5. Fooladmand, H. Evaluation of Blaney-Criddle equation for estimating evapotranspiration in south of Iran. *Afr. J. Agric. Res.* **2011**, *6*, 3103–3109.
6. Jhajharia, D.; Ali, M.; Barma, D.; Durbude, D.G.; Kumar, R. Assessing Reference Evapotranspiration by Temperature-based Methods for Humid Regions of Assam. *J. Indian Water Resour. Soc.* **2009**, *29*, 1–8.
7. Mehdi, H.M.; Morteza, H. Calibration of Blaney-Criddle equation for estimating reference evapotranspiration in semiarid and arid regions. *Disaster Adv.* **2014**, *7*, 12–24.
8. Pandey, P.K.; Dabral, P.P.; Pandey, V. Evaluation of reference evapotranspiration methods for the northeastern region of India. *Int. Soil Water Conserv. Res.* **2016**, *4*, 52–63. [CrossRef]
9. Rahimikhoob, A.; Hosseinzadeh, M. Assessment of Blaney-Criddle Equation for Calculating Reference Evapotranspiration with NOAA/AVHRR Data. *Water Resour. Manag.* **2014**, *28*, 3365–3375. [CrossRef]
10. Zhang, L.; Cong, Z. Calculation of reference evapotranspiration based on FAO-Blaney-Criddle method in Hetao Irrigation district. *Trans. Chin. Soc. Agric. Eng.* **2016**, *32*, 95–101. [CrossRef]
11. Abd El-wahed, M.; Ali, T. Estimating reference evapotranspiration using modified Blaney-Criddle equation in arid region. *Bothalia J.* **2015**, *44*, 183–195.
12. El-Nashar, W.Y.; Hussien, E.A. Estimating the potential evapo-transpiration and crop coefficient from climatic data in Middle Delta of Egypt. *Alex. Eng. J.* **2013**, *52*, 35–42. [CrossRef]
13. Ramanathan, K.C.; Saravanan, S.; Adityakrishna, K.; Srinivas, T.; Selokar, A. Reference Evapotranspiration Assessment Techniques for Estimating Crop Water Requirement. *Int. J. Eng. Technol.* **2019**, *8*, 1094–1100. [CrossRef]
14. Tzimopoulos, C.; Mpallas, L.; Papaevangelou, G. Estimation of Evapotranspiration Using Fuzzy Systems and Comparison With the Blaney-Criddle Method. *J. Environ. Sci. Technol.* **2008**, *1*, 181–186. [CrossRef]
15. Schwalm, C.R.; Huntinzger, D.N.; Michalak, A.M.; Fisher, J.B.; Kimball, J.S.; Mueller, B.; Zhang, Y. Sensitivity of inferred climate model skill to evaluation decisions: A case study using CMIP5 evapotranspiration. *Environ. Res. Lett.* **2013**, *8*, 24028. [CrossRef]
16. Ferreira, L.B.; da Cunha, F.F.; de Oliveira, R.A.; Fernandes Filho, E.I. Estimation of reference evapotranspiration in Brazil with limited meteorological data using ANN and SVM–A new approach. *J. Hydrol.* **2019**, *572*, 556–570. [CrossRef]
17. Yu, H.; Wen, X.; Li, B.; Yang, Z.; Wu, M.; Ma, Y. Uncertainty analysis of artificial intelligence modeling daily reference evapotranspiration in the northwest end of China. *Comput. Electron. Agric.* **2020**, *176*, 105653. [CrossRef]
18. Shabani, S.; Samadianfard, S.; Sattari, M.T.; Mosavi, A.; Shamshirband, S.; Kmet, T.; Várkonyi-Kóczy, A.R. Modeling pan evaporation using Gaussian process regression K-nearest neighbors random forest and support vector machines; comparative analysis. *Atmosphere* **2020**, *11*, 66. [CrossRef]
19. Mohammadi, B.; Mehdizadeh, S. Modeling daily reference evapotranspiration via a novel approach based on support vector regression coupled with whale optimization algorithm. *Agric. Water Manag.* **2020**, *237*, 106145. [CrossRef]
20. Granata, F.; Di Nunno, F. Forecasting evapotranspiration in different climates using ensembles of recurrent neural networks. *Agric. Water Manag.* **2021**, *255*, 107040. [CrossRef]
21. Doorenbos, J.; Pruitt, W.O. *Guidelines for Predicting Crop Water Requirements*; XF2006236315; FAO: Rome, Italy, 1977; 24p.
22. Allen, R.; Pruitt, W. Rational Use of the FAO Blaney-Criddle Formula. *J. Irrig. Drain. Eng.* **1986**, *112*, 139–155. [CrossRef]
23. Allen, R.G.; Jensen, M.E.; Wright, J.L.; Burman, R.D. Operational Estimates of Reference Evapotranspiration. *Agron. J.* **1989**, *81*, 650–662. [CrossRef]
24. Jensen, M.E.; Burman, R.D.; Allen, R.G. *Evapotranspiration and Irrigation Water Requirements: A Manual*; American Society of Civil, Engineers Committee on Irrigation Water Requirements: New York, NY, USA, 1990.
25. Trajkovic, S.; Stankovic, M.; Todorovic, B. Estimation of FAO Blaney-Criddle b factor by RBF network. *J. Irrig. Drain. Eng.* **2000**, *126*, 268–270. [CrossRef]
26. Frevert, D.K.; Hill, R.W.; Braaten, B.C. Estimation of FAO evapotranspiration coefficients. *J. Irrig. Drain. Eng.* **1983**, *109*, 265–270. [CrossRef]
27. Allen, R.G.; Pruitt, W.O. FAO-24 reference evapotranspiration factors. *J. Irrig. Drain. Eng.* **1991**, *117*, 758–773. [CrossRef]
28. Ambas, V.; Evanggelos, B. The Estimation of b Factor of the FAO24 Blaney—Cridlle Method with the Use of Weighted Least Squares. 2010. Available online: https://ui.adsabs.harvard.edu/abs/2010EGUGA..1213424V/abstract (accessed on 31 July 2022).
29. Solomatine, D.; See, L.M.; Abrahart, R.J. Data-Driven Modelling: Concepts, Approaches and Experiences. In *Practical Hydroinformatics: Computational Intelligence and Technological Developments in Water Applications*; Springer: Berlin/Heidelberg, Germany, 2008; pp. 17–30.
30. Breiman, L. Random forests. *Mach. Learn.* **2001**, *45*, 5–32. [CrossRef]

31. Akay, H. Spatial modeling of snow avalanche susceptibility using hybrid and ensemble machine learning techniques. *Catena* **2021**, *206*, 105524. [CrossRef]
32. Breiman, L. *Random Forests—Random Features*; Statistics Department, University of California: Berkeley, CA, USA, 1999.
33. Quinlan, J.R. Learning with continuous classes. In Proceedings of the 5th Australian Joint Conference on Artificial Intelligence, Hobart, TAS, Australia, 16–18 November 1992; pp. 343–348.
34. Vapnik, V.N. *The Nature of Statistical Learning Theory*; Springer: Berlin/Heidelberg, Germany, 1995.
35. Xie, W.; Li, X.; Jian, W.; Yang, Y.; Liu, H.; Robledo, L.F.; Nie, W. A Novel Hybrid Method for Landslide Susceptibility Mapping-Based GeoDetector and Machine Learning Cluster: A Case of Xiaojin County, China. *ISPRS Int. J. Geo-Inf.* **2021**, *10*, 93. [CrossRef]
36. Xie, W.; Nie, W.; Saffari, P.; Robledo, L.F.; Descote, P.; Jian, W. Landslide hazard assessment based on Bayesian optimization–support vector machine in Nanping City, China. *Nat. Hazards* **2021**, *109*, 931–948. [CrossRef]

atmosphere

MDPI

Article

Applicability of a CEEMD–ARIMA Combined Model for Drought Forecasting: A Case Study in the Ningxia Hui Autonomous Region

Dehe Xu [1], Yan Ding [1,*], Hui Liu [1], Qi Zhang [2] and De Zhang [3]

1 College of Surveying and Geo-Informatics, North China University of Water Resources and Electric Power, Zhengzhou 450046, China; xudehe@ncwu.edu.cn (D.X.); liuhui_cehui@ncwu.edu.cn (H.L.)
2 Faculty of Geographical Science, Beijing Normal University, Beijing 100875, China; zqqz@mail.bnu.edu.cn
3 State Key Laboratory of Geo-Information Engineering, Xi'an 710054, China; zhangde01@infu.ac.cn
* Correspondence: x201911548804@stu.ncwu.edu.cn

Abstract: In the context of global warming, the increasing frequency of drought events has caused negative impacts on agricultural productivity and societal activities. However, the drought occurrences have not been well predicted by any single model, and precipitation may show nonstationary behavior. In this study, 60 years of monthly precipitation data from 1960 to 2019 for the Ningxia Hui Autonomous Region were analyzed. The standard precipitation index (SPI) was used to classify drought events. This study combined the strengths of autoregressive integrated moving average (ARIMA) and complementary ensemble empirical mode decomposition (CEEMD) to predict drought. First, based on the precipitation dataset, the SPI at timescales of 1, 3, 6, 9, 12, and 24 months was calculated. Then, each of these SPI time series was predicted using the ARIMA model and the CEEMD–ARIMA combined model. Finally, the models' performance was compared using statistical metrics, namely, root-mean-square error (RMSE), mean absolute error (MAE), Kling–Gupta efficiency (KGE), Willmott index (WI), and Nash–Sutcliffe efficiency (NSE). The results show that the following: (1) Compared with the ARIMA forecast value, the prediction results of the CEEMD–ARIMA model were in good agreement with the SPI values, indicating that the combined model outperformed the single model. (2) Two different models obtained the lowest accuracy for the SPI1 prediction and the highest accuracy for the SPI24 prediction. (3) The CEEMD–ARIMA model achieved higher prediction accuracy than the ARIMA model at each time scale. The most precise model during the test phase was the CEEMD–ARIMA model at SPI24 at Xiji Station, with error measures of MAE = 0.076, RMSE = 0.100, NSE = 0.994, KGE = 0.993, and WI = 0.999. Such findings will be essential for government to make decisions.

Keywords: CEEMD–ARIMA combined model; ARIMA model; drought prediction; SPI

check for updates

Citation: Xu, D.; Ding, Y.; Liu, H.; Zhang, Q.; Zhang, D. Applicability of a CEEMD–ARIMA Combined Model for Drought Forecasting: A Case Study in the Ningxia Hui Autonomous Region. *Atmosphere* **2022**, *13*, 1109. https://doi.org/10.3390/atmos13071109

Academic Editors: Jinping Liu, Quoc Bao Pham, Arfan Arshad and Masoud Jafari Shalamzari

Received: 8 June 2022
Accepted: 11 July 2022
Published: 14 July 2022

1. Introduction

Drought, which frequently occurs around the world, causes tremendous losses to agricultural production and economic operation [1]. For instance, the 2014 California drought was a record-breaking event that cost the United States USD 2.2 billion [2]. Drought is one of the costliest disasters that humankind faces all over the world [3]. With climate changes and temperature increases, droughts are becoming more and more frequent. Quantitative studies on drought will help countries to avoid damage caused by climate disasters in the future. Improved drought-monitoring ability has obvious significance in city development, which could help in the creation of drought management agencies.

Droughts are generally categorized into four types: meteorological droughts, agricultural droughts, hydrological droughts, and socioeconomic droughts [4]. Meteorological drought initiates when precipitation presents with volumes below normal in a particular place. Such anomalies go on to affect agriculture and hydrology (agricultural and hydrological droughts, respectively) [5]. This study focuses on meteorological drought. An effective method of meteorological drought detection is to use drought indices [6]. Drought indices can be used to quantitatively evaluate the influence range of a drought. In recent years, a variety of indices have been used to study drought, such as the Palmer drought severity index (PDSI) [7], standardized precipitation index (SPI) [8], reconnaissance drought index (RDI) [9], and temperature vegetation drought index (TVDI) [10]. Among them, the SPI is widely used in drought research at home and abroad due to its variable time scale, and only precipitation data are used for calculation [6,11,12]. Based on these two major advantages of the SPI—and particularly its ability to describe drought on multiple time scales—Oliveira-Júnior et al. [13] evaluated the drought severity in northern and northwestern Rio de Janeiro State (SRJ) regions from 1967 to 2013. Wu et al. [14] and Xu et al. [15] used the SPI to investigate the characteristics of meteorological drought in China. Łabędzki [16] adopted the SPI to estimate meteorological drought frequency in the central part of Poland from 1861 to 2005. Therefore, the SPI was used in this study due to its wide acceptability and advantages in drought research. Drought prediction provides an early warning to decision makers in disaster management. However, it is often notably challenging to obtain proper forecasts, because of the complexity of measuring the precision of a time series [17,18]. In recent years, numerous models have been used in drought forecasting, such as the autoregressive integrated moving average (ARIMA), artificial neural network (ANN), and support-vector regression (SVR) [18–22]. Among them, ARIMA is widely used in drought prediction because of its flexibility and richer information on time-related changes [23]. Nevertheless, the forecasting accuracy of a single model cannot meet the needs of drought prediction. Moreover, the precipitation data have nonlinear and nonstationary characteristics. Therefore, hybrid models are used in drought research to improve the prediction accuracy. Empirical mode decomposition (EMD) has apparent advantages in the processing of nonlinear and nonstationary signal time–frequency sequences. Özger et al. [24] used EMD for decomposing self-calibrated Palmer drought severity index (sc-PDSI) time series into their sub-bands on drought prediction, but this decomposition method has the problem of mode aliasing. As a further improvement of EMD, ensemble empirical mode decomposition (EEMD) effectively reduces the occurrence of mode aliasing. Libanda et al. [25] used it to understand consecutive dry days, but in this decomposition method, Gaussian white noise was added to the original signal, and its influence on the results could not be ignored. Therefore, since CEEMD has achieved great results in many fields based on advantages in processing signals [26–28], and it can effectively reduce the residual white noise and process nonlinear and nonstationary signals, a new drought prediction method was proposed, combining the ARIMA model and complementary ensemble empirical mode decomposition (CEEMD).

As mentioned above, with the nonlinear and nonstationary characteristics of precipitation data, it is important to accurately predict the occurrence of drought. The main objectives of the present study are as follows: to (1) quantify the precipitation situation by multi-timescale SPI, (2) develop the ARIMA model and then propose the hybrid model by combining the strengths of ARIMA with CEEMD, and (3) evaluate the efficiency of the ARIMA model and the CEEMD–ARIMA model according to the evaluated indices.

2. Study Area

The Ningxia Hui Autonomous Region extends from 104°17′ E to 107°39′ E and from 35°14′ N to 39°23′ N, with altitudes mostly above 1000 m (Figure 1). Helan Mountain is located in the northwest of the province. As a natural barrier, this mountain reduces intrusion from Tengger Desert quicksand and cold northwest winds into the Ningxia Hui Autonomous Region. The southern part of the province is the Liupan mountainous area. As the wettest region in Ningxia Hui Autonomous Region, it has a humid climate and dense jungles. The area from the Helan Mountains to the Weining Plain has an arid climate, and the area from the Weining Plain to the Liupan mountainous area has a semiarid climate.

The Liupan mountainous area consists of jungle, wet, and pleasantly cool areas. The climate characteristics of Ningxia vary widely from north to south on spatial and temporal scales. The annual average temperature decreases from the north to the south in Ningxia, while the annual precipitation shows the opposite tendency. The mean annual temperature is between 5.3 and 9.9 °C, with the southern part below 7 °C, the central part above 7 °C, and the northern part above 8 °C. The mean annual precipitation is between 150 mm and 600 mm, and the average annual water surface evaporation in Ningxia is 1250 mm.

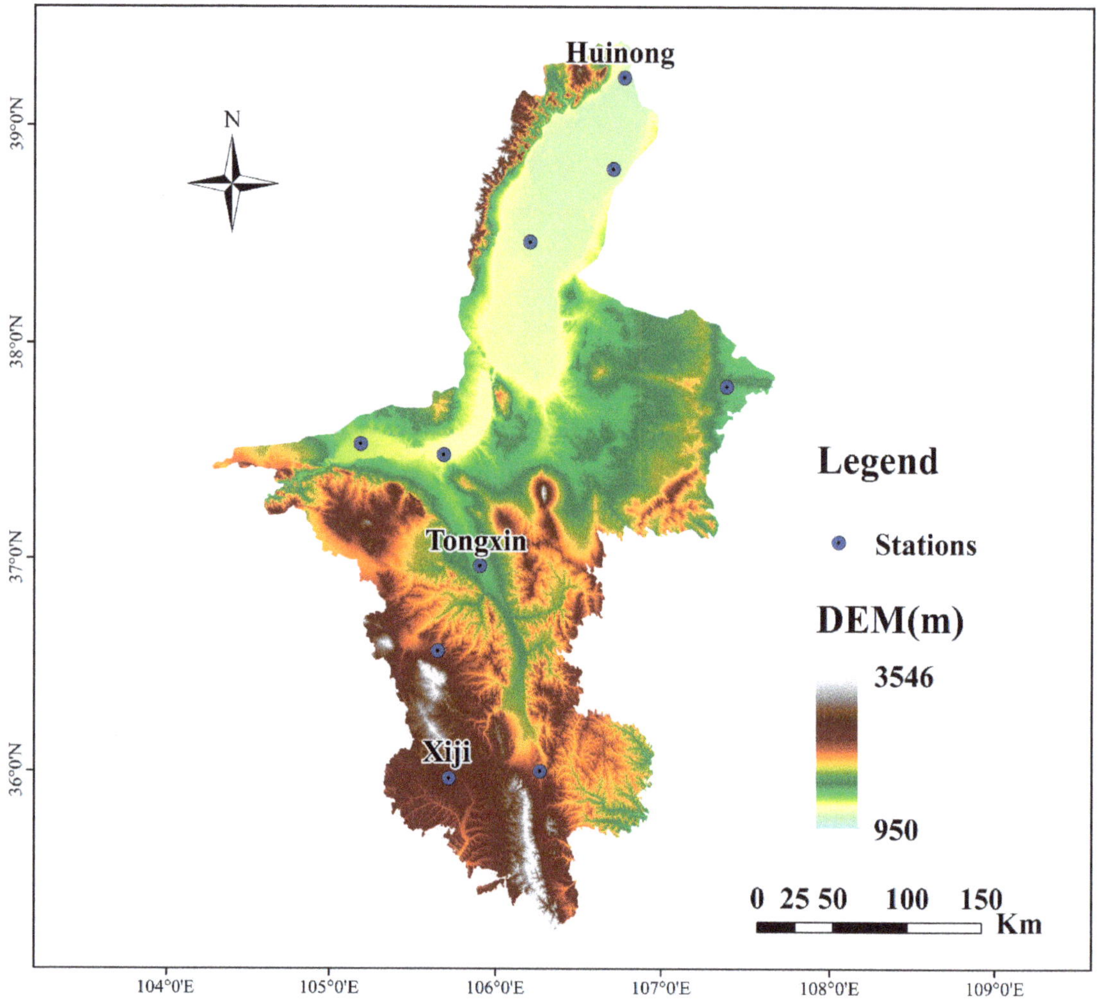

Figure 1. The distribution of meteorological stations in the Ningxia Hui Autonomous Region.

3. Materials and Methods

3.1. Data Sources

Monthly precipitation datasets from 10 meteorological stations were used in this study. The datasets from January 1960 to December 2019 were obtained from the Ningxia Hui Autonomous Region weather station in the National Meteorological Data Center (http://data.cma.cn/ accessed on 13 March 2020). The elevation data were obtained from the Geospatial Data Cloud (http://www.gscloud.cn/search accessed on 28 June 2021). Due to the terrain inclines from the southwest to the northeast in the Ningxia Hui Autonomous Region, there is a large spatial difference in climate characteristics from the south to the north. Since this paper focuses on the applicability of the combined model in drought prediction, three meteorological stations were selected as representative stations of the southern, northern, and central Ningxia Hui Autonomous Region. Table 1 shows the information of three representative stations. The observed and predicted values of 10 sites are visualized by the empirical Bayesian kriging interpolation method of ArcGIS.

Table 1. Information about meteorological stations in the sample.

Station Number	Station Name	Longitude/°E	Latitude/°N	Altitude/m
53519	Huinong	106.46	39.13	1092.5
53810	Tongxin	105.54	36.58	1339.3
53903	Xiji	105.43	35.58	1916.5

3.2. Research Methods

3.2.1. SPI

The SPI was developed by McKee et al. [29] to quantify precipitation on different time scales. It can be calculated based solely on precipitation. The time scale of the SPI is variable. Short-timescale SPI could reflect the water supply of crops. On a 1-month time scale, the SPI can be used to reflect short-term precipitation conditions. The SPI on a 3-month time scale can be used to analyze seasonal changes in precipitation. The calculated SPI data for February, May, August, and November reflect winter, spring, summer, and autumn drought conditions, respectively. The 6-month time scale SPI can be used to reflect mid-term precipitation conditions. On a 9-month time scale, the SPI can be used to characterize groundwater level changes over a longer period of time. The 12-month time scale SPI and 24-month time scale SPI are good indicators of long-term drought conditions [30]. The SPIs of the 1-, 3-, 6-, 9-, 12-, and 24-month time scales are denoted as SPI1, SPI3, SPI6, SPI9, SPI12, and SPI24, respectively. The computation procedure of SPI followed the method of Lloyd-Hughes and Saunders [31]:

$$\mathrm{SPI} = B\left(t - \frac{c_0 + c_1 t + c_2 t^2}{1 + d_1 t + d_2 t^2 + d_3 t^3}\right) \quad (1)$$

where B is the positive and negative coefficient of probability density; for $B = -1$, $t = \sqrt{ln\frac{1}{G(x)^2}}$, and for $B = 1$, $t = \sqrt{ln\frac{1}{(1-G(x))^2}}$, where $G(x)$ is a cumulative probability [32]. The constants c_0 = 2.515517, c_1 = 0.802853, c_2 = 0.010328, d_1 = 1.432788, d_2 = 0.189269, and d_3 = 0.001308.

The classification of drought based on SPI values is shown in Table 2 [31].

Table 2. Drought classification based on SPI.

SPI Value	Category
SPI > −0.5	No drought
−1.0 < SPI ≤ −0.5	Mild drought
−1.5 < SPI ≤ −1.0	Moderate drought
−2.0 < SPI ≤ −1.5	Severe drought
SPI ≤ −2.0	Extreme drought

3.2.2. ARIMA Model

The ARIMA model developed by Box and Jenkins [33] includes three basic types: autoregressive (AR) models, moving average (MA) models, and the combined AR and MA (ARMA) models. AR, MA, and ARMA can be used when the data are stationary [19]. Nonstationary and nonwhite noise sequences can be predicted by the ARIMA model. First, a stationary time series is obtained from a nonstationary series through d-order difference. Then, the ARMA model is used for prediction. The formula of the ARIMA (p, d, q) model is written as follows:

$$X_t = \omega_1 X_{t-1} + \omega_2 X_{t-2} + \cdots + \omega_p X_{t-p} + h_t - \theta_1 h_{t-1} - \theta_2 h_{t-2} - \cdots - \theta_q h_{t-q} \quad (2)$$

where X_t is a time-series value, and $\omega_i(i = 1, 2, \cdots, p)$ and $\theta_j(j = 1, 2, \cdots, q)$ are the autoregressive coefficient and moving average coefficient, respectively. h_t is a white noise sequence, $h_t \sim N(0, \sigma^2)$.

The modeling process of the ARIMA model is as follows:

First, each time series should go through stationary testing. In this paper, the augmented Dickey–Fuller test (ADF) is used for judgment. If it is a nonstationary time series, the d-order difference of the series is determined. Then, the value range of the model order should be determined. The value range of P and Q is determined according to the autocorrelation function (ACF) and partial autocorrelation function (PACF) of the data. After that, the Akaike information criterion (AIC) and Bayesian information criterion (BIC) are used to determine the order of the model. The formulae of the AIC and BIC are as follows:

$$AIC(p, q) = N \ln \sigma^2(p, q) + 2(p + q + 1) \quad (3)$$

$$BIC(p, q) = N \ln \sigma^2(p, q) + (p + q + 1) \ln N \quad (4)$$

where N is the number of parameters. In different combinations of p and q, the combination corresponding to the minimum value of AIC and BIC is selected to obtain the optimal ARIMA model. The datasets were divided by grid search and cross-validation; 80% of the data were selected as the training set for model prediction, and 20% of the data were selected as the test set.

3.2.3. CEEMD

As proposed by Yeh et al. [34], CEEMD has apparent advantages in the processing of nonlinear and nonstationary signal time–frequency sequences. It can adaptively decompose the original sequence into several intrinsic mode function (IMF) components, with different scales that are mutually independent and a residual trend quantity. The steps are as follows:

A group of white noise includes positive noise and negative noise. $P(t)$ is the original sequence, with n groups of auxiliary white noise added to the positive noise sequence Y_1 and negative noise sequence Y_2. Now, the total number of sequences obtained is $2n$,

$$\begin{bmatrix} Y_1 \\ Y_2 \end{bmatrix} = \begin{bmatrix} 1 & 1 \\ 1 & -1 \end{bmatrix} \begin{bmatrix} P \\ N \end{bmatrix} \quad (5)$$

where N is an auxiliary sequence. The obtained sequences are decomposed by EMD to obtain m IMF components, and each group of components is denoted as $C_{ij}^{+}(t)$ and $C_{ij}^{-}(t)$ $(i = 1, \ldots, n, j = 1, \ldots, m)$. $C_{ij}^{+}(t)$ and $C_{ij}^{-}(t)$ of the IMF components in each group are averaged to obtain the IMF_j value.

$$\mathrm{IMF}_j = \frac{1}{2n} \sum_{i=1}^{n} \left(C_{ij}^{+}(t) + C_{ij}^{-}(t) \right) \quad (6)$$

Take the decomposed IMF components as the final result. The original sequence is decomposed into:

$$P_{(t)} = \sum_{j=1}^{m} \mathrm{IMF}_{j(t)} + R_{(t)} \quad (7)$$

where $R_{(t)}$ is a residual trend quantity.

3.2.4. CEEMD–ARIMA Combined Model

The original sequences with strong volatility are decomposed by CEEMD to obtain a set of IMF components with low volatility, improving the predictability of the sequences predicted by the ARIMA model. The CEEMD and ARIMA models are combined using Python to form the CEEMD–ARIMA model. The steps are as follows:

Step 1: The SPI sequence is imported into the CEEMD for decomposition to obtain IMF1, IMF2 ... , IMFn, and residual trends from high frequency to low frequency.

Step 2: The sequences decomposed by CEEMD are imported into the ARIMA model. The stationarity of each component is tested through the ARIMA model. After the order and prediction are determined, the prediction result is obtained. The predicted results are denoted as $P1$, $P2$, and $Pn + 1$.

Step 3: Finally, sum $P1$, $P2$, and ... , $Pn + 1$ as the combined model's predicted results.

$$P = \sum_{i=1}^{n+1} P_i \tag{8}$$

The modeling process of the CEEMD–ARIMA model is as shown in Figure 2.

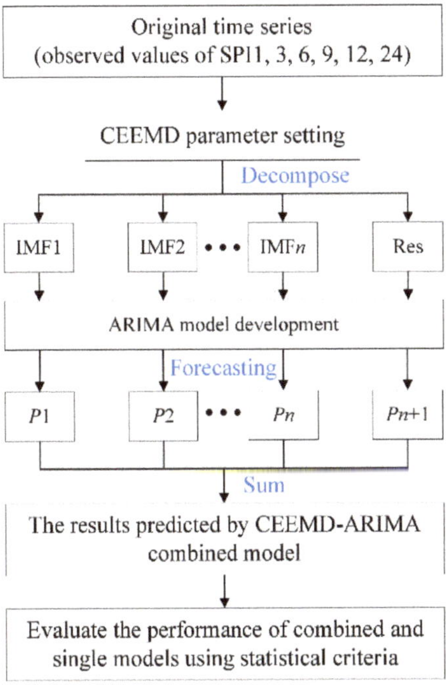

Figure 2. Workflow of the CEEMD–ARIMA combined model.

3.2.5. Evaluation Index

To evaluate the performance of the ARIMA model and the CEEMD–ARIMA model, statistical criteria such as root-mean-square error (RMSE), mean absolute error (MAE), Kling–Gupta efficiency (KGE), Willmott index (WI), and Nash–Sutcliffe efficiency (NSE) were used. A model with the lowest RMSE and MAE and the highest KGE, WI, and NSE was selected and proposed as an appropriate model. The formulae are as follows [35–38]:

$$RMSE = \sqrt{\frac{1}{N}\sum_{i=1}^{N}(y_i - \hat{y}_i)^2} \tag{9}$$

$$MAE = \frac{1}{N}\sum_{i=1}^{N}|y_i - \hat{y}_i| \tag{10}$$

$$KGE = 1 - \sqrt{(CC-1)^2 + (\alpha-1)^2 + (\beta-1)^2} \tag{11}$$

$$WI = \left|1 - \left[\frac{\sum_{i=1}^{N}(y_i - \hat{y}_i)^2}{\sum_{i=1}^{N}(|y_i - \overline{y}| + |\hat{y}_i - \overline{y}|)^2}\right]\right| \tag{12}$$

$$NSE = 1 - \frac{\sum_{i=1}^{N}(y_i - \hat{y}_i)^2}{\sum_{i=1}^{N}(y_i - \overline{y})^2} \tag{13}$$

where y_i is the observed value, $\overline{y}$ is the average value of y_i, $\hat{y}_i$ is the forecasted value, and N is the total data size of y_i. CC, α, and β in the KGE index illustrate the correlation coefficient, the standard deviation ratio, and the average ratio of the observed and forecasted data, respectively.

4. Results

The calculation of the SPI and the fitting of the ARIMA model were both accomplished on the Python 3.7 platform.

4.1. SPI Values at Different Time Scales

The research applicability of the CEEMD–ARIMA model in drought is mainly through the prediction of the SPI on the time scales of 1, 3, 6, 9, 12, and 24 months. Monthly precipitation data from 10 meteorological stations in the Ningxia Hui Autonomous Region during 1960–2019 were used to calculate the SPI. The calculated SPI results characterize drought conditions, as shown in Table 2. Huinong, Tongxin, and Xiji were taken as examples to demonstrate multiple-timescale SPIs, and the calculated SPIs of the sample stations are shown in Figure 3. Through the Mann–Kendall trend test, the SPI12 and SPI24 sequences of Tongxin Station and the SPI9, SPI12, and SPI24 sequences of Xiji Station have a decreasing trend. The other sequences have no trend.

4.2. The ARIMA Modeling and Prediction

The first 80% of the calculated SPI data were used as observation training data, and the last 20% of the data were used as prediction comparison data. That is, the data from 1960 to 2007 were used as the training set, and the data from 2008 to 2019 were used as the test set. The stability of the 80% training data should be judged before prediction. If the data are a stable series, then $d = 0$ in the ARIMA model can be used for prediction; if not, then $d \neq 0$. Through the ADF test, the p-values of all SPI sequences of the sample sites were less than 0.05 (Table 3). Therefore, all SPI sequences were stationary time series.

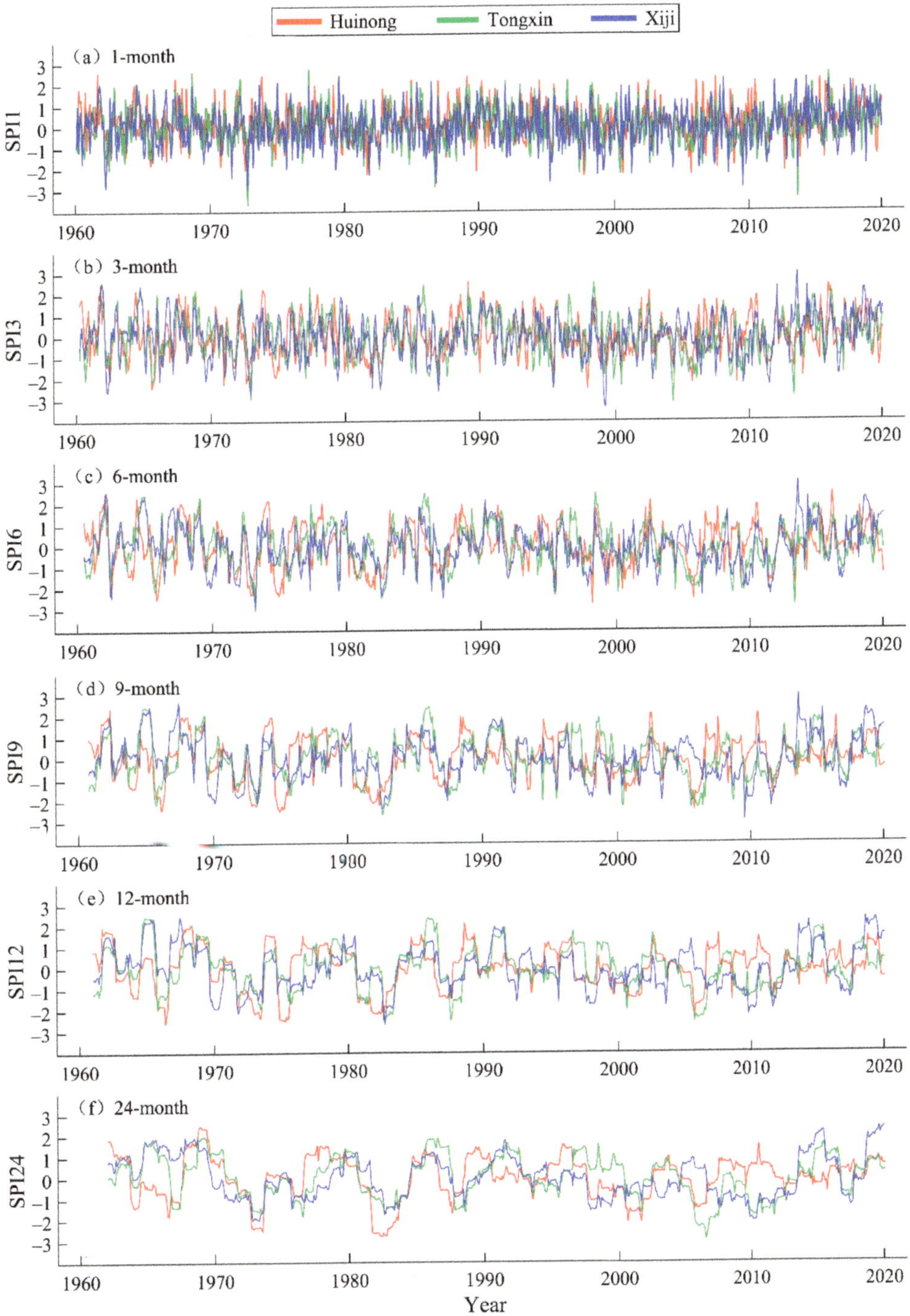

Figure 3. Observed SPI values at different time scales of the example stations.

Table 3. Unit root test of SPI original sequence.

Example Stations	SPI Series	ADF	Critical Value			p-Value
			1%	5%	10%	
Huinong	SPI1	−20.0550	−3.4418	−2.8666	−2.5694	0.0000
	SPI3	−9.6732	−3.4419	−2.8666	−2.5695	1.2610×10^{-16}
	SPI6	−6.9028	−3.4420	−2.8667	−2.5695	1.2693×10^{-9}
	SPI9	−5.3241	−3.4423	−2.8668	−2.5696	4.8806×10^{-6}
	SPI12	−4.7455	−3.4423	−2.8668	−2.5696	6.9075×10^{-5}
	SPI24	−4.1882	−3.4423	−2.8668	−2.5696	0.0007
Tongxin	SPI1	−21.6155	−3.4418	−2.8666	−2.5694	0.0000
	SPI3	−9.6077	−3.4419	−2.8666	−2.5695	1.8469×10^{-16}
	SPI6	−6.7922	−3.4420	−2.8667	−2.5695	2.3486×10^{-9}
	SPI9	−4.9104	−3.4423	−2.8668	−2.5696	3.3288×10^{-5}
	SPI12	−4.4071	−3.4423	−2.8668	−2.5696	0.0003
	SPI24	−3.7087	−3.4423	−2.8668	−2.5696	0.0040
Xiji	SPI1	−22.0945	−3.4418	−2.8666	−2.5694	0.0000
	SPI3	−10.7739	−3.4419	−2.8666	−2.5695	2.3469×10^{-19}
	SPI6	−7.3216	−3.4420	−2.8667	−2.5695	1.1900×10^{-10}
	SPI9	−4.1113	−3.4423	−2.8668	−2.5696	0.0009
	SPI12	−3.4578	−3.4422	−2.8668	−2.5696	0.0091
	SPI24	−3.3257	−3.4423	−2.8668	−2.5696	0.0138

Because the SPI series of the three sites were stationary time series, the ARMA model was selected for the prediction. ACF and PACF were used to rank the ARMA model, and p- and q-values corresponding to the minimum AIC and BIC values were selected. The model ranking results of each sequence are shown in Table 4. The optimal model of the SPI series at various time scales was applied to predict the SPI series of the three stations from 2008 to 2019.

Table 4. Model order based on SPI values of six time scales.

Example Stations	SPI Series	Model Select	AIC	BIC	Model Order Estimation
Huinong	SPI1	ARMA	1826.071	1839.804	ARMA (1, 0)
	SPI3	ARMA	1631.778	1650.079	ARMA (0, 2)
	SPI6	ARMA	1398.692	1412.404	ARMA (1, 0)
	SPI9	ARMA	1026.739	1045.006	ARMA (1, 0)
	SPI12	ARMA	538.884	579.946	ARMA (5, 2)
	SPI24	ARMA	64.999	87.725	ARMA (3, 0)
Tongxin	SPI1	ARMA	1937.225	1950.959	ARMA (1, 0)
	SPI3	ARMA	1593.929	1612.230	ARMA (0, 2)
	SPI6	ARMA	1302.638	1343.776	ARMA (5, 2)
	SPI9	ARMA	957.282	970.982	ARMA (1, 0)
	SPI12	ARMA	536.069	586.256	ARMA (7, 2)
	SPI24	ARMA	43.954	62.136	ARMA (2, 0)
Xiji	SPI1	ARMA	2012.614	2026.347	ARMA (0, 1)
	SPI3	ARMA	1628.778	1647.078	ARMA (0, 2)
	SPI6	ARMA	1453.959	1472.242	ARMA (2, 0)
	SPI9	ARMA	1061.371	1075.071	ARMA (1, 0)
	SPI12	ARMA	575.482	616.544	ARMA (5, 2)
	SPI24	ARMA	31.131	62.949	ARMA (3, 2)

4.3. The CEEMD–ARIMA Combined Model

Multiscale SPI was decomposed by CEEMD. After several parameters were modified and compared, when the Gaussian white noise logarithm was 100, the total number of modes (not including the trend) was 7, and the standard deviation of the original time series was multiplied by 0.2. CEEMD had the best decomposition effect. Seven IMF components and one trend item were obtained by the CEEMD decomposition of the SPI3 sequence of Xiji Station (Figure 4). The trend term represents the general trend of a sequence over time. As shown in Figure 4, the fluctuation range of the IMF component obtained by decomposition is smaller than that of the original sequence, and with the gradual progress of decomposition, the fluctuation of the component tends to be smooth. Therefore, the predictability of the subsequence obtained after decomposition is higher than that of the original sequence.

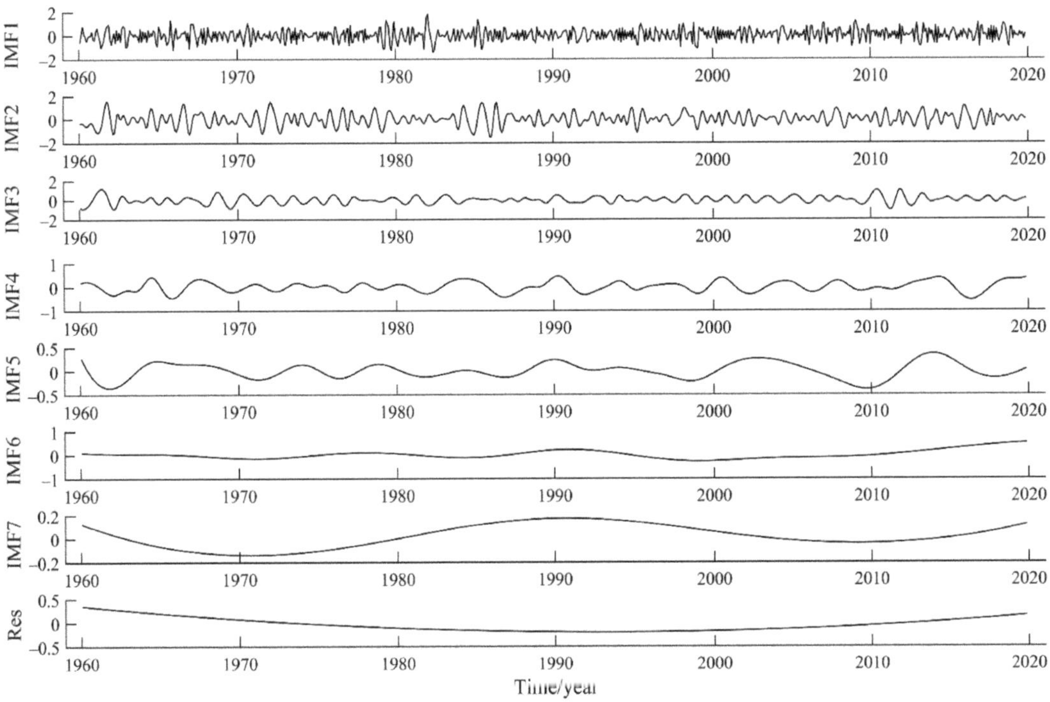

Figure 4. The CEEMD decomposition results of the SPI3 sequence at Xiji Station.

A stationary test was carried out for the components decomposed by CEEMD. IMF1~IMF6 were stationary time series; therefore, the ARMA model was selected for prediction. IMF7 and RES were nonstationary time series, and the ARIMA model was used for prediction after the stationary series was obtained by data difference. AIC and BIC were used to determine the order of the model, and the ranking results are shown in Table 5.

In this paper, the data from 1960 to 2007 were used as observation training data. Therefore, only this part of the SPI sequence was decomposed by CEEMD and then predicted by the ARIMA model, and the sum of the predicted results of each component was used as the final prediction result of the SPI sequence. The prediction comparison plot includes the actual calculated SPI values and the predicted SPI values of the CEEMD–ARIMA model and the ARIMA model, as shown in Figures 5–7 for Huinong, Tongxin, and Xiji sample stations, respectively.

Table 5. Model order based on IMFs of the SPI3 sequence at Xiji Station.

SPI Series	Decompose Results	Model Select	Model Order Estimation
SPI3	IMF1	ARMA	ARMA (1, 1)
	IMF2	ARMA	ARMA (2, 5)
	IMF3	ARMA	ARMA (4, 2)
	IMF4	ARMA	ARMA (4, 5)
	IMF5	ARMA	ARMA (4, 6)
	IMF6	ARMA	ARMA (2, 1)
	IMF7	ARIMA	ARIMA (4, 1, 1)
	Res	ARIMA	ARIMA (3, 1, 1)

Figures 5a, 6a and 7a show that there is a big difference between the predicted value of the ARIMA model and the actual value of SPI at the 1-month time scale, and the predicted result of the CEEMD–ARIMA combination model at the SPI1 time scale is better than that of the ARIMA model. The poor stationarity of the 1-month time series resulted in bad prediction results of the ARIMA model. As the time scale increases and the data stationarity improves, the ARIMA prediction results become closer and closer to the actual situation. The prediction of the ARIMA model in the combined model relies on the stable basis provided by the CEEMD. At the time scales of 1 month and 3 months, the predicted value of the CEEMD–ARIMA model is different from the actual value. At the time scales of 6, 9, 12, and 24 months, the predicted values of the combined model are less different. According to the comparison diagram of SPI3 and SPI6 in Figures 5–7, the prediction results of the combined model in the extreme drought year (SPI ≤ -2) are closer to the actual situation, indicating that the combined model is more suitable for the study of extreme drought than the single model. With increasing time scale, the difference between the predicted value of the model and the actual calculated value tends to decrease. The two models′ predicted values of SPI12 and SPI24 were very close to the actual values, and the combined model was closer than the single model.

In 2009, due to high temperature and drier weather, a severe large-scale drought occurred in Ningxia, which spread from the central region. From the predicted values of the combined model shown in Figures 5–7, drought first appeared near Tongxin Station, and then around Huinong Station and Xiji Station, which is consistent with the record in the China Meteorological Network. The combined model predicts that there would be no drought or mild drought at Huinong Station and Tongxin Station in 2016, but drought at Xiji Station, which is also consistent with the record. In July 2017, large-scale precipitation occurred in Ningxia, with heavy rain in some areas. In 2019, the precipitation in the whole region was relatively high, with annual precipitation of 341.7 mm, and the southern mountainous region was rainy for eight consecutive months. The prediction results of the combined model for drought in these periods were nearly consistent with the actual situation. This indicates that the prediction of the combined model has a high consistency with the actual situation, illustrating that the combined model is suitable for studying drought prediction.

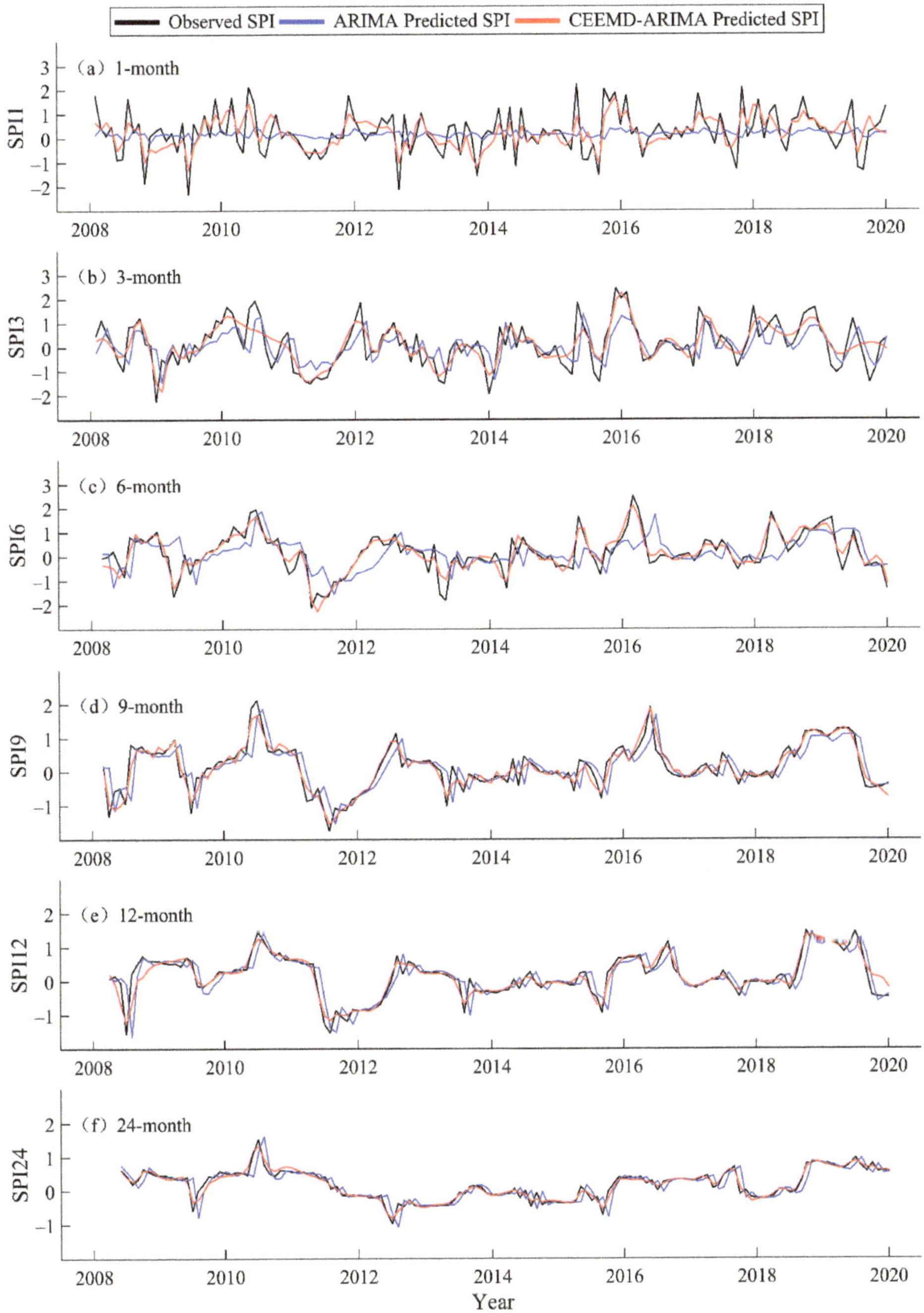

Figure 5. Comparison of the predicted and observed values of the SPIs on the 1-month (**a**), 3-month (**b**), 6-month (**c**), 9-month (**d**), 12-month (**e**), and 24-month (**f**) time scales of the ARIMA model and the CEEMD–ARIMA model at Huinong Station (2008–2019).

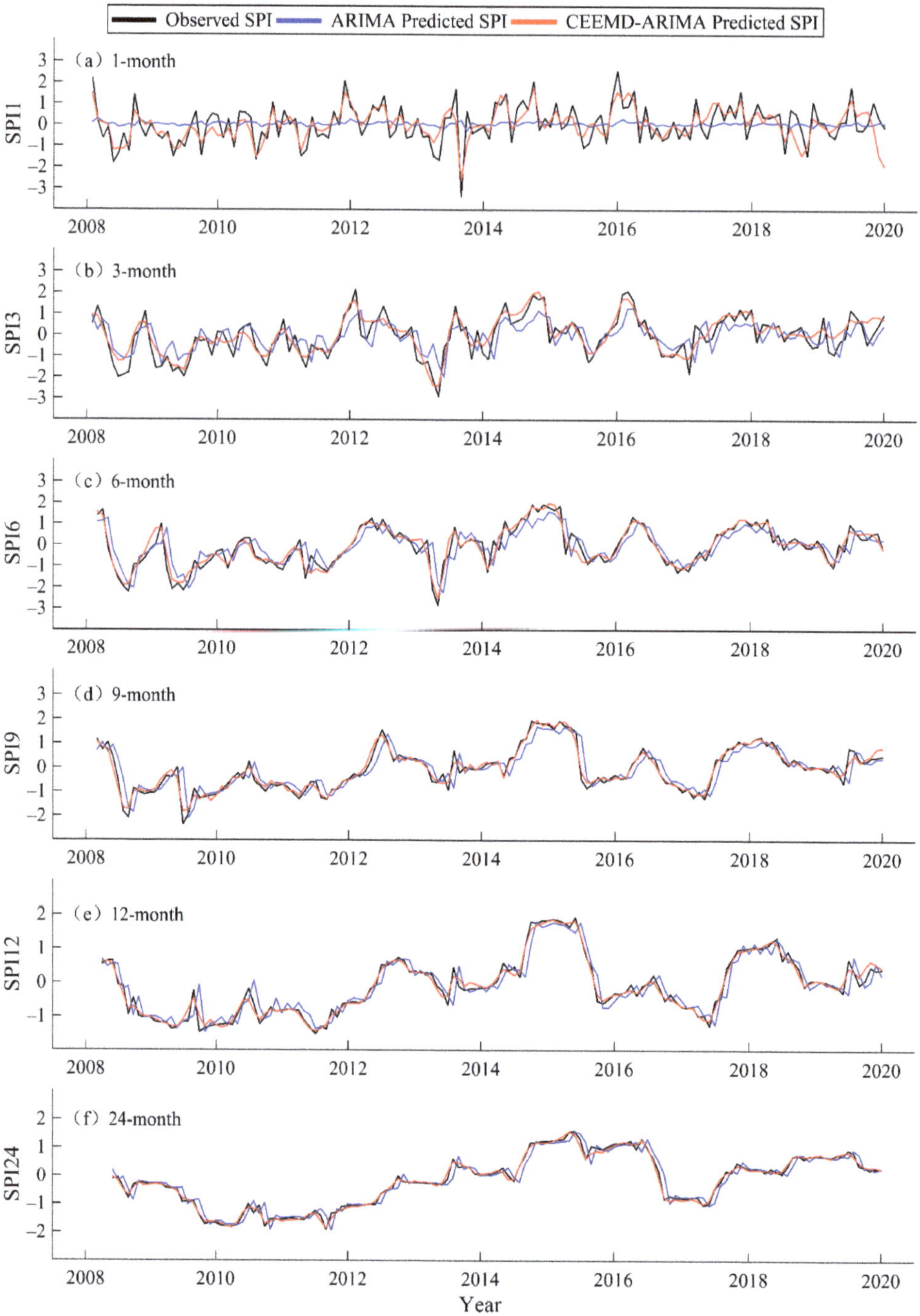

Figure 6. Comparison of the predicted and observed values of the SPIs on the 1-month (**a**), 3-month (**b**), 6-month (**c**), 9-month (**d**), 12-month (**e**), and 24-month (**f**) time scales of the ARIMA model and the CEEMD–ARIMA model at Tongxin Station (2008–2019).

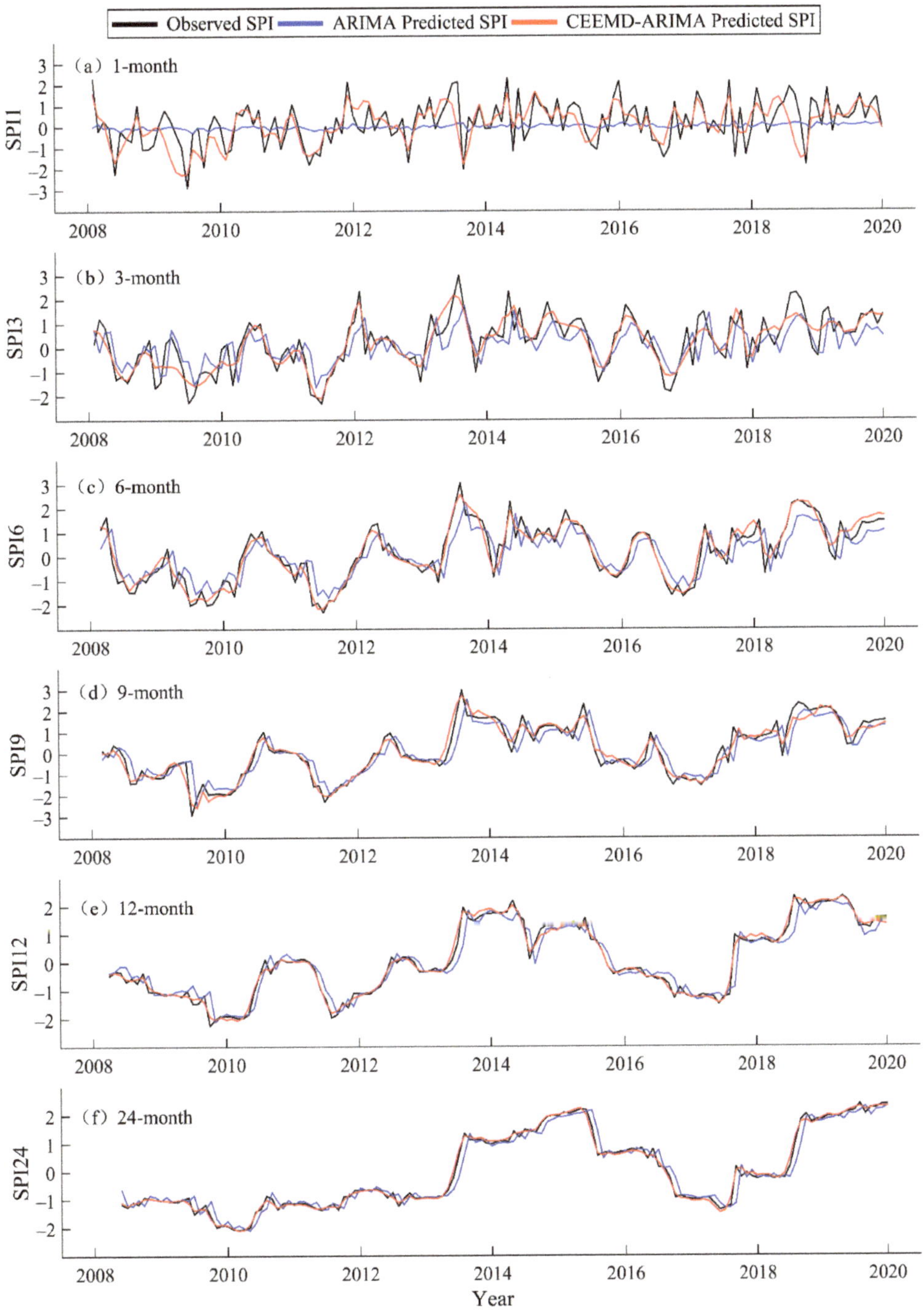

Figure 7. Comparison of the predicted and observed values of the SPIs on the 1-month (**a**), 3-month (**b**), 6-month (**c**), 9-month (**d**), 12-month (**e**), and 24-month (**f**) time scales of the ARIMA model and the CEEMD–ARIMA model at Xiji Station (2008–2019).

The prediction results of the two models were evaluated using MAE, RMSE, NSE, KGE, and WI. With increasing time scale, the MAE and RMSE values of the two models decreased, while the NSE, KGE, and WI generally increased (Table 6), indicating that the prediction accuracy of the two models gradually improved with increasing time scale, and reached the maximum at the 24-month time scale. For example, the ARIMA model implemented at SPI1 had a WI = 0.150, and at SPI24 had a WI = 0.911, at Huinong Station. The evaluation index values of the two models were compared. At all time scales, the MAE and RMSE values of the CEEMD–ARIMA model were lower than those of the ARIMA model, and the NSE, KGE, and WI values were higher than those of the ARIMA model, illustrating the higher prediction accuracy of the combined model, which is more suitable for the prediction of multiscale SPI. At the 1-month time scale, the prediction accuracy of the combined model was much higher than that of the single model, and the prediction accuracy of SPI9, SPI12, and SPI24 was slightly higher than that of the single model. The most precise model during the test phase was the CEEMD–ARIMA model at SPI24 at Xiji Station with MAE = 0.076 RMSE = 0.100, NSE = 0.994, KGE = 0.993, and WI = 0.999. With increasing time scale, the fluctuation of the SPI series tended to be flat, and the fitting degree of the ARIMA model to the calculated SPI value was gradually improved.

Table 6. The statistical criteria of the ARIMA and CEEMD–ARIMA models.

Example Stations	SPI Series	Model	Training					Testing				
			MAE	RMSE	NSE	KGE	WI	MAE	RMSE	NSE	KGE	WI
Huinong	SPI1	ARIMA	0.634	0.850	−31.759	−3.881	0.204	0.667	0.892	−48.453	−4.992	0.150
		CEEMD–ARIMA	0.459	0.580	−0.020	0.440	0.830	0.465	0.596	−0.058	0.420	0.817
	SPI3	ARIMA	0.535	0.708	−0.544	0.284	0.750	0.549	0.723	−0.663	−0.250	0.730
		CEEMD–ARIMA	0.393	0.502	0.497	0.654	0.894	0.407	0.526	0.448	0.632	0.886
	SPI6	ARIMA	0.429	0.609	0.363	0.643	0.867	0.440	0.618	−0.013	0.452	0.783
		CEEMD–ARIMA	0.244	0.312	0.860	0.886	0.962	0.250	0.321	0.808	0.861	0.954
	SPI9	ARIMA	0.304	0.434	0.711	0.816	0.934	0.315	0.460	0.384	0.671	0.850
		CEEMD–ARIMA	0.143	0.188	0.927	0.893	0.981	0.150	0.199	0.906	0.876	0.977
	SPI12	ARIMA	0.219	0.348	0.883	0.921	0.972	0.226	0.363	**0.604**	0.783	0.896
		CEEMD–ARIMA	0.125	0.186	0.925	0.927	0.982	0.129	0.194	**0.884**	0.923	**0.972**
	SPI24	ARIMA	0.149	0.233	0.939	0.953	0.985	0.157	0.248	**0.670**	0.831	0.911
		CEEMD–ARIMA	0.067	0.087	0.957	0.978	0.990	0.069	0.090	0.954	0.972	0.989
Tongxin	SPI1	ARIMA	0.711	0.909	−87.660	−7.274	0.127	0.724	0.918	−100.523	−8.116	0.115
		CEEMD–ARIMA	0.452	0.557	0.415	0.133	0.879	0.466	0.574	0.374	0.130	0.868
	SPI3	ARIMA	0.578	0.729	−0.286	0.360	0.783	0.606	0.740	−0.395	0.133	0.758
		CEEMD–ARIMA	0.343	0.416	0.787	0.377	0.952	0.349	0.424	0.750	0.369	0.944
	SPI6	ARIMA	0.437	0.588	0.489	0.704	0.890	0.467	0.626	0.357	0.499	0.859
		CEEMD–ARIMA	0.207	0.275	0.934	0.553	0.985	0.224	0.296	0.894	0.541	0.974
	SPI9	ARIMA	0.323	0.472	0.731	0.791	0.938	0.325	0.482	0.632	0.783	0.915
		CEEMD–ARIMA	0.138	0.181	0.960	0.804	0.991	0.142	0.187	0.952	0.797	0.988
	SPI12	ARIMA	0.235	0.336	0.873	0.916	0.969	0.239	0.341	0.823	0.853	0.957
		CEEMD–ARIMA	0.090	0.122	0.984	0.967	0.996	0.096	0.130	0.976	0.962	0.994
	SPI24	ARIMA	0.159	0.247	0.937	0.956	0.985	0.172	0.253	0.921	0.944	0.980
		CEEMD–ARIMA	0.062	0.079	0.996	0.975	0.999	0.065	0.083	0.992	0.972	0.998
Xiji	SPI1	ARIMA	0.782	0.961	−116.898	−10.640	0.237	0.825	1.036	−126.675	**−36.326**	0.224
		CEEMD–ARIMA	0.570	0.706	0.269	0.205	0.846	0.584	0.739	0.256	0.182	0.831
	SPI3	ARIMA	0.574	0.731	−0.313	0.370	0.774	0.649	0.820	−0.487	−0.528	0.752
		CEEMD–ARIMA	0.391	0.481	0.717	0.794	0.939	0.407	0.508	0.689	0.776	0.930
	SPI6	ARIMA	0.492	0.657	0.332	0.529	0.877	0.547	0.670	0.313	0.262	0.869
		CEEMD–ARIMA	0.235	0.297	0.930	0.842	0.984	0.247	0.309	0.923	0.835	0.981
	SPI9	ARIMA	0.346	0.490	0.711	0.768	0.936	0.412	0.576	0.696	0.482	0.933
		CEEMD–ARIMA	0.211	0.279	0.948	0.934	0.988	0.221	0.291	0.940	0.923	0.985
	SPI12	ARIMA	0.229	0.354	0.890	0.888	0.980	0.245	0.377	0.890	0.625	0.974
		CEEMD–ARIMA	0.102	0.136	0.987	0.937	0.997	0.107	0.141	0.987	0.921	0.997
	SPI24	ARIMA	0.158	0.233	0.950	0.753	0.989	0.188	0.285	0.949	**0.514**	0.988
		CEEMD–ARIMA	0.069	0.087	0.995	0.994	0.999	0.076	0.100	0.994	0.993	0.999

The actual calculated SPI values, the ARIMA-predicted values, and the CEEMD–ARIMA-predicted values of 10 sites in 2019 were visualized by the empirical Bayesian kriging interpolation method in ArcGIS. The SPI at different time scales is suitable for different analyses. In this paper, SPI3 was selected to show the drought situation of spring, summer, autumn, and winter in the Ningxia Hui Autonomous Region, which can be used to analyze the seasonal variation in drought. As shown in Figure 8, the prediction of the CEEMD–ARIMA model was closer to the actual situation than that of the ARIMA model, and the predicted results were consistent with the actual approximation. In the summer of 2019, the precipitation in southern Ningxia Hui Autonomous Region was abnormal and excessive. The instability of precipitation data led to a big difference between the ARIMA-predicted values and the observed SPI values. Based on the advantage of CEEMD in nonstationary signal processing, the precision of the combined model is good, and is consistent with actual states.

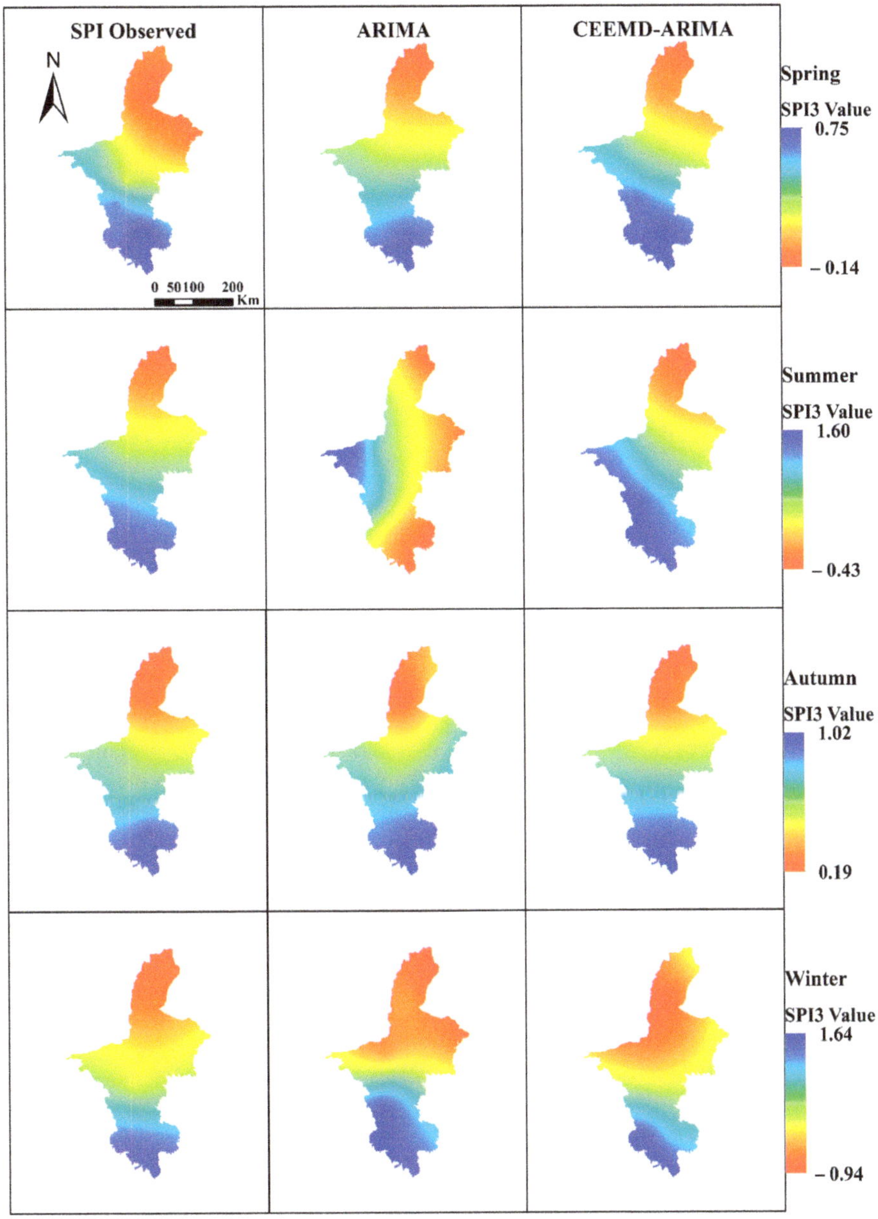

Figure 8. Kriging interpolation results of the actual calculated values and the predicted values of the ARIMA model and the CEEMD–ARIMA model.

5. Discussion

Recent studies have highlighted the superiority of EMD, EEMD, and CEEMD in forecasting aspects [39–42]. Ali et al. [43] forecasted rainfall at a monthly time scale. They resolved the non-stationarity challenges faced by rainfall forecasting models via CEEMD. Their study, which used a hybrid model for forecasting rainfall, achieved a WI value of 0.966. Their findings indicated that the CEEMD effectively avoids the non-stationarity in rainfall forecasting. Previous studies in drought prediction have used single models. For example, using the ARIMA model, Shatanawi et al. [20] predicted 3 out of 4 actual moderate droughts at Amman and Mafraq Stations. Similarly, Liu et al. [6] used the ARMA model to predict SPI9; the results showed that the prediction results of ARMA model were closer to the observed values in Longkou Station (the average relative error is 20.39%). However, when considering the entire stations, the highest standard error was as much as 43.69%. This result showed that more emphasis should be given to studying larger areas, which will be essential in regional drought management to make decisions. In this study, data stability is considered, along with the applicability of the CEEMD–ARIMA model over a large area.

The non-stationarity of data affects the drought prediction results of the model. The prediction result of the ARIMA model for SPI1 was significantly different from the actual situation. The ARIMA model had lower prediction accuracy at short time scales and higher prediction accuracy at long time scales, depending on the characteristics of the ARIMA model. As an overall linear autoregressive model, the prediction of the ARIMA model tends to become stable gradually with increasing test set time. In this study, the data volume of the 1-month time scale was larger than that of the 3-, 6-, 9-, 12-, and 24-month timescales, and the data series tended to be strictly stationary (i.e., the sequence distribution structure does not change over time). Therefore, the lowest prediction accuracy of the ARIMA model was obtained at the 1-month time scale. With increasing time scale, the amount of time-series data decreased, and the data series tended to be weakly stationary (i.e., the expectation, variance, and covariance of the stochastic process were constant; that is, the future value was related to the past value). Therefore, the fitting accuracy of the ARIMA model gradually improves as the time scale increases. At the same time, with increasing time scale, the SPI sequence obtains more information from the original sequence, and the fit of the predicted value with the actual calculated value becomes increasingly better.

Some studies have compared and analyzed the signal decomposition methods of EMD, EEMD, and CEEMD, and the analysis results show that the effects of different decomposition methods are very good, and that CEEMD can control residual noise at a relatively low level [34,44,45]. The large residual auxiliary noise of the defective EEMD cannot be avoided, and influences the experimental results. However, the influence from CEEMD can be ignored. Therefore, to stabilize the SPI sequence, in this study, CEEMD was used to extract the local features of the original sequence at different scales. Drought prediction based on CEEMD decomposition provides a stable premise for the ARIMA model. Therefore, the CEEMD–ARIMA model has a high prediction accuracy. If the data stationarity is poor, the prediction accuracy of the CEEMD–ARIMA is be reduced, as determined by the characteristics of the ARIMA model. In this case, although the prediction accuracy of the CEEMD–ARIMA model has a significant improvement over the single model, as shown in Figures 5a, 6a and 7a, its predictive effect is still poor. In August 2016, local rainstorms and short-term heavy precipitation occurred in the central and northern parts of Ningxia, precipitation in the southern mountainous area was rare, and the whole region presented a rare flood in the north and drought in the south. However, the combined model predicted conditions consistent with the record, indicating that CEEMD–ARIMA is suitable for use in drought research.

One limitation of this study is the diversity of factors that contribute to drought. In the arid zone of central Ningxia, the average annual precipitation is only 183.1 mm, and it is concentrated between June and August. However, with low precipitation, transpiration is high—close to 2000 mm in arid areas. Therefore, considering the influence of evapotranspiration is of great significance for accurately judging the arid zones' drought status. In a subsequent study on the drought situation in Ningxia Hui Autonomous Region, it was necessary not only to judge the drought situation based on precipitation, but also to consider the influence of evapotranspiration. This can be achieved by dividing the study area and then selecting different drought indices to analyze the drought in each area.

6. Conclusions

In this paper, the multiscale SPI was calculated based on precipitation data from 10 stations in the Ningxia Hui Autonomous Region. Combining CEEMD in the signal processing field and the ARIMA model in the machine learning field to predict the SPI, through comparative analysis of the prediction results, the following conclusions were obtained:

(1) As an effective nonlinear and nonstationary time-series decomposition method, CEEMD can extract the change trend of the SPI series and describe the characteristics of drought trends under climate change. Using CEEMD to decompose the SPI sequence of the Ningxia Hui Autonomous Region, seven IMF components and one trend item were obtained. The fluctuation of the component quantity became smoother than that of the original sequence, providing a basis for model prediction.

(2) The ARIMA model had the lowest prediction accuracy on the 1-month time scale and the highest on the 24-month time scale. At the same time scales, the prediction accuracy of the CEEMD–ARIMA model was higher than that of the ARIMA model. According to the visual display of the forecast results of the 3-month time scale, in the seasons of spring, summer, autumn, and winter, the drought conditions predicted by CEEMD–ARIMA were more consistent with the actual conditions.

(3) The drought prediction of CEEMD–ARIMA was approximately consistent with the China Meteorological Network records, indicating that the combined model is suitable for drought prediction. The original sequence was decomposed by CEEMD, and then the decomposed sequence was predicted by the ARIMA model. Finally, the predicted values of each component were added together to obtain the final prediction result. The final prediction result had high precision. According to the prediction results, the CEEMD–ARIMA model obtains higher prediction accuracy than the ARIMA model at multiple time scales, meaning that the combined model can better fit the SPI sequence at different time scales.

Author Contributions: Data curation, D.Z.; methodology, Y.D.; project administration, D.X.; resources, H.L.; software, H.L.; validation, Q.Z.; visualization, Q.Z.; writing—original draft, D.X.; writing—review and editing, Y.D. All authors have read and agreed to the published version of the manuscript.

Funding: This work was partially funded by State Key Laboratory of Geo-Information Engineering (No. SKLGIE2019-Z-4-2) and the Henan Provincial Science and Technology Research Project (No.212102310052, No.222102320021).

Institutional Review Board Statement: Not applicable.

Informed Consent Statement: Not applicable.

Data Availability Statement: The datasets from January 1960 to December 2019 were obtained from the Ningxia Hui Autonomous Region weather station in the National Meteorological Data Center (http://data.cma.cn/accessed on 13 March 2020). The elevation data were obtained from the Geospatial Data Cloud (http://www.gscloud.cn/search accessed on 28 June 2021).

Conflicts of Interest: The authors declare no conflict of interest.

References

1. Fang, O.; Zhang, Q.B.; Vitasse, Y.; Zweifel, R.; Cherubini, P. The Frequency and Severity of Past Droughts Shape the Drought Sensitivity of Juniper Trees on the Tibetan Plateau. *For. Ecol. Manag.* **2021**, *486*, 118968. [CrossRef]
2. Xu, Y.; Zhang, X.; Wang, X.; Hao, Z.; Singh, V.P.; Hao, F. Propagation from Meteorological Drought to Hydrological Drought under the Impact of Human Activities: A Case Study in Northern China. *J. Hydrol.* **2019**, *579*, 124147. [CrossRef]

3. Khan, R.; Gilani, H. Global Drought Monitoring with Drought Severity Index (DSI) Using Google Earth Engine. *Theor. Appl. Climatol.* **2021**, *146*, 411–427. [CrossRef]
4. Esfahanian, E.; Nejadhashemi, A.P.; Abouali, M.; Adhikari, U.; Zhang, Z.; Daneshvar, F.; Herman, M.R. Development and Evaluation of a Comprehensive Drought Index. *J. Environ. Manag.* **2017**, *185*, 31–43. [CrossRef] [PubMed]
5. Araneda-Cabrera, R.J.; Bermúdez, M.; Puertas, J. Benchmarking of Drought and Climate Indices for Agricultural Drought Monitoring in Argentina. *Sci. Total Environ.* **2021**, *790*, 148090. [CrossRef]
6. Liu, Q.; Zhang, G.; Ali, S.; Wang, X.; Wang, G.; Pan, Z.; Zhang, J. SPI-Based Drought Simulation and Prediction Using ARMA-GARCH Model. *Appl. Math. Comput.* **2019**, *355*, 96–107. [CrossRef]
7. Zhang, J.; Sun, F.; Lai, W.; Lim, W.H.; Liu, W.; Wang, T.; Wang, P. Attributing Changes in Future Extreme Droughts Based on PDSI in China. *J. Hydrol.* **2019**, *573*, 607–615. [CrossRef]
8. Yuan, X.; Jian, J.; Jiang, G. Spatiotemporal Variation of Precipitation Regime in China from 1961 to 2014 from the Standardized Precipitation Index. *ISPRS Int. J. Geo-Inf.* **2016**, *5*, 194. [CrossRef]
9. Asadi Zarch, M.A.; Sivakumar, B.; Sharma, A. Droughts in a Warming Climate: A Global Assessment of Standardized Precipitation Index (SPI) and Reconnaissance Drought Index (RDI). *J. Hydrol.* **2015**, *526*, 183–195. [CrossRef]
10. Bai, J.J.; Yu, Y.; Di, L. Comparison between TVDI and CWSI for Drought Monitoring in the Guanzhong Plain, China. *J. Integr. Agric.* **2017**, *16*, 389–397. [CrossRef]
11. Shi, B.; Zhu, X.; Hu, Y.; Yang, Y. Drought Characteristics of Henan Province in 1961-2013 Based on Standardized Precipitation Evapotranspiration Index. *J. Geogr. Sci.* **2017**, *27*, 311–325. [CrossRef]
12. Sivakumar, V.L.; Ramalakshmi, M.; Krishnappa, R.R.; Manimaran, J.C.; Paranthaman, P.K.; Priyadharshini, B.; Periyasami, R.K. An Integration of Geospatial Technology and Standard Precipitation Index (SPI) for Drought Vulnerability Assessment for a Part of Namakkal District, South India. *Mater. Today Proc.* **2020**, *33*, 1206–1211. [CrossRef]
13. De Oliveira-Júnior, J.F.; de Gois, G.; de Bodas Terassi, P.M.; da Silva Junior, C.A.; Blanco, C.J.C.; Sobral, B.S.; Gasparini, K.A.C. Drought Severity Based on the SPI Index and Its Relation to the ENSO and PDO Climatic Variability Modes in the Regions North and Northwest of the State of Rio de Janeiro—Brazil. *Atmos. Res.* **2018**, *212*, 91–105. [CrossRef]
14. Wu, J.; Chen, X.; Yao, H.; Zhang, D. Multi-Timescale Assessment of Propagation Thresholds from Meteorological to Hydrological Drought. *Sci. Total Environ.* **2021**, *765*, 144232. [CrossRef] [PubMed]
15. Xu, Y.; Zhang, X.; Hao, Z.; Singh, V.P.; Hao, F. Characterization of Agricultural Drought Propagation over China Based on Bivariate Probabilistic Quantification. *J. Hydrol.* **2021**, *598*, 126194. [CrossRef]
16. Łabędzki, L. Estimation of Local Drought Frequency in Central Poland Using the Standardized Precipitation Index SPI. *Irrig. Drain.* **2007**, *56*, 67–77. [CrossRef]
17. Hao, Z.; Hao, F.; Xia, Y.; Singh, V.P.; Hong, Y.; Shen, X.; Ouyang, W. A Statistical Method for Categorical Drought Prediction Based on NLDAS-2. *J. Appl. Meteorol. Climatol.* **2016**, *55*, 1049–1061. [CrossRef]
18. Khan, M.M.H.; Muhammad, N.S.; El-Shafie, A. Wavelet Based Hybrid ANN-ARIMA Models for Meteorological Drought Forecasting. *J. Hydrol.* **2020**, *590*, 125380. [CrossRef]
19. Han, P.; Wang, P.X.; Zhang, S.Y.; Zhu, D.H. Drought Forecasting Based on the Remote Sensing Data Using ARIMA Models. *Math. Comput. Model.* **2010**, *51*, 1398–1403. [CrossRef]
20. Shatanawi, K.; Rahbeh, M.; Shatanawi, M. Characterizing, Monitoring and Forecasting of Drought in Jordan River Basin. *J. Water Resour. Prot.* **2013**, *5*, 1192–1202. [CrossRef]
21. Fung, K.F.; Huang, Y.F.; Koo, C.H. Coupling Fuzzy–SVR and Boosting–SVR Models with Wavelet Decomposition for Meteorological Drought Prediction. *Environ. Earth Sci.* **2019**, *78*, 1–18. [CrossRef]
22. Xu, D.; Zhang, Q.; Ding, Y.; Huang, H. Application of a Hybrid Arima–Svr Model Based on the Spi for the Forecast of Drought—A Case Study in Henan Province, China. *J. Appl. Meteorol. Climatol.* **2020**, *59*, 1239–1259. [CrossRef]
23. Han, P.; Wang, P.; Tian, M.; Zhang, S.; Liu, J.; Zhu, D. Application of the ARIMA Models in Drought Forecasting Using the Standardized Precipitation Index. *IFIP Adv. Inf. Commun. Technol.* **2013**, *392*, 352–358. [CrossRef]
24. Özger, M.; Başakın, E.E.; Ekmekcioğlu, Ö.; Hacısüleyman, V. Comparison of Wavelet and Empirical Mode Decomposition Hybrid Models in Drought Prediction. *Comput. Electron. Agric.* **2020**, *179*, 105851. [CrossRef]
25. Libanda, B.; Nkolola, N.B. An Ensemble Empirical Mode Decomposition of Consecutive Dry Days in the Zambezi Riparian Region: Implications for Water Management. *Phys. Chem. Earth* **2022**, *126*, 103147. [CrossRef]
26. Tang, L.; Dai, W.; Yu, L.; Wang, S. A Novel CEEMD-Based Eelm Ensemble Learning Paradigm for Crude Oil Price Forecasting. *Int. J. Inf. Technol. Decis. Mak.* **2015**, *14*, 141–169. [CrossRef]
27. Niu, M.; Wang, Y.; Sun, S.; Li, Y. A Novel Hybrid Decomposition-and-Ensemble Model Based on CEEMD and GWO for Short-Term PM2.5 Concentration Forecasting. *Atmos. Environ.* **2016**, *134*, 168–180. [CrossRef]
28. Li, J.; He, K.; Tan, M.; Cheng, X. An Adaptive CEEMD-ANN Algorithm and Its Application in Pneumatic Conveying Flow Pattern Identification. *Flow Meas. Instrum.* **2021**, *77*, 101860. [CrossRef]
29. Mckee, T.B.; Doesken, N.J.; Kleist, J. The relationship of drought frequency and duration to time scales. In Proceedings of the 8th Conference on Applied Climatology, Anaheim, CA, USA, 17–22 January 1993; pp. 179–184.
30. Belayneh, A.; Adamowski, J.; Khalil, B.; Ozga-Zielinski, B. Long-Term SPI Drought Forecasting in the Awash River Basin in Ethiopia Using Wavelet Neural Networks and Wavelet Support Vector Regression Models. *J. Hydrol.* **2014**, *508*, 418–429. [CrossRef]

31. Lloyd-Hughes, B.; Saunders, M.A. A Drought Climatology for Europe. *Int. J. Climatol.* **2002**, *22*, 1571–1592. [CrossRef]
32. Javed, T.; Li, Y.; Rashid, S.; Li, F.; Hu, Q.; Feng, H.; Chen, X.; Ahmad, S.; Liu, F.; Pulatov, B. Performance and Relationship of Four Different Agricultural Drought Indices for Drought Monitoring in China's Mainland Using Remote Sensing Data. *Sci. Total Environ.* **2021**, *759*, 143530. [CrossRef]
33. Box, G.E.P.; Jenkins, G.M.; Reinsel, G.C.; Ljung, G.M. *Time Series Analysis: Forecasting and Control*; Holden-Day: San Francisco, CA, USA, 1976.
34. Yeh, J.R.; Shieh, J.S.; Huang, N.E. Complementary Ensemble Empirical Mode Decomposition: A Novel Noise Enhanced Data Analysis Method. *Adv. Adapt. Data Anal.* **2010**, *2*, 135–156. [CrossRef]
35. Mehdizadeh, S.; Ahmadi, F.; Kozekalani Sales, A. Modelling Daily Soil Temperature at Different Depths via the Classical and Hybrid Models. *Meteorol. Appl.* **2020**, *27*, 1–15. [CrossRef]
36. Esmaeili-Gisavandani, H.; Farajpanah, H.; Adib, A.; Kisi, O.; Riyahi, M.M.; Lotfirad, M.; Salehpoor, J. Evaluating Ability of Three Types of Discrete Wavelet Transforms for Improving Performance of Different ML Models in Estimation of Daily-Suspended Sediment Load. *Arab. J. Geosci.* **2022**, *15*, 1–13. [CrossRef]
37. Ahmadi, F.; Mehdizadeh, S.; Nourani, V. Improving the Performance of Random Forest for Estimating Monthly Reservoir Inflow via Complete Ensemble Empirical Mode Decomposition and Wavelet Analysis. *Stoch. Environ. Res. Risk Assess.* **2022**, 1–16. [CrossRef]
38. Adib, A.; Zaerpour, A.; Kisi, O.; Lotfirad, M. A Rigorous Wavelet-Packet Transform to Retrieve Snow Depth from SSMIS Data and Evaluation of Its Reliability by Uncertainty Parameters. *Water Resour. Manag.* **2021**, *35*, 2723–2740. [CrossRef]
39. Liu, M.D.; Ding, L.; Bai, Y.L. Application of Hybrid Model Based on Empirical Mode Decomposition, Novel Recurrent Neural Networks and the ARIMA to Wind Speed Prediction. *Energy Convers. Manag.* **2021**, *233*, 113917. [CrossRef]
40. Sun, H.; Zhai, W.; Wang, Y.; Yin, L.; Zhou, F. Privileged Information-Driven Random Network Based Non-Iterative Integration Model for Building Energy Consumption Prediction. *Appl. Soft Comput.* **2021**, *108*, 107438. [CrossRef]
41. Wu, C.; Wang, J.; Chen, X.; Du, P.; Yang, W. A Novel Hybrid System Based on Multi-Objective Optimization for Wind Speed Forecasting. *Renew. Energy* **2020**, *146*, 149–165. [CrossRef]
42. Zhang, X.; Wu, X.; He, S.; Zhao, D. Precipitation Forecast Based on CEEMD-LSTM Coupled Model. *Water Supply* **2021**, *21*, 4641–4657. [CrossRef]
43. Ali, M.; Prasad, R.; Xiang, Y.; Yaseen, Z.M. Complete Ensemble Empirical Mode Decomposition Hybridized with Random Forest and Kernel Ridge Regression Model for Monthly Rainfall Forecasts. *J. Hydrol.* **2020**, *584*, 124647. [CrossRef]
44. Zhang, Y.; Yan, B.; Aasma, M. A Novel Deep Learning Framework: Prediction and Analysis of Financial Time Series Using CEEMD and LSTM. *Expert Syst. Appl.* **2020**, *159*, 113609. [CrossRef]
45. Rezaei, H.; Faaljou, H.; Mansourfar, G. Stock Price Prediction Using Deep Learning and Frequency Decomposition. *Expert Syst. Appl.* **2021**, *169*, 114332. [CrossRef]

atmosphere

MDPI

Article

Innovative Trend Analysis of High-Altitude Climatology of Kashmir Valley, North-West Himalayas

Ishfaq Gujree [1,2], Ijaz Ahmad [1,3,*], Fan Zhang [1,2] and Arfan Arshad [4]

1 State Key Laboratory of Tibetan Plateau Earth System, Resources and Environment (TPESRE), Institute of Tibetan Plateau Research, Chinese Academy of Sciences, Beijing 100101, China; ishfaqgujree@itpcas.ac.cn (I.G.); zhangfan@itpcas.ac.cn (F.Z.)
2 University of Chinese Academy of Sciences, Beijing 100086, China
3 Centre of Excellence in Water Resources Engineering, University of Engineering and Technology, Lahore 54890, Pakistan
4 Department of Biosystems and Agricultural Engineering, Oklahoma State University, Stillwater, OK 74075, USA; aarshad@okstate.edu
* Correspondence: ijaz.ahmad@cewre.edu.pk

Abstract: This paper investigates the annual and seasonal variations in the minimum and maximum air temperature (T_{min} and T_{max}) and precipitation over Kashmir valley, Northwestern Himalayas from 1980–2019 by using the innovative trend analysis (ITA), Mann-Kendall (MK), and Sen's slope estimator methods. The results indicated that the annual and seasonal T_{min} and T_{max} are increasing for all the six climatic stations, whereas four of them exhibit significant increasing trends at (α = 0.05). Moreover, this increase in T_{min} and T_{max} was found more pronounced at higher altitude stations, i.e., Pahalgam (2650 m asl) and Gulmarg (2740 m asl). The annual and seasonal precipitation patterns for all climatic stations showed downward trends. For instance, Gulmarg station exhibited a significant downward trend for the annual, spring, and winter seasons (α = 0.05). Whereas, Qazigund showed a significant downward trend for the annual and spring seasons (α = 0.05). The overall analysis revealed that the increased T_{min} and T_{max} trends during the winter season are one of the reasons behind the early onset of melting of snow and the corresponding spring season. Furthermore, the observed decreased precipitation trends could result in making the region vulnerable towards drier climatic extremes. Such changes in the region's hydro-meteorological processes shall have severe implications on the delicate ecological balance of the fragile environment of the Kashmir valley.

Keywords: climate change; innovative trend analysis; Jhelum basin; Kashmir Himalayas; Mann Kendall test

check for updates

Citation: Gujree, I.; Ahmad, I.; Zhang, F.; Arshad, A. Innovative Trend Analysis of High-Altitude Climatology of Kashmir Valley, North-West Himalayas. *Atmosphere* **2022**, *13*, 764. https://doi.org/10.3390/atmos13050764

Academic Editors: Carlos E. Ramos Scharrón and Alexey V. Eliseev

Received: 11 March 2022
Accepted: 4 May 2022
Published: 9 May 2022

1. Introduction

Climate change is a natural phenomenon, though extensive research indicates that the anthropogenic activities in the 20th century are one of the major reasons for the temperature increase [1–3]. Increasing air temperatures and fluctuating precipitation patterns have gained a lot of attention in recent years because of their importance in understanding the climate change of any region [4,5]. An average increase in air temperature of 0.74 °C has been reported worldwide over the next 100 years [6]. In contrast to global predictions, the estimations of regional climate change rates differ due to differing methodology and datasets used to estimate future climate change [7–9].

Precipitation and temperature are considered key climatic variables affecting the spatiotemporal patterns of regional water resources availability [10,11]. Numerous studies have shown that assessing the implications of climate change on regional economic development, agriculture, and human society requires measuring fluctuations in regional air temperature and precipitation [12–15]. The most significant parameters in hydrometeorology are to evaluate a region's climate and estimate the consequences of changing climate, which are the air temperature and precipitation [16].

In recent years, substantial research has been carried out to quantify the climate change consequences by identifying the precipitation and temperature trends at various spatiotemporal scales in order to regulate regional resources of water and related hazards [17]. Significant warming trends with a magnitude of 0.16 °C per decade in air temperature were reported during the 20th century over the Tibetan Plateau [18–20]. According to [21].

According to Shrestha and Devkota (2010), in most regions of the Hind-Kush Himalayan (HKH), the warming rate is higher in the winter compared to other seasons [22]. In the last few decades, annual and winter precipitation has increased over the Tibetan Plateau and Indus basin. However, in these regions, an incoherent spatial pattern was witnessed for long-term precipitation variability [23]. Several studies have used methodologies to evaluate the temperature and precipitation variations, such as Mann-Kendall (MK) test, Sen's slope estimator, linear regression (LR), and Spearman's rho (SR) tests [24]. However, MK test is considered as the most common method used and has been applied in many regions worldwide to detect the changes in the hydro-meteorological variables. The MK test was employed by [25] to examine annual daily maximum precipitation trends and showed a significant increase. Pingale et al. (2014) examined the spatiotemporal mean and extreme rainfall and temperature trends using the MK test and Sen's slope estimator and found equally positive and negative trends for Rajasthan state urban centres [26]. Using LR and the SR tests [27] evaluated the significant upward trends for heatwaves and air temperatures in northwest Mexico. Gemmer et al. (2011) for observing the spatiotemporal characteristics for trends of rainfall used the MK test and found that while some stations showed annual trends in rainfall, monthly rainfall time series showed significant positive and negative trends in entire China's Zhujiang River Basin [28]. In another study, the results of the precipitation and temperature trend analysis were used to manage the scarce water resources of such regions for future water resource management development [29].

Few studies have been carried out in the Kashmir valley by using traditional statistical tests for trend analysis in hydro-meteorological data. Gujree et al. (2017), using Sen's slope estimator and MK test analyzed the spatial variability of precipitation and temperature extremes [30]. They showed that areas in plains exhibited an upward trend in T_{max} extremes, while in the near future the mountain areas may showcase more extreme events in T_{min} and precipitation. Shafiq et al. (2019) assessed the changing trends of precipitation and air temperature variables using the non-parametric tests in the Kashmir valley at various elevation zones [31]. Dad et al. (2021) examined the significance of trends and estimated the magnitude of trends in air temperature and precipitation on annual, seasonal, and monthly scales for all six meteorological stations positioned throughout the Kashmir valley employed non-parametric method. Ahmad et al. (2021) used non-parametric tests to assess the trend significance of air temperature and precipitation for the whole Kashmir valley [32]. Zaz et al. (2019) used statistical tests, such as Student's test, cumulative deviation, MK, and LR to examine the annual and seasonal precipitation and temperature changes in the six meteorological stations of the Kashmir valley [33].

However, traditional trend analysis methods can only detect the monotonic trends through pure statistical calculations and cannot identify the trends in different subcategories of the time series [34]. The innovative trend analysis (ITA) technique had been widely utilized to check trends predicted by existing methods and identify unseen trends in high, medium, and low-value categories utilizing springy graphical tools [17,34]. ITA is an intuitive and straightforward method that can be applied irrespective of distribution assumptions to identify trends in various time series subcategories [35]. In many parts of the world ITA has been used for investigating the hidden trends in hydro-meteorological variables. Ay and Kisi [36] carried out ITA-based trend analysis at six different provinces of Turkey for monthly precipitation and observed significantly upward trends at Trabzon and Samsun regions along with other four regions that were found insignificant. Similarly, Elouissi et al. [35] used this ITA method for assessing monthly precipitation to conduct trend analysis for 25 stations and observed a downward trend towards the northern parts and an upward trend towards the southern parts of the Macta watershed, Algeria. Tosunoglu and

Kisi [37] analyzed drought variables for nine stations by means of the modified ITA and MK tests. Moreover, the results showed that the MK test depicted trendless results for the investigated stations, whereas the modified MK showed a significantly decreasing trend at 10% significance level. The results for the Coruh River basin in Turkey, on the other hand, were consistent. Wu and Qian [38] assessed the 14 stations for annual and seasonal rainfall trends using the ITA, MK, and linear regression methods at Shanxi Province, China. They concluded that their results were good agreement across the tests and perfect covenant among tests showing significant trends.

Kashmir valley, the north-western part of the Himalayas, showed pronounced indicators of climate change. The Himalaya has complete control over the meteorological and hydrological conditions in Kashmir's valley. Even a little alteration in their climate has the potential to have severe effects for people's socioeconomic existence. Previous studies have used monotonous statistical techniques predominantly (MK-based tests) to understand key climatic indicators (air temperature and precipitation). However, studies have shown that it does not significantly address the reasons behind the changing hydrological regime of the Kashmir Valley [28–30,32,39–41]. In the present study, ITA based trend analysis method is explored to understand the climatic variability in high, medium, and low values of precipitation and air temperature over the last few decades in the Jhelum basin (Kashmir valley). As previously stated, this technique has been used all around the world to uncover hidden trends in Hydro-meteorological variables. Specifically, the present study aims to evaluate the regional climatic variability of the Kashmir valley by analyzing the time-series of air temperature and precipitation data between 1980 and 2019 using an ITA-based approach. It also aims to assess whether this technique is more reliable in terms of revealing better insights on the climatic variability of the region compared to MK test and Sen's slope approach. The main objectives of the study include: (i) Spatio -Temporal variations for air temperature and precipitation for Kashmir valley. (ii) Detection and quantification of trends in air temperature and precipitation. (iii) Comparision between the different trend analysis approaches.

2. Materials and Methods

2.1. Study Area

Kashmir Valley on the south is bordered by the Pir Panjal range and by the western Himalayan peaks on the north side [42]. The Himalayan complex has a pervasive influence on the valley's geographic entity. The total area of the region is about 15,948 km^2. The oval-bowl shaped valley extends from latitudes, 32°22′–34°43′ N and longitudes, 73°52′ 75°42′ E with an elevation range of 1300–1800 masl as shown in Figure 1. It is traversed by the Jhelum river, one of the Indus basin tributary. The weather in the Kashmir valley keeps on fluctuating owing to elevational differences [30].

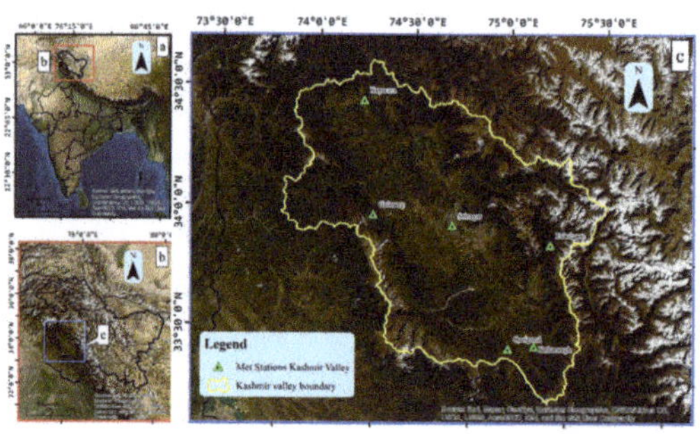

Figure 1. Location of study area (**a**) Indian political map; (**b**) Jammu & Kashmir map; (**c**) Kashmir valley map.

Summer monsoons originating in the Indian Ocean and Central Siberia's winter air masses are separated by the mountain ranges of the valley that act as a barrier to them [29]. Westerly troughs moving during the winter at higher altitudes enter the west and northwest of the valley, while the greater Himalayas obstruct their influx. The climate is unpleasant above the tops of surrounding mountains due to micro-level variations but generally warm and temperate over the valley. With more than 105 glaciers, the Kashmir valley is an important watershed of the upper Indus basin (UIB) [43]. Based on mean temperature and precipitation, the Kashmir valley's climate is characterized as sub-Mediterranean type with four seasons, spring (March–May), summer (June–August), autumn (September–November), and winter (December–February), [44]. The winters are usually cold and unpleasant, and summers are scorching, while spring is usually wet and autumn is dry. The valley's annual temperature ranges from −10 to 35 °C. Winter precipitation is coupled with western disturbances, dominates the rainfall pattern in the valley [45], while snowfall occurs primarily in the winter and early spring [46,47].

2.2. Datasets

The Himalayas (Greater), valley floor (Jhelum), and the Pir Panjal divide the Kashmir valley into three physiographic regions. Six well-distributed meteorological stations with varying mean sea levels, namely Gulmarg station (2740 m), Pahalgam station (2600 m), Kokarnag station (2000 m), Srinagar station (1600 m), Kupwara station (1670 m), and Qazigund station (1650 m), were chosen to represent the entire valley for analyzing the spatiotemporal variations in climatic variables (Table 1 and Figure 1). These six stations' topographic setting is characterized into two groups: (1) stations located on the plains (Qazigund, Kokarnag, Srinagar, and Kupwara) and (2) stations located in the mountainous areas (Pahalgam and Gulmarg) [32]. The data used were collected from the IMD-Srinagar and IMD-Pune centers for the period of 40 years (1980–2019) of six ground-based meteorological stations. This time series was deemed adequate for trend analysis to observe the fluctuations in different time scales at various climatic variables in the region. In order to understand climatic fluctuations over the region, inter and intra-annual trend analysis was carried out. The double-mass curve analysis method was utilized to cross-examine and check the data's homogeneity and consistency, which might have occurred due to instrumentation error [48].

Table 1. List of data ranges, basic geographic characteristics, and variables for those stations used in this study.

S.No.	Met Stations	Latitude	Longitude	Resolution	Time Period	Variables
1	Srinagar	34.05	74.80	Monthly	1980–2019	T_{max}, T_{min}, Precp
2	Gulmarg	34.06	74.39	Monthly	1980–2019	T_{max}, T_{min}, Precp
3	Kupwara	34.53	74.27	Monthly	1980–2019	T_{max}, T_{min}, Precp
4	Phalgham	34.02	75.33	Monthly	1980–2019	T_{max}, T_{min}, Precp
5	Qazigund	33.60	75.17	Monthly	1980–2019	T_{max}, T_{min}, Precp
6	Kukarnagh	33.59	75.30	Monthly	1980–2019	T_{max}, T_{min}, Precp

In prior studies on meteorological time series data, different statistical approaches (parametric and non-parametric) were used for determining whether the observed values of a hydro-meteorological time series are increasing, decreasing, or trendless. However, parametric methods with many restricted measures such as normal distribution and serially independent data are considered more powerful than non-parametric approaches, which is hardly true when it comes to meteorological time series data [49]. Non-parametric approaches, on the other hand, have been employed to identify trends in hydro-meteorological time series data since they don't need data to be distributed normally; nonetheless, this is the necessity for data to be free of serial correlation. A pre-whitening method was used to remove the serial correlation prior to using the MK test on the meteorological time series data [50]. Using the pre-whitening method on time series data, on the other hand may compromise the uniqueness of the data and erase a trend component that is truly present [51,52]. As a result, [53] suggested ITA technique that does not require such preprocessing in addition to having broad applications.

The results of the ITA approach are compared with those of MK and Sen's slope tests to assess the approach's trustworthiness. The annual and seasonal precipitation and air temperature time series were analyzed using ITA method for six stations across the valley. The monthly data were averaged for temperature and precipitation to develop the seasonal and annual precipitation time series [49,54]. Further, the air temperature and precipitation time series trends were analyzed at 10 percent, 5 percent, and 1 percent levels of significance using the ITA, MK, and Sen's slope approaches. A significance threshold of 10 percent level was used to establish a significant trend.

2.3. MK and Sen's Slope Tests

The non-parametric MK test is one of the robust statistical trend method for hydro-meteorological time series to detect the monotonic trends because of outlier's insensitivity and normal distribution [55–57].

The MK statistics, S is known as:

$$S = \sum_{k=1}^{n-1} \sum_{j=k+1}^{n} \text{sgn}(Y_j - Y_k) \tag{1}$$

$$\text{sgn}(Y_j - Y_k) = \begin{cases} if(Y_j - Y_k) < 0; & then \quad -1 \\ if(Y_j - Y_k) = 0; & then \quad 0 \\ if(Y_j - Y_k) > 0; & then \quad 1 \end{cases} \tag{2}$$

Here, Y_k and Y_j are successive data points of time-series with period k and j, n defines the no. of points, sgn represents the fn. taking the values of 1, 0, and −1; if >, $Y_j = Y_k$ and $Y_j < Y_k$, respectively. +ve values of S define the upward trend, and −ve values of S denote a decreasing trend in the hydro-meteorological time series [58]. The size of the sample for which $n > 10$, the test has to be escorted through a normal distribution ($\sigma^2 = 1$) and average ($\mu = 0$) with variance (Var) and probability (E) as presented below:

$$E[S] = 0 \tag{3}$$

$$Var(S) = \frac{n(n-1)(2n+5) - \sum_{p=1}^{q} t_p(t_p - 1)(2t_p + 5)}{18} \tag{4}$$

where q is the taut groups signifying observations having the common value, excluding unique rank numbers position, t_p defines the no. of data points of the pth group, symbol (Σ) describes all the tied groups summation. $Var(S)$ is the variance after manipulating from Equation (4), the test statistics standardized value (Z_{MK}) is evaluated by means of the eqn. below:

$$Z_{MK} = \begin{cases} \frac{S-1}{\sqrt{VAR(S)}}, & if \quad S > 0 \\ 0, & if \quad S = 0 \\ \frac{S+1}{\sqrt{VAR(S)}}, & if \quad S < 0 \end{cases} \tag{5}$$

Normal distribution with variance is followed by the regular Z_{MK} values follow a "1" and means "0," and is employed for calculating the variational weight. It is employed for checking the null theory, H_0. If Z_{MK} is bigger than $Z_{\alpha/2}$, consequently, the data series shows trends that are significant. Such a calculated estimation of Z_{MK} is matched with the two-tailed test table for normal distribution pertaining to α confidence level = 10%. However, for tests that is two-tailed, the null theory (H_0) is settled for zero (no) trend if the calculated estimation of Z_{MK} falls from—$Z1\text{-}_{\alpha/2}$ through $Z1\text{-}_{\alpha/2}$, and so, H_1 is excluded. In our study, the meteorological time series data trends are assessed for the levels of significance of 1%, 5%, and 10%.

The non-parametric Sen's slope estimator test is used for assessing the trend's weight in time series data [59,60]. The slope for n number of pairs of data-values is assessed by means of the equation given below.

$$b_i = \ median \ \left[\frac{Y_j - Y_k}{j-k}\right] \ \ \forall(k < j) \tag{6}$$

where Y_j and Y_k defines data points at time j and k. The n number of values that are median of bi actually depicts Sen's slope of trend. The +ve values of bi denote an increasing trend, while −ve values reveal the downward trend. Here, n = odd number, consequently the slope of the trend using Sen's method is calculated as:

$$Q_{med} = b_{[(n+1)/2]} \tag{7}$$

where, n = even number, now the trend slope using Sen's method is estimated as:

$$Q_{med} = \frac{1}{2}\left(b_{[n/2]} + b_{[(n+2)/2]}\right) \tag{8}$$

Finally, a two-tailed test is used to verify Q_{med} at desired confidence interval, and the real trend magnitude of the slope can be assessed through a non-parametric test [61].

2.4. Innovative Trend Analysis (ITA) Method

Most studies have employed the innovative trend analysis (ITA) approach in conjunction with many further trend analysis approaches to find disparities in climatological, meteorological, and hydrological data time series around the world due to its advantages over other non-parametric approaches. The trustworthiness of ITA is proven, however, by matching its results to those with the MK test results. The initial stage in this strategy is to divide hydro-meteorological time series data into 2 equal halves and position each one in increasing order separately. The second stage involves, the first 1/2 of the sub-series (X_i; i = 1, 2 ... n/2) positioned at X-axis, with the second 1/2 (X_j; j = n/2 + 1, n/2 + 2 ... n) is positioned at Y-axis of cartesian coordinate system, as illustrated in Figure 2. Both the axes (vertical & horizontal) necessarily have the same range. A series of clusters can be used to describe the domain variance of each sub-series (subgroups). This type of graph provides a quick visual inspection of the nature of time-series trends. Each subgroup's range can be resolute qualitatively or numerically. Data values on the scatter plot may be collected on the 45°-1:1 linear line. The hydro-meteorological time series has no trend. Otherwise, data values accumulating at the area of triangulation below or above the linear line specify an upward trend or a downward trend within the time series, respectively [53]. On computing the average difference between the X_j and X_i values at every point, the increasing or downward longitudinal trend in the time series can be evaluated. The horizontal and vertical distance from the linear line can be used to calculate this absolute difference. When comparing the amplitude of two subseries' trends, however, these average disparities should be normalized. The first half of the time series is used to determine the trend change. As a result, the indicator of trend is derived by dividing the mean difference between the linear line and the time series' first 1/2. On multiplying by ten the ITA trend indicator has represented the scale of the Sen's slope estimator and MK test at a 10% significance level as shown in the equation below:

$$D = \frac{1}{n}\sum_{i=1}^{n} \frac{10\left(Y_j - Y_i\right)}{\mu} \tag{9}$$

Here, D denotes the indicator of trend, n the number for data points in each sub-series, Y_i and Y_j denote the 1st and 2nd sub-series data points, respectively μ denotes the first subseries average. However, the +ve or −ve values of D represent an increasing or decreasing trend, respectively. If the value of observational data in the original time-series are found odd, at that time the initial results might be omitted earlier when dividing it into two equal halves so as to make the recent records are fully utilized.

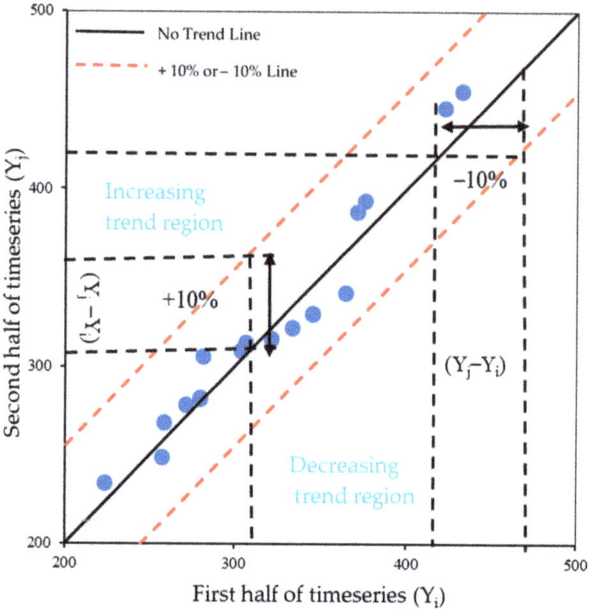

Figure 2. Illustration of upward, downward, and trendless portions in the ITA method.

3. Results

3.1. Spatio-Temporal Variations of T_{max}, T_{min} and Precipitation for Kashmir Valley Stations

Seasonal T_{max}, T_{min}, and precipitation time series at six stations were investigated for spatiotemporal changes in Kashmir valley are presented in Figure 3. The average observed monthly T_{max}, T_{min}, and precipitation at six dissimilar stations was combined to signify the overall temporal distribution of T_{max}, T_{min}, and precipitation throughout the Kashmir valley. Due to the northern side of the Himalayas, the valley (Kashmir) is influenced by several climate regimes such as westerly disturbances, monsoonal effects, and orographic fluctuations, making it a complicated region. The annual T_{max}, T_{min}, mean temperature and precipitation in the valley stations were approximately 20 °C, 7.6 °C, 13.8 °C, and 723.8 mm for Srinagar, 19.3 °C, 6.4 °C, 12.8 °C and 1212.7 mm for Qazigund, 16.6 °C, 3.1 °C, 9.8 °C and 1288.9 mm for Pahalgam, 20.1 °C, 6.3 °C, 13.2 °C and 1081.2 mm for Kupwara 18.1 °C, 4.1 °C, 11.1 °C, and 1080.2 mm for Kukarnagh 11.7 °C, 2.4 °C, 7.0 °C and 1485.1 mm for Gulmarg station for the period of forty years (1980–2019) as shown in Table 2. Furthermore, the precipitation was mainly concentrated in the spring and winter in all stations across the valley, with summer precipitation contributing a good portion also.

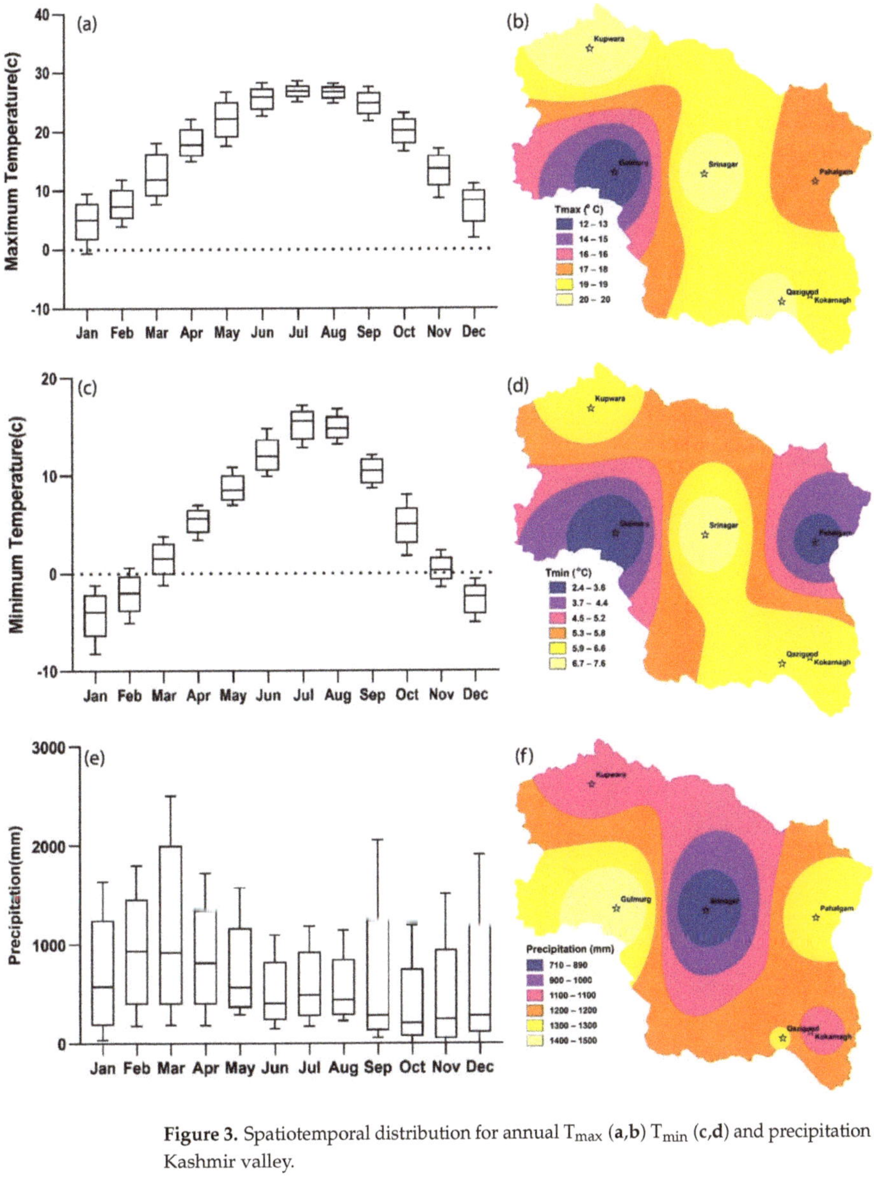

Figure 3. Spatiotemporal distribution for annual T_{max} (**a**,**b**) T_{min} (**c**,**d**) and precipitation (**e**,**f**) for entire Kashmir valley.

The mean monthly T_{max}, T_{min} and precipitation over the Kashmir valley are presented in Table 2. The annual precipitation over the Kashmir valley was governed by two climatic systems, the Indian summer monsoons (ISM) and the Western Disturbances. Over three-quarters of precipitation (71.54%) account for Western Disturbances from October to May, with the peak monthly precipitation occurring in Mar through May of the spring season. However, the residual 28.46% of rainfall falls between June and September, with a cluster of highest monthly precipitation in September Figure 3e,f, which is attributed to the Indian Summer Monsoon (ISM). Kashmir valley is influenced by mid-latitude westerlies considerably and is captured by the northern part represented by the two IMD stations,

Gulmarg and Kupwara, while the south side of the valley was influenced by ISM and was captured by three IMD stations (Kukarnagh and Pahalgam and Qazigund) [30,62].

Table 2. T_{max}, T_{min}, and Mean temperature and precipitation mean in Kashmir valley stations over a multi-year period.

Stations Name	Seasons	T_{max}	T_{min}	Mean-Temperature	Precipitation
		1980–2019	1980–2019	1980–2019	1980–2019
Srinagar	Annual	20.0	7.6	13.8	723.8
	Spring	20.1	7.7	13.9	281.0
	Summer	29.3	17.0	23.2	173.0
	Autumn	21.7	6.7	14.2	93.0
	Winter	8.8	−1.1	3.9	172.5
Qazigund	Annual	19.3	6.4	12.8	1212.7
	Spring	19.4	6.3	12.8	135.9
	Summer	27.7	15.3	21.5	82.5
	Autumn	21.4	5.8	13.6	43.8
	Winter	8.7	−1.9	3.4	129.1
Pahalgam	Annual	16.6	3.1	9.8	1288.9
	Spring	16.6	2.9	9.7	463.9
	Summer	25.0	11.2	18.1	300.1
	Autumn	18.6	3.2	10.9	181.6
	Winter	6.1	−4.9	0.6	332.6
Kupwara	Annual	20.1	6.3	13.2	1081.2
	Spring	19.8	6.2	13.0	442.4
	Summer	29.5	15.3	22.4	207.7
	Autumn	22.5	5.6	14.1	138.9
	Winter	8.6	−1.9	3.3	281.4
Kukarnagh	Annual	18.1	4.1	11.1	1080.2
	Spring	18.4	6.4	12.4	394.3
	Summer	27.0	14.9	21.0	259.3
	Autumn	19.9	6.8	13.3	151.2
	Winter	7.1	−2.1	2.5	267.7
Gulmarg	Annual	11.7	2.4	7.0	1485.1
	Spring	10.8	1.9	6.4	176.3
	Summer	20.0	10.7	15.3	104.3
	Autumn	13.2	3.1	8.2	63.3
	Winter	2.6	−6.0	−1.7	439.6

3.2. Annual and Seasonal T_{max} Variations over Time

The annual and seasonal T_{max} over the Kashmir valley stations are examined using the MK test and their results are summarized in Table 3. Annual T_{max} showed significant increasing trends at Srinagar, Pahalgam, Kupwara, and Kukarnagh stations ($\alpha = 0.01$, $\alpha = 0.05$) whereas Qazigund and Gulmarg stations also exhibited an increasing trend but statistically insignificant trends. Spring T_{max} showed significantly increasing trends at Kupwara ($\alpha = 0.01$), Srinagar, Qazigund, Pahalgam, and Kukarnagh stations ($\alpha = 0.05$) and Gulmarg ($\alpha = 0.1$). The summer T_{max} showed a decreasing trend at Srinagar, Qazigund, Pahalgam, and Gulmarg stations, while Kupwara and Kukarnagh stations exhibited increasing trends. Autumn T_{max} showed a significantly increasing trend for Srinagar and Pahalgam stations ($\alpha = 0.05$, $\alpha = 0.1$) and a significantly decreasing trend for the Qazigund station ($\alpha = 0.05$). Winter T_{max} indicates a significantly increasing trend for all the stations, Srinagar ($\alpha = 0.01$) Pahalgam and Kupwara ($\alpha = 0.001$) Kukarnagh ($\alpha = 0.05$) Qazigund and Gulmarg station ($\alpha = 0.1$) as presented in Figure 4.

Table 3. Summarized results for the seasonal T_{max} time series using ITA method statistic D, MK test statistic Z and Sen's slope estimator β.

S.No.	Station Name	Annual			Spring			Summer			Autumn			Winter		
		ITA	Zmk	Sen Slope (β)	ITA	Zmk	Sen Slope (β)	ITA	Zmk	Sen Slope (β)	ITA	Zmk	Sen Slope (β)	ITA	Zmk	Sen Slope (β)
1	Srinagar	0.57	2.92 **	0.04	0.88	2.23 *	0.05	0.04	−0.20	0.00	0.37	2.34 *	0.04	2.17	2.81 **	0.06
2	Qazigund	0.22	1.13	0.01	0.70	2.16 *	0.04	−0.06	−0.62	0.00	−0.13	−2.11 *	−0.03	1.32	1.85 +	0.04
3	Pahalgam	0.56	2.80 **	0.04	0.94	2.21 *	0.05	−0.13	−1.00	−0.01	0.36	1.68 +	0.03	3.45	3.93 ***	0.08
4	Kupwara	0.58	3.18 **	0.04	1.06	2.62 **	0.07	0.20	1.06	0.02	0.15	1.06	0.02	2.04	3.30 ***	0.06
5	Kukarnagh	0.60	2.57 *	0.04	0.96	2.20 *	0.06	0.07	1.25	0.01	0.27	0.73	0.01	2.75	2.57 *	0.06
6	Gulmarg	0.25	0.85	0.01	1.55	1.71 +	0.05	−0.20	−0.83	−0.02	−0.50	−0.52	−0.01	2.44	1.78 +	0.04

*** if trend at $\alpha = 0.001$ level of significance. ** if trend at $\alpha = 0.01$ level of significance. * if trend at $\alpha = 0.05$ level of significance. + if trend at $\alpha = 0.1$ level of significance. If the cell is blank, the significance level is greater than 0.1.

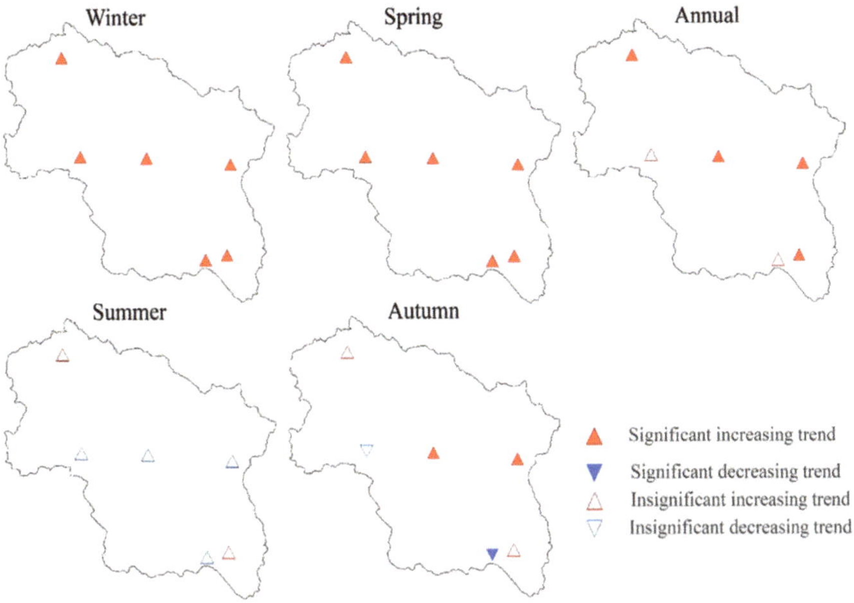

Figure 4. Using the MK approach, seasonal trends in T_{max} over Kashmir Valley.

The annual and seasonal trends of T_{max} based on Innovative Trend Analysis (ITA) over the Kashmir valley stations are summed-up in Table 3 and Figure 5. The trends for the annual T_{max} showed positive values of ITA statistic D dominated statistics, showing mostly significant increasing trends. At Srinagar, Kupwara, Kukarnagh, Pahalgam and Qazigund stations, significantly increasing and decreasing trends for the annual T_{max} data points were observed falling above 10% range from the 1:1 line, In comparison, Gulmarg station's majority of temperature data points fall in +10% range showing increasing trend with few points fall in −10% range from the 1:1 line during the period of forty years (1980–2019). The trends for T_{max} for the spring season statistics exhibited by significant (positive) values of ITA D, which is evidence of an increasing trend, are summarized in Table 3.

Srinagar, Qazigund, Pahalgam, Kupwara, and Kukarnagh stations showed increasing and decreasing trends for T_{max} data points, falling above 10% range from 1:1 line depicting a significantly positive trend. In comparison, Gulmarg station exhibited the increasing and decreasing trend for T_{max} data points which falls on +10% range from the 1:1 line. Summer T_{max} trends exhibited positive values of ITA statistic D increasing trend for three stations (Srinagar, Kupwara, and Kukarnagh) and negative values for three stations (Qazigund, Pahalgam, and Gulmarg). The combination of T_{max} data points falls within the range of 10% from the 1:1 line, which exhibits the decreasing trend for Srinagar and Qazigund stations. In contrast, Kupwara Pahalgam stations showed a decreasing trend for T_{max} data points falling in the 10% range. The results suggest that T_{max} trends for the autumn season showed four stations with significant positive values (Srinagar, Pahalgam, Kupwara, and Kukarnagh) and two stations with significant negative values (Qazigund and Gulmarg). In autumn, most T_{max} data points for Srinagar, Pahalgam, Kupwara, and Kukarnagh stations fall on the +10% range from the 1:1 line with an increasing trend. In comparison, Qazigund and Gulmarg showed T_{max} data points falling over the −10% range with an insignificant decreasing trend. The T_{max} for the winter season showed that all the station's ITA statistics for D exhibit positive values with a significant increasing trend. Moreover, an increasing and decreasing trends combination for T_{max} data points that fall above the 10% range from the 1:1 line exhibit a significantly positive trend for all the stations except Gulmarg station. The latter exhibited that the T_{max} data points falling on the +10% range of 1:1 linear line with an upward trend.

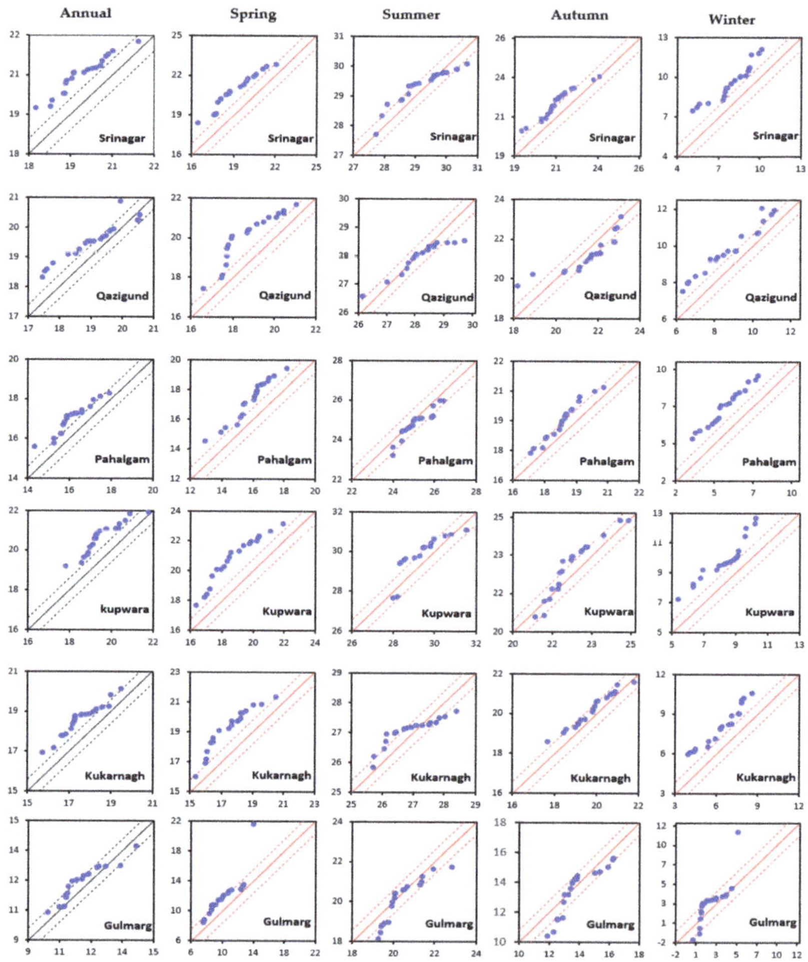

Figure 5. ITA method results for different seasons T_{max} at 6 stations.

Results depict an overall increase in the annual temperature is largely attributed to a mean T_{max}. It is clear for Figure 5 that it is reasonable to conclude that the climate in the Kashmir valley is fluctuating towards a system subjugated by high temperatures; as a result, the spring and winter seasons are changing in their durations.

3.3. Annual and Seasonal T_{min} Variations

Significant increasing trend based on Mann Kendall (MK) were noticed at Srinagar, Pahalgam ($\alpha = 0.01$), Qazigund ($\alpha = 0.001$) and Kupwara ($\alpha = 0.05$) stations whereas Kukarnagh and Gulmarg stations displayed an increasing but insignificant trend in the annual T_{min} as summarized in Table 4. Spring T_{min} exhibited a significantly increasing trend for Kupwara, Srinagar, Pahalgam, and Kukarnagh stations ($\alpha = 0.05$, $\alpha = 0.01$) while Qazigund and Gulmarg exhibited an increasing but insignificant trend. The summer T_{min} indicates a significantly increasing trend for Pahalgam Srinagar, Qazigund, Kupwara, and Kukarnagh ($\alpha = 0.01$), whereas for Gulmarg exhibited decreasing but an insignificant trend. Autumn T_{min} exhibited a significantly increasing trend for Srinagar, Pahalgam, Kupwara, Kukarnagh, and Gulmarg ($\alpha = 0.001$), with Qazigund showing decreasing but insignificant trend. Winter T_{min} exhibits a significant increasing trend for Kupwara ($\alpha = 0.001$), Gulmarg ($\alpha = 0.01$), and Pahalgam ($\alpha = 0.05$) while Qazigund and Kukarnagh exhibit an increasing trend with Srinagar showing a significantly decreasing trend as presented in Figure 6.

Table 4. Summarized results for the seasonal T_{min} time series using ITA method statistic D, MK test statistic Z and Sen's slope estimator β.

S.No.	Station Name	Annual			Spring			Summer			Autumn			Winter		
		ITA	Zmk	Sen Slope (β)	ITA	Zmk	Sen Slope (β)	ITA	Zmk	Sen Slope (β)	ITA	Zmk	Sen Slope (β)	ITA	Zmk	Sen Slope (β)
1	Srinagar	0.58	2.83 **	0.02	0.80	2.50 *	0.02	0.20	1.43	0.01	0.96	3.55 ***	0.04	−1.47	−0.15	0.00
2	Qazigund	0.14	0.85 ***	0.00	0.21	0.43	0.00	0.11	0.94	0.01	−0.59	−0.10	0.00	−1.61	0.62	0.01
3	Pahalgam	2.72	3.66 **	0.04	1.62	2.14 *	0.03	1.27	2.91 **	0.05	2.01	3.45 ***	0.03	−1.23	2.18 *	0.04
4	Kupwara	0.62	1.27	0.04	1.10	1.62	0.07	0.31	0.97	0.02	0.53	1.06	0.02	−0.63	−0.48	0.06
5	Kukarnagh	0.77	2.27	0.02	1.41	2.18 *	0.03	0.21	0.66	0.01	0.60	1.29	0.01	−4.44	1.20	0.03
6	Gulmarg	0.40	0.97	0.01	7.00	1.32	0.03	−0.81	−0.90	−0.02	−2.14	1.06	0.02	−1.45	2.62 **	0.05

*** if trend at $\alpha = 0.001$ level of significance. ** if trend at $\alpha = 0.01$ level of significance. * if trend at $\alpha = 0.05$ level of significance. If the cell is blank, the significance level is greater than 0.1.

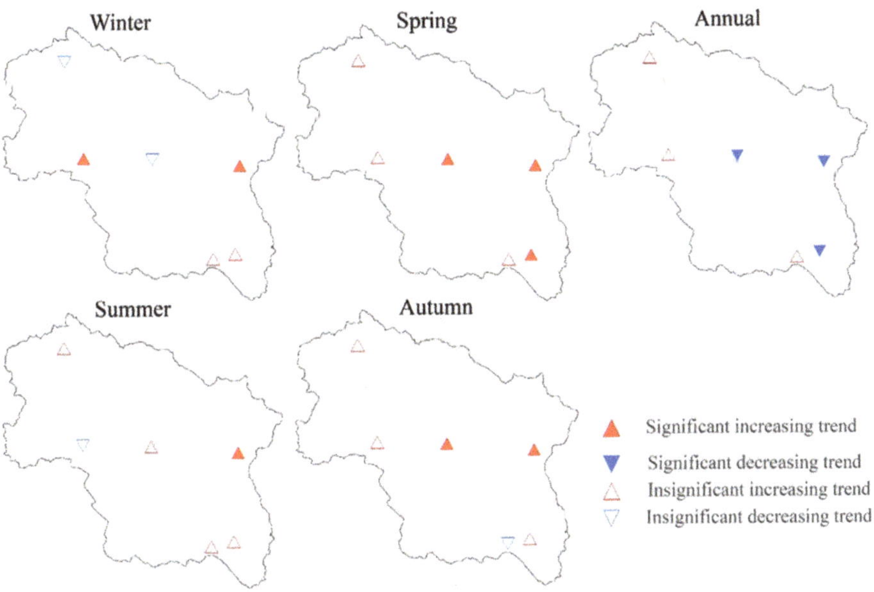

Figure 6. Using the MK approach, seasonal trends in T_{min} over Kashmir Valley.

Annual (annual) and seasonal T_{min} trends based on Innovative Trend Analysis for Kashmir valley are summed up in Table 4 and Figure 7. The trends for annual T_{min} for all the stations showed that the statistics for ITA D were dominated by positive values evident of increasing trend. The combination of increasing and decreasing trends for T_{min} data points falls above the 10% range for Srinagar and Pahalgam stations. However, Qazigund, Kupwara, Kukarnagh, and Gulmarg's T_{min} data points fall on a +10% range from the 1:1 line. The trends for spring T_{min} for all the stations showed that the statistics were significantly positive dominated. Moreover, the combination of upward and downward trends for T_{min} data points falls on the +10% range for Qazigund and Gulmarg stations. In contrast, Srinagar, Pahalgam, Kupwara, and Kukarnagh have T_{min} data points that mostly fall >10% range from the 1:1 line. The trends for summer T_{min} for all the stations showed that the statistics for ITA D were positively dominated, which exhibits the insignificant increasing trend except for the Gulmarg station, which exhibits a negative trend. However, Pahalgam station exhibits a significant increasing trend, as summarized in Table 4. The combination of increasing and decreasing trends for T_{min} data points falls on +10% range for Srinagar, Qazigund, Kupwara, and Kukarnagh stations whereas, Gulmarg station is at −10% with Pahalgam station as exception data points falls >10% range from 1:1 line. The trends for autumn T_{min} for Srinagar, Pahalgam, Kupwara, and Kukarnagh showed positive values, with significantly increasing trends for Srinagar and Pahalgam. Further, the combination of increasing and decreasing T_{min} data points for Qazigund, Kupwara, Kukarnagh, and Gulmarg stations falls within the 10% range.

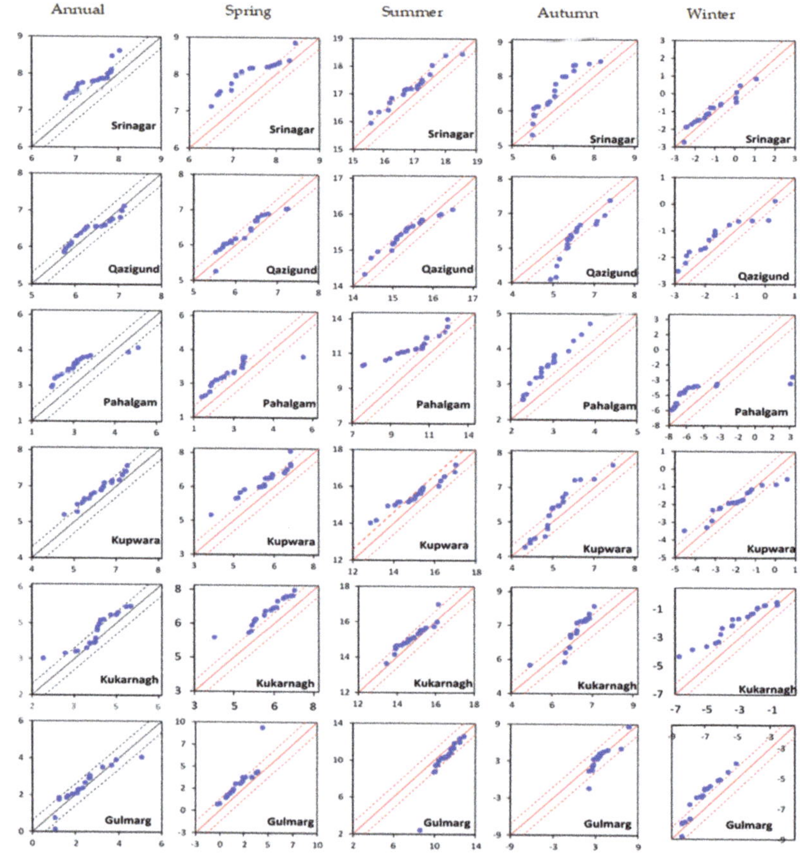

Figure 7. ITA method results for different seasons T_{min} at six stations.

In comparison, Srinagar and Pahalgam station's T_{min} data points fall above the 10% range from the 1:1 line. The trends for winter T_{min} for all the stations showed negative values for ITA of D-dominated statistics. Most of them were significantly decreasing trends, which are summarized in Table 4. The combination of increasing and decreasing for T_{min} data points for Srinagar, Qazigund, and Kupwara stations falls in the +10% range, while Pahalgam, Kukarnagh, and Gulmarg stations T_{min} data points fall above the 10% range from 1:1 line. The overall results suggest that T_{min} is in comparison with the T_{max}. It may be noted that the spring and winter seasons are warming in this region than the autumn and summer seasons. This anomaly contributes towards a reduction in the snow depth/cover and shrinking of glaciers.

3.4. Annual and Seasonal Precipitation Variations over Time

Table 5 represents the annual (annual) and seasonal trends using Mann Kendall (MK) test over Kashmir valley. Annual precipitation exhibited decreasing trend but was significant for the Gulmarg station ($\alpha = 0.05$), while the rest of the stations exhibited insignificant decreasing trends. Spring precipitation indicated a significantly decreasing trend for Gulmarg ($\alpha = 0.01$), Qazigund, Pahalgam ($\alpha = 0.05$), and Kupwara ($\alpha = 0.1$), while Srinagar and Kukarnagh stations exhibited decreasing insignificant trend. Summer precipitation for Srinagar, Kupwara, and Gulmarg stations displayed a downward trend while Qazigund, Pahalgam, and Kukarnagh stations exhibited an increasing significant trendMeanwhile, autumn precipitation showed an increasing trend for all stations except Kupwara, which exhibited an insignificant decreasing trend. Winter precipitation exhibited a significantly decreasing trend for Gulmarg ($\alpha = 0.05$) while Qazigund, Pahalgam, Kupwara, and Kukarnagh exhibited decreasing trends except for Srinagar station, which showed an increasing but significant trend as presented in Figure 8.

Table 5. Summarized results for the seasonal precipitation time series using ITA method statistic D, MK test statistic Z and Sen's slope estimator β.

S.No.	Station Name	Annual			Spring			Summer			Autumn			Winter		
		ITA	Zmk	Sen Slope (β)	ITA	Zmk	Sen Slope (β)	ITA	Zmk	Sen Slope (β)	ITA	Zmk	Sen Slope (β)	ITA	Zmk	Sen Slope (β)
1	Srinagar	−0.78	−0.36	−1.08	−2.09	−1.27	−2.02	−0.48	−0.34	−0.22	1.88	0.15	0.15	−0.30	0.08	0.09
2	Qazigund	−1.64	−1.18	−5.59	−2.72	−2.53 *	−5.31	−0.46	0.66	0.60	−2.55	0.10	0.18	−1.14	−1.55	−2.96
3	Pahalgam	−0.61	−0.42	−1.29	−2.20	−2.28 *	−4.73	0.26	0.77	0.70	1.27	1.26	1.65	0.23	−0.26	−0.60
4	Kupwara	−1.03	−1.57	−5.57	−2.28	−1.85 +	−3.71	−1.45	−1.35	−1.28	0.61	−0.17	−0.19	1.03	−0.17	−0.32
5	Kukarnagh	−0.58	−0.20	−0.68	−0.90	−1.50	−2.99	−0.68	0.70	0.77	0.69	0.27	0.31	−0.44	−0.61	−1.09
6	Gulmarg	−2.42	−2.34 *	−12.30	−3.36	−2.64 **	−2.37	−0.90	−0.69	−0.40	0.51	0.45	0.27	−3.10	−2.41 *	−5.69

** if trend at $\alpha = 0.01$ level of significance. * if trend at $\alpha = 0.05$ level of significance. + if trend at $\alpha = 0.1$ level of significance. If the cell is blank, the significance level is greater than 0.1.

The annual (annual) and seasonal precipitation trends results based on Innovative Trend Analysis (ITA) for Kashmir valley stations are presented in Figure 9 and Table 5. The annual precipitation trends for all the stations showed negative values for ITA of D dominated statistics, with Qazigund, Kupwara, and Gulmarg stations showing significantly decreasing trends, as summarized in Table 5. The grouping of increasing/decreasing trends for precipitation data points falls <10% range for Qazigund and Gulmarg stations. However, Srinagar, Kupwara, and Kukarnagh showed that the precipitation data points fall in the −10% range, with Pahalgam station showing a majority of data points below and some on +10% range from the 1:1 line. The trends for spring precipitation for all the stations showed that statistics for ITA of D values were negatively dominated, while Srinagar, Qazigund, Pahalgam, Kupwara, and Gulmarg stations were showing significant decreasing trends as summarized in Table 5 and presented in Figure 9, while Kukarnagh station showed an increasing trend but not significant. The increasing and decreasing trends combination for precipitation data points falls in <10% range for Srinagar, Qazigund, Pahalgam, Kupwara, and Gulmarg stations, with Kukarnagh station majority of data points −10% and some points fall in +10% range from the 1:1 line. The trends for summer precipitation for all the stations showed that negative values dominated statistics (decreasing trend), but all insignificant, as summarized in Table 5. The increasing and decreasing trends combination for precipitation data points falls in −10% range for Srinagar, Qazigund, Pahalgam, Kukarnagh, and Gulmarg stations with Kupwara station, showing that majority of precipitation data points ranges below from the 1:1 line and some are present on no trend line. The trends for autumn precipitation for all the stations showed that statistics of ITA of D were positively dominated, with Qazigund station as an exception showing negative values (decreasing) trend but insignificant. The increasing and decreasing trends combination for data points precipitation falls between 10% range on the 1:1 line for all observatories. The trends for the winter precipitation of all stations showed negative values of ITA for D dominated statistics with Pahalgam and Kupwara as exceptions showing positive (increasing trend), but all insignificant except for Gulmarg station, which is showing significant trend as summarized in Table 5. The increasing and decreasing trends combination for data points of precipitation falling between the 10% range for all stations except Gulmarg station, showing a majority of precipitation data points below (<) and some points are at −10% range from the 1:1 line. The results for overall precipitation suggest drought conditions for Kashmir valley; however, decreases in spring and winter precipitation are in agreement with the temperature suggesting seasonal inconsistency and strong inter-station, which results in a visible shift in precipitation pattern.

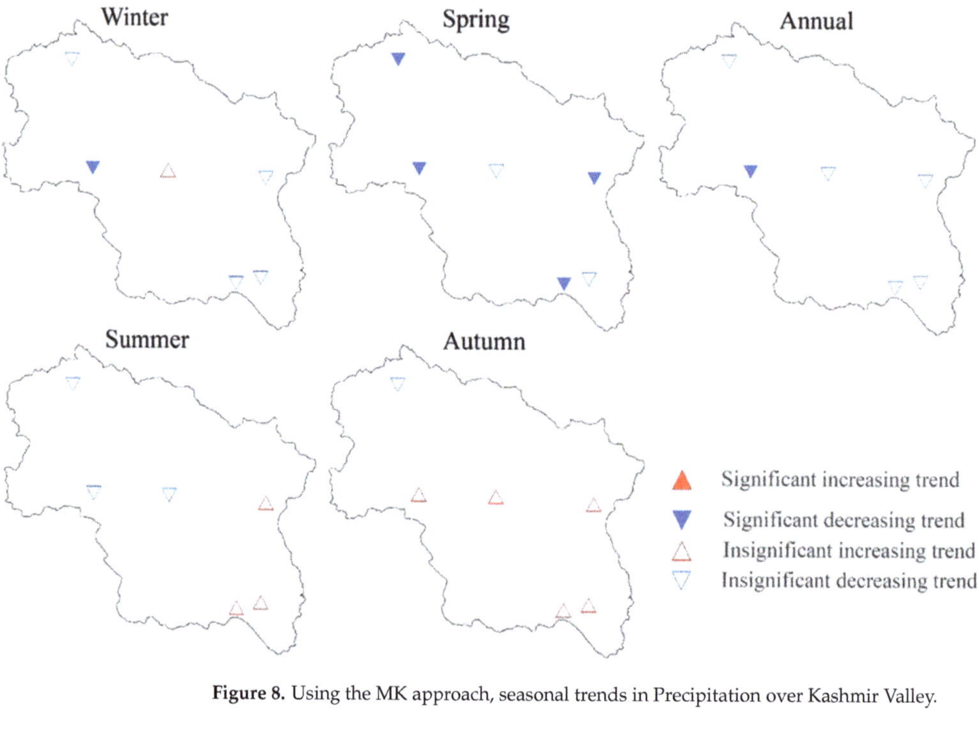

Figure 8. Using the MK approach, seasonal trends in Precipitation over Kashmir Valley.

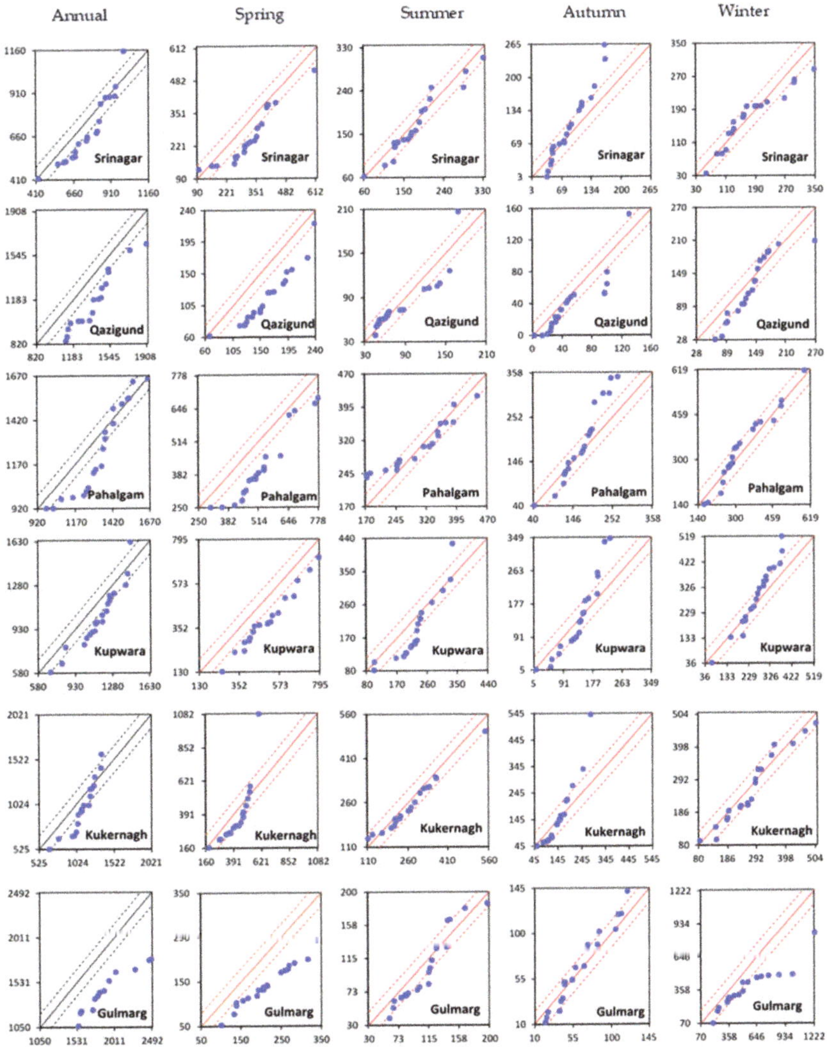

Figure 9. ITA method results for different seasons precipitation at 6 stations.

3.5. Comparison of ITA, MK, and Sen Slope Estimation Approach Results

On 40 years data annual and seasonal time series, MK and Sen's slope estimator tests have been used for assessing the reliability of the ITA method. Except for the Gulmarg station, winter season of T_{min} and precipitation, nine (09) time series on a seasonal time-scale prove significant trends at dissimilar observatories that are sign-steady under the above-mentioned approaches. In order to match the ITA results for all significant and insignificant trends, scatter plots were generated in between the Zmk of the MK with statistic D of the ITA technique and Sen's slope estimator with statistic D of the ITA method as depicted in Figure 10a,b. The results demonstrate 80% widely held of the scatter points fall in between the 1st and 3rd quadrants, indicating all these methods are in general agreement. The other 20% of the points fall in between the 2nd and 3rd quadrants, with same sign fluctuation; nevertheless, the trends are insignificant, and their S, i.e., the scale of Sen's slope is minimal. The Figure 10a,b also showed a high agreement between the

ITA statistic D and the MK statistic Zmk and Sen's slope estimator. As a result of this comparison, it is evident that ITA results are consistent across all trend detection approaches, indicating that it is a reliable and successful tool. The reason for this is its ability to assess meteorological and hydrological data patterns from graphical representations at low to high values. When compared to the MK test, which has various limitations such as seasonal cycle, normality of distribution, serially independent data, and time-series length, the ITA technique has universal applicability [38,49,53].

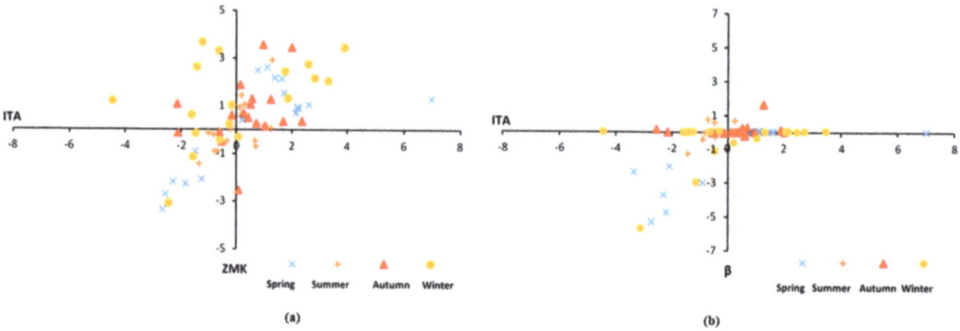

Figure 10. Scatter-plot results of (**a**) Zmk, MK test compared to D of ITA approach, (**b**) β, Sen's slope estimator, compared to D of ITA approach.

4. Discussion

The observed annual and seasonal T_{min} and T_{max} increase for six observational stations in the valley (Kashmir) has profound implications on the various land system and socio-economic processes operating in the region. These findings have substantiated the results of many previous studies on trend analysis for hydro-meteorological data from the Kashmir region. According to Kumar and Jain [42], T_{max} in the region (Kashmir) showed a significant increase (+0.04 to +0.05 °C/year) and T_{min} in the region (Jammu) (+0.030 to + 0.080 °C/year) had been detected when analyzed over the period 1980–2012. According to Singh et al. [63], Western Himalayas showed an increase in air temperatures, which are in line with the observed temperature trend for the Northwest Himalayas and the present study. [30] analyzed station-wise spatiotemporal variation of observed temperature in the Kashmir valley and found that T_{max} and T_{min} in the Valley showed an increase by 0.0350 °C and 0.0220 °C, respectively.

Shafiq, et al. [29] concluded that minimum temperatures in various topographic zones of the Kashmir valley increased at a relatively consistent rate of around 0.020 °C/year, with the peak rate of increase in the mountains (0.020 °C/year). Further, Zaz et al. [32], found that the annual mean temperature has increased by 0.8 degrees Celsius over 37 years (from 1980 to 2016), with T_{max} (0.97 °C) increasing faster than T_{min} (0.76 °C). The study also found that the Pahalgam (1.13 °C) and Gulmarg (1.04 °C) being at high altitudes have significantly increasing trends in temperature consistent with the results of the present study. This changing temperature pattern shall have serious environmental consequences, affecting food security and ecological sustainability in the region and water availability and other natural resources. Moreover, in all observational stations in the valley, the annual and seasonal precipitation patterns were found to have downward trends, consistent with the results of the previous studies. Zaz, et al. [32] studied the long and short-term precipitation patterns using LR, MK, Student's *t*-tests, and cumulative deviation; however, the results from the study showed that spring precipitation is decreasing. Shafiq, et al. [29] examined the annual and season-wise precipitation in Kashmir Valley from 1980 to 2016 and found a −5.1 mm/year decline in annual precipitation. The long-term upward trend results for precipitation at Pahalgam and Gulmarg stations for two seasons (spring, summer), which are also compatible with the [32]. It had been observed that the northern stations of the

Kashmir valley (Jhelum basin) recommend a moderately stronger monsoonal influence; meanwhile, two stations within the Kashmir valley (Jhelum basin) revealed a considerable increase in precipitation.

An obvious shift in precipitation regime can be seen from the north to the south of the basin, consistent with earlier research findings in this region which can be linked to the Western disturbances' movements (WD). The seasonal precipitation in the region is controlled by large-scale circulations from western disturbances (WDs), and it affects the supply and inventory of water resources of the Kashmir valley (Jhelum basin) [64]. It is hence concluded that, in winter, precipitation gets increased, but in spring, it decreases, and this phenomenon linked to specific variations in the WD precipitation trends. This observed winter precipitation increase is inconsistent with observations and future projections of the incursions of western disturbances, combined with the drying of the spring season in the region. This also designates a less erratic future WD regime [32]. However, the beginning of snowmelt from the Jhelum basin is expected to change as the changes in the western disturbances become more common in the future, as projected by the climate models. Understanding the Kashmir valley's climatic variability is crucial for improved management and use of the valley's available water resources where the population is dependent on agriculture activities. As the Himalayan hydrology is controlled by snow and glaciers, the changing climate of the region is bound to have drastic impacts on its water resources. Climate change impacts are evident in the Kashmir Himalayas. Natural calamities such as landslides and floods posing a significant threat to the people and infrastructure of this region which are climate change consequences [65,66]. Due to increased climatic variability associated with the ongoing changing climate, the region has recently witnessed one of the worst flooding disasters of the century [32,67]. Moreover, changing precipitation patterns have already impacted the recharge of groundwater, water supply in the streams during lean periods, and agriculture. The increase in the trends of T_{max} and T_{min} in the valley, as observed in the present study, has led to the early onset of spring and snowmelt, thus affecting the region's ecological setup. Moreover, as observed in the present study, decreasing trends in precipitation affect the water supply and management scenarios.

Hence assessing the trends in hydro-meteorological parameters such as T_{max}, T_{min} and Precipitation have helped to understand the variability associated with the changing climate in the fragile Kashmir valley in North-Western Himalayas [68]. It is because of the variability of trends at different stations, further extensive research on the climate of this region at the micro-level are needed. In order to check how climate may get changed, trend analysis is used to assess the hydro-metrological data, which may impact river discharge in the Kashmir Himalayas. However, the trend analysis will help us to know how river discharge will behave under projected climate scenarios for kashmir valley under different models like BCC-CSM2-MR, CNRM-CM6-1 and IPSL-CM6A-LR.

5. Conclusions

The present study addressed the climatic variability assessment for the Jhelum basin (Kashmir valley) using different statistical methods. The investigation of the annual and seasonal variations of the temperature and precipitation using the MK, Sen's slope estimator, and ITA approaches revealed some interesting findings related to the changing climatic scenario in the Kashmir valley. The rising temperature in two seasons (spring and winter) was credited for this major warming trend. T_{max} was observed to be significantly increasing at spring and winter seasons as well as annually, in almost all the stations except the Gulmarg, which exhibited an increasing but non-significant trend. The mean annual T_{min} followed the same significant trends as T_{max}, barring Kukarnagh and Gulmarg stations which exhibited non-significant increasing trends. T_{min} in the spring season also shows a significant positive trend for all stations except Qazigund and Gulmarg stations across the 1980–2019 time period. This rise in temperature across the valley, coupled with a considerable surge in spring and winter temperatures, reveals that less snowfall in the winter season results in decreased snow cover/depth. However, even at a small increase in temperature, a good amount of snowmelt is happening on the glaciers, clearing the way for early springs. Furthermore, annual precipitation showed decreasing trends for all the stations across the Kashmir valley. A significant decreasing trend in precipitation was detected for Qazigund, Pahalgam, Kupwara stations for the spring season, whereas for Gulmarg station, spring and winter seasons exhibited significantly decreasing trends since 1980–2019. This changing temperature and precipitation patterns in the region might have catastrophic implications for agriculture, hydropower, and drinking water supplies, affecting the region's food security and ecological sustainability.

However, understanding the climatic variability has various intrinsic uncertainties, which may arise from the secondary source data and area characteristics, particularly in mountainous in the Himalayan mountainous regions like Kashmir valley. Unavailability of the long-term time series data for air temperature and precipitation hinders the more accurate assessment of the region's climate variability, owing to only six meteorological stations representing 15,500 km^2 area. However, the major findings from this research confirm well with the large-scale land system changes taking place in the form of increased melting rates of snow and glacier resources in the valley.

Author Contributions: Conceptualization, I.G., I.A. and F.Z.; methodology, I.G. and I.A. and F.Z.; formal analysis, I.G., I.A. and A.A.; investigation, I.G., A.A. and F.Z.; resources, F.Z.; data curation, I.G. and A.A.; writing—original draft preparation, I.G. and I.A.; writing—review and editing, F.Z., I.A.; supervision, F.Z. and I.A.; project administration, F.Z.; funding acquisition, F.Z., I.G. All authors have read and agreed to the published version of the manuscript.

Funding: This research was financially supported by the National Natural Science Foundation of China (Grant No. 42125104).

Institutional Review Board Statement: Not applicable.

Informed Consent Statement: Not applicable.

Data Availability Statement: The data used in this study is available from the first author upon reasonable request.

Acknowledgments: Special acknowledgements are due to Indian Metrological Department (IMD), Srinagar and Pune for providing the data to conduct this research work, respectivly. Ishfaq Gujree and Ijaz Ahmad was supported by the CAS-TWAS President's PhD Fellowship Programme and the CAS-PIFI program, respectively. Ishfaq Gujree would like to specially thank to M.S Bhat and Muhammad Muslim for supporting to gather the data and providing Lab support, respectively.

Conflicts of Interest: The authors declare that no competing interest exists.

References

1. Rahmat, A.; Zaki, M.K.; Effendi, I.; Mutolib, A.; Yanfika, H.; Listiana, I. Effect of global climate change on air temperature and precipitation in six cities in Gifu Prefecture, Japan. *J. Phys. Conf. Ser.* **2019**, *1155*, 012070. [CrossRef]
2. Arora, N.K.; Fatima, T.; Mishra, I.; Verma, M.; Mishra, J.; Mishra, V. Environmental sustainability: Challenges and viable solutions. *Environ. Sustain.* **2018**, *1*, 309–340. [CrossRef]
3. Aamir, M.; Rai, K.K.; Dubey, M.K.; Zehra, A.; Tripathi, Y.N.; Divyanshu, K.; Samal, S.; Upadhyay, R.S. Impact of Climate Change on Soil Carbon Exchange, Ecosystem Dynamics, and Plant–Microbe Interactions. In *Climate Change and Agricultural Ecosystems*; Elsevier: Amsterdam, The Netherlands, 2019; pp. 379–413.
4. Le Quéré, C.; Andrew, R.M.; Canadell, J.G.; Sitch, S.; Korsbakken, J.I.; Peters, G.P.; Manning, A.C.; Boden, T.A.; Tans, P.P.; Houghton, R.A.; et al. Global carbon budget. *Earth Syst. Sci. Data* **2016**, *8*, 605–649. [CrossRef]
5. Cui, L.; Wang, L.; Lai, Z.; Tian, Q.; Liu, W.; Li, J. Innovative trend analysis of annual and seasonal air temperature and rainfall in the Yangtze River Basin, China during 1960–2015. *J. Atmos. Sol. Terr. Phys.* **2017**, *164*, 48–59. [CrossRef]
6. Higashino, M.; Stefan, H.G. Trends and correlations in recent air temperature and precipitation observations across Japan (1906–2005). *Theor. Appl. Climatol.* **2020**, *140*, 517–531. [CrossRef]
7. Thuiller, W. Patterns and uncertainties of species' range shifts under climate change. *Glob. Chang. Biol.* **2004**, *10*, 2020–2027. [CrossRef]
8. Walsh, K.J.; McBride, J.L.; Klotzbach, P.J.; Balachandran, S.; Camargo, S.J.; Holland, G.; Knutson, T.R.; Kossin, J.P.; Lee, T.C.; Sobel, A.; et al. Tropical cyclones and climate change. *Wiley Interdiscip. Rev. Clim. Chang.* **2016**, *7*, 65–89. [CrossRef]

9. Ren, Y.-Y.; Ren, G.-Y.; Sun, X.-B.; Shrestha, A.B.; You, Q.-L.; Zhan, Y.-J.; Rajbhandari, R.; Zhang, P.-F.; Wen, K.-M. Observed changes in surface air temperature and precipitation in the Hindu Kush Himalayan region over the last 100-plus years. *Adv. Clim. Chang. Res.* **2017**, *8*, 148–156. [CrossRef]
10. Gedefaw, M.; Yan, D.; Wang, H.; Qin, T.; Girma, A.; Abiyu, A.; Batsuren, D. Innovative trend analysis of annual and seasonal rainfall variability in Amhara regional state, Ethiopia. *Atmosphere* **2018**, *9*, 326. [CrossRef]
11. Shyam, G.M.; Taloor, A.K.; Singh, S.K.; Kanga, S. Sustainable water management using rainfall-runoff modeling: A geospatial approach. *Groundw. Sustain. Dev.* **2021**, *15*, 100676. [CrossRef]
12. Piao, S.L.; Ciais, P.; Huang, Y.; Shen, Z.H.; Peng, S.S.; Li, J.S.; Zhou, L.P.; Liu, H.Y.; Ma, Y.C.; Ding, Y.H.; et al. The impacts of climate change on water resources and agriculture in China. *Nature* **2010**, *467*, 43–51. [CrossRef] [PubMed]
13. Chen, J.; Wu, X.; Finlayson, B.L.; Webber, M.; Wei, T.; Li, M.; Chen, Z. Variability and trend in the hydrology of the Yangtze River, China: Annual precipitation and runoff. *J. Hydrol.* **2014**, *513*, 403–412. [CrossRef]
14. Wang, R.; Chen, J.; Chen, X.; Wang, Y. Variability of precipitation extremes and dryness/wetness over the southeast coastal region of China, 1960–2014. *Int. J. Climatol.* **2017**, *37*, 4656–4669. [CrossRef]
15. Yue, S.; Pilon, P.; Phinney, B.; Cavadias, G. The influence of autocorrelation on the ability to detect trend in hydrological series. *Hydrol. Processes* **2002**, *16*, 1807–1829. [CrossRef]
16. Sen, Z. Partial trend identification by change-point successive average methodology (SAM). *J. Hydrol.* **2019**, *571*, 288–299. [CrossRef]
17. Dong, Z.; Jia, W.; Sarukkalige, R.; Fu, G.; Meng, Q.; Wang, Q. Innovative Trend Analysis of Air Temperature and Precipitation in the Jinsha River Basin, China. *Water* **2020**, *12*, 3293. [CrossRef]
18. Kang, S.; Xu, Y.; You, Q.; Flügel, W.-A.; Pepin, N.; Yao, T. Review of climate and cryospheric change in the Tibetan Plateau. *Environ. Res. Lett.* **2010**, *5*, 015101. [CrossRef]
19. You, Q.; Min, J.; Kang, S. Rapid warming in the Tibetan Plateau from observations and CMIP5 models in recent decades. *Int. J. Climatol.* **2016**, *36*, 2660–2670. [CrossRef]
20. Liu, X.; Chen, B. Climatic warming in the Tibetan Plateau during recent decades. *Int. J. Climatol. A J. R. Meteorol. Soc.* **2000**, *20*, 1729–1742. [CrossRef]
21. Shrestha, A.B.; Devkota, L.P. *Climate Change in the Eastern Himalayas: Observed Trends and Model Projections*; International Centre for Integrated Mountain Development (ICIMOD): Kathmandu, Nepal, 2010.
22. Ren, G. Climate changes of China's mainland over the past half century. *Acta Meteorol. Sin.* **2005**, *63*, 942–955.
23. Westra, S.; Alexander, L.V.; Zwiers, F.W. Global increasing trends in annual maximum daily precipitation. *J. Clim.* **2013**, *26*, 3904–3918. [CrossRef]
24. Pingale, S.M.; Khare, D.; Jat, M.K.; Adamowski, J. Spatial and temporal trends of mean and extreme rainfall and temperature for the 33 urban centers of the arid and semi-arid state of Rajasthan, India. *Atmos. Res.* **2014**, *138*, 73–90. [CrossRef]
25. Martinez-Austria, P.F.; Bandala, E.R.; Patiño-Gómez, C. Temperature and heat wave trends in northwest Mexico. *Phys. Chem. Earth Parts A/B/C* **2016**, *91*, 20–26. [CrossRef]
26. Gemmer, M.; Fischer, T.; Jiang, T.; Su, B.; Liu, L.L. Trends in precipitation extremes in the Zhujiang River basin, South China. *J. Clim.* **2011**, *24*, 750–761. [CrossRef]
27. Xu, M.; Kang, S.; Wu, H.; Yuan, X. Detection of spatio-temporal variability of air temperature and precipitation based on long-term meteorological station observations over Tianshan Mountains, Central Asia. *Atmos. Res.* **2018**, *203*, 141–163. [CrossRef]
28. Gujree, I.; Wani, I.; Muslim, M.; Farooq, M.; Meraj, G. Evaluating the variability and trends in extreme climate events in the Kashmir Valley using PRECIS RCM simulations. *Modeling Earth Syst. Environ.* **2017**, *3*, 1647–1662. [CrossRef]
29. Shafiq, M.U.; Rasool, R.; Ahmed, P.; Dimri, A. Temperature and precipitation trends in Kashmir Valley, north western Himalayas. *Theor. Appl. Climatol.* **2019**, *135*, 293–304. [CrossRef]
30. Dad, J.M.; Muslim, M.; Rashid, I.; Rashid, I.; Reshi, Z.A. Time series analysis of climate variability and trends in Kashmir Himalaya. *Ecol. Indic.* **2021**, *126*, 107690. [CrossRef]
31. Ahmad, T.; Pandey, A.C.; Kumar, A. Long-term precipitation monitoring and its linkage with flood scenario in changing climate conditions in Kashmir valley. *Geocarto Int.* **2021**, 1–26. [CrossRef]
32. Zaz, S.N.; Romshoo, S.A.; Krishnamoorthy, R.T.; Viswanadhapalli, Y. Analyses of temperature and precipitation in the Indian Jammu and Kashmir region for the 1980–2016 period: Implications for remote influence and extreme events. *Atmos. Chem. Phys.* **2019**, *19*, 15–37. [CrossRef]
33. Sabzevari, A.A.; Zarenistanak, M.; Tabari, H.; Moghimi, S. Evaluation of precipitation and river discharge variations over southwestern Iran during recent decades. *J. Earth Syst. Sci.* **2015**, *124*, 335–352. [CrossRef]
34. Sen, Z. Trend identification simulation and application. *J. Hydrol. Eng.* **2014**, *19*, 635–642. [CrossRef]
35. Elouissi, A.; Şen, Z.; Habi, M. Algerian rainfall innovative trend analysis and its implications to Macta watershed. *Arab. J. Geosci.* **2016**, *9*, 303. [CrossRef]
36. Ay, M.; Kisi, O. Investigation of trend analysis of monthly total precipitation by an innovative method. *Theor. Appl. Climatol.* **2015**, *120*, 617–629. [CrossRef]
37. Tosunoglu, F.; Kisi, O. Trend analysis of maximum hydrologic drought variables using Mann–Kendall and Şen's innovative trend method. *River Res. Appl.* **2017**, *33*, 597–610. [CrossRef]

38. Wu, H.; Qian, H. Innovative trend analysis of annual and seasonal rainfall and extreme values in Shaanxi, China, since the 1950s. *Int. J. Climatol.* **2017**, *37*, 2582–2592. [CrossRef]
39. Romshoo, S.; Zaz, S.; Ali, N. Recent Climate Variability in Kashmir Valley, India and its Impact on Streamflows of the Jhelum River. *J. Res. Dev.* **2018**, *17*, 1–22.
40. Zaz, S.N.; Romshoo, S.A. Recent variation in temperature trends in Kashmir Valley (India). *J. Himal. Ecol. Sustain. Dev.* **2013**, *8*, 42–63.
41. Ahsan, S.; Bhat, M.S.; Alam, A.; Ahmed, N.; Farooq, H.; Ahmad, B. Assessment of trends in climatic extremes from observational data in the Kashmir basin, NW Himalaya. *Environ. Monit. Assess.* **2021**, *193*, 649. [CrossRef]
42. Kumar, V.; Jain, S.K. Trends in seasonal and annual rainfall and rainy days in Kashmir Valley in the last century. *Quat. Int.* **2010**, *212*, 64–69. [CrossRef]
43. Romshoo, S.A.; Dar, R.A.; Rashid, I.; Marazi, A.; Ali, N.; Zaz, S.N. Implications of shrinking cryosphere under changing climate on the streamflows in the Lidder catchment in the Upper Indus Basin, India. *Arct. Antarct. Alp. Res.* **2015**, *47*, 627–644. [CrossRef]
44. Bagnouls, F. *Bioclimatic Types of South-East Asia*; Institute Francais de Pondichery: Pondichery, India, 1959.
45. Dar, R.A.; Romshoo, S.A.; Chandra, R.; Ahmad, I. Tectono-geomorphic study of the Karewa Basin of Kashmir Valley. *J. Asian Earth Sci.* **2014**, *92*, 143–156. [CrossRef]
46. Iqbal, M.J.; Ilyas, K. Influence of Icelandic Low pressure on winter precipitation variability over northern part of Indo-Pak Region. *Arab. J. Geosci.* **2013**, *6*, 543–548. [CrossRef]
47. Archer, D.R.; Fowler, H.J. Spatial and temporal variations in precipitation in the Upper Indus Basin, global teleconnections and hydrological implications. *Hydrol. Earth Syst. Sci.* **2004**, *8*, 47–61. [CrossRef]
48. Gao, P.; Li, P.; Zhao, B.; Xu, R.; Zhao, G.; Sun, W.; Mu, X. Use of double mass curves in hydrologic benefit evaluations. *Hydrol. Processes* **2017**, *31*, 4639–4646. [CrossRef]
49. Ahmad, I.; Zhang, F.; Tayyab, M.; Anjum, M.N.; Zaman, M.; Liu, J.; Saddique, Q. Spatiotemporal analysis of precipitation variability in annual, seasonal and extreme values over upper Indus River basin. *Atmos. Res.* **2018**, *213*, 346–360. [CrossRef]
50. Von Storch, H. Misuses of Statistical Analysis in Climate Research. In *Analysis of Climate Variability*; Springer: Berlin/Heidelberg, Germany, 1999; pp. 11–26.
51. Yue, S.; Pilon, P.; Cavadias, G. Power of the Mann–Kendall and Spearman's rho tests for detecting monotonic trends in hydrological series. *J. Hydrol.* **2002**, *259*, 254–271. [CrossRef]
52. Douglas, E.; Vogel, R.; Kroll, C. Trends in floods and low flows in the United States: Impact of spatial correlation. *J. Hydrol.* **2000**, *240*, 90–105. [CrossRef]
53. Sen, Z. Innovative trend analysis methodology. *J. Hydrol. Eng.* **2012**, *17*, 1042–1046. [CrossRef]
54. Ahmad, I.; Tang, D.; Wang, T.; Wang, M.; Wagan, B. Precipitation trends over time using Mann-Kendall and spearman's rho tests in swat river basin, Pakistan. *Adv. Meteorol.* **2015**, *2015*, 431860. [CrossRef]
55. Mann, H. Non-Parametric Tests against Trend. *Econmetrica J. Econom. Soc.* **1945**, *13*, 245–259. [CrossRef]
56. Kendall, K. Thin-film peeling-the elastic term. *J. Phys. D Appl. Phys.* **1975**, *8*, 1449. [CrossRef]
57. Hamed, K.H. Trend detection in hydrologic data: The Mann–Kendall trend test under the scaling hypothesis. *J. Hydrol.* **2008**, *349*, 350–363. [CrossRef]
58. Helsel, D.R.; Hirsch, R.M. *Statistical Methods in Water Resources*; Elsevier: Amsterdam, The Netherlands, 1992; Volume 49.
59. Sen, P.K. Estimates of the regression coefficient based on Kendall's tau. *J. Am. Stat. Assoc.* **1968**, *63*, 1379–1389. [CrossRef]
60. Ohlson, J.A.; Kim, S. Linear valuation without OLS: The Theil-Sen estimation approach. *Rev. Account. Stud.* **2015**, *20*, 395–435. [CrossRef]
61. Partal, T.; Kahya, E. Trend analysis in Turkish precipitation data. *Hydrol. Processes Int. J.* **2006**, *20*, 2011–2026. [CrossRef]
62. Jeelani, G.; Deshpande, R. Isotope fingerprinting of precipitation associated with western disturbances and Indian summer monsoons across the Himalayas. *J. Earth Syst. Sci.* **2017**, *126*, 108. [CrossRef]
63. Singh, D.; Sharma, V.; Juyal, V. Observed linear trend in few surface weather elements over the Northwest Himalayas (NWH) during winter season. *J. Earth Syst. Sci.* **2015**, *124*, 553–565. [CrossRef]
64. Meraj, G.; Romshoo, S.A.; Yousuf, A.; Altaf, S.; Altaf, F. Assessing the influence of watershed characteristics on the flood vulnerability of Jhelum basin in Kashmir Himalaya. *Nat. Hazards* **2015**, *77*, 153–175. [CrossRef]
65. Meraj, G.; Farooq, M.; Singh, S.K.; Islam, M.; Kanga, S. Modeling the sediment retention and ecosystem provisioning services in the Kashmir valley, India, Western Himalayas. *Modeling Earth Syst. Environ.* **2021**, 1–26. [CrossRef]
66. Rather, M.A.; Meraj, G.; Farooq, M.; Shiekh, B.A.; Kumar, P.; Kanga, S.; Singh, S.K.; Sahu, N.; Tiwari, S.P. Identifying the Potential Dam Sites to Avert the Risk of Catastrophic Floods in the Jhelum Basin, Kashmir, NW Himalaya, India. *Remote Sens.* **2022**, *14*, 1538. [CrossRef]
67. Altaf, S.; Meraj, G.; Romshoo, S.A. Morphometry and land cover based multi-criteria analysis for assessing the soil erosion susceptibility of the western Himalayan watershed. *Environ. Monit. Assess.* **2014**, *186*, 8391–8412. [CrossRef] [PubMed]
68. Altaf, F.; Meraj, G.; Romshoo, S.A. Morphometric analysis to infer hydrological behaviour of Lidder watershed, Western Himalaya, India. *Geogr. J.* **2013**, *2013*, 178021. [CrossRef]

atmosphere

Article

Crop Yield, Nitrogen Recovery, and Soil Mineral Nitrogen Accumulation in Extremely Arid Oasis Cropland under Long-Term Fertilization Management

Shimin Li [1,2], Xihe Wang [3], Changlin Kou [2], Jinling Lv [2,*] and Jianhua Gao [1,*]

1. College of Geography and Environmental Science, Henan University, Kaifeng 475004, China; mbaism@126.com
2. Institute of Plant Nutrition, Resources and Environmental Sciences, Henan Academy of Agricultural Sciences, Zhengzhou 450002, China; koucl@126.com
3. Soil and Fertilizer Institute, Xinjiang Academy of Agricultural Sciences, Urumqi 830000, China; wxh810701@163.com

* Correspondence: lvjinling2008@163.com (J.L.); jhgao@henu.edu.cn (J.G.)

Abstract: Crop yield stability and soil mineral nitrogen (Nmin) have rarely been evaluated from a long-term perspective in the extremely arid cropland regions of China. Therefore, a nationwide experiment aimed to optimize fertilizer application and increase productivity and nitrogen use efficiency in gray desert soils was initiated in 1990. Eight combinations of chemical fertilizers (CK, N, NK, NP, and NPK), straw return (NPKS), and manure amendments (NPKM and NPKM+) were tested for 24 years on spring wheat, winter wheat, and maize. The results displayed that the yield of three crops from balanced fertilizer treatments (NPK, NPKS, NPKM, and NPKM+) did not differ significantly after 24 years; however, reliable yield stability due to lower coefficient of variation (CV) and higher nitrogen harvest index (NHI) were recorded for manure amendment treatments. Compared to NPKM, NHI was lower for the NPKM+ treatment, but crop yield and stability did not improve, suggesting that the appropriate choice for manure amendment is important for guaranteeing food security in extremely arid regions. Balanced fertilizer treatments resulted in lower Nmin residual in the 300 cm soil profile, compared to unbalanced fertilizer treatments. The NPKS treatment gave the lowest value. In the 0–100 cm soil profile, Nmin was higher in NPKM than in the NPK treatment, suggesting that straw or manure amendment can effectively maintain Nmin in the topsoil undercurrent cropland management in arid areas. The NPKM treatment had the highest crop nitrogen recovery rate and the lowest nitrogen losses, further illustrating that manure amendment has higher N retention potential. Overall, although Nmin residues are relatively high in these regions, balanced fertilizer treatments, especially NPKM and NPKS, are the optimum strategies in extremely arid regions.

Keywords: extremely arid regions; long-term experiment; yield stability; nitrogen harvest index; soil mineral nitrogen; ^{15}N-labeled urea; gray desert soil

check for updates

Citation: Li, S.; Wang, X.; Kou, C.; Lv, J.; Gao, J. Crop Yield, Nitrogen Recovery, and Soil Mineral Nitrogen Accumulation in Extremely Arid Oasis Cropland under Long-Term Fertilization Management. *Atmosphere* **2022**, *13*, 754. https://doi.org/10.3390/atmos13050754

Academic Editors: Jinping Liu, Quoc Bao Pham, Arfan Arshad and Masoud Jafari Shalamzari

Received: 28 March 2022
Accepted: 4 May 2022
Published: 7 May 2022

Publisher's Note: MDPI stays neutral with regard to jurisdictional claims in published maps and institutional affiliations.

1. Introduction

Ensuring high crop yields while maintaining yield stability and reducing environmental risks have always been challenging issues in modern agriculture [1,2]. However, due to the intensification of global climate change [3–5], extreme meteorological conditions frequently occur, and under such conditions, the stability of food has increasingly become a hot issue of concern. For example, severe drought can cause plant water deficit, thereby affecting nitrogen (N) transport and metabolism, ultimately affecting crop yield, while excess rainfall may limit soil N retention and lead to mineral nutrients loss [2,6,7]; in this case, the crop yield is inevitably affected. At the same time, due to the continuous improvement of cropland management technology and the improvement of crop varieties, whether the traditional fertilization method can meet the current technical needs and ensure the continued stability of crop yield is also worth further discussion [8]. Furthermore, the developments in energy, transportation, and livestock and poultry farming have all significantly increased environmental N content [9,10]; for example, Liu et al. (2014) found that nitrogen deposition in the North China Plain has markedly increased over the past 20 years, with dry and wet N deposition reaching up to 80 kg N ha^{-1} yr^{-1} in Northwest China [11], and even 30 kg ha^{-1} yr^{-1}, respectively, in extremely arid oasis areas [12,13], which inevitably affects the N input/output balance of the cropland ecosystem.

Excessive N use is well known to cause severe environmental problems, such as groundwater pollution and eutrophication of streams, rivers (nitrate leaching), and terrestrial ecosystems (NH_3 volatilization) [14] and global-warming-related (NO and N_2O) gas emissions [7,15,16]. Moreover, soil acidification [17–19] and depletion of soil structure [20,21] have also been reported after the improper application of inorganic fertilizers in the absence of organic manure inputs. Therefore, it is very important to evaluate yield stability, migration, and distribution of mineral nitrogen (Nmin) in different croplands from a long-term perspective under typical environmental conditions.

Over the last few decades, arid and extremely arid regions have become increasingly important for grain production in China [21,22], leading to a continuous overuse of chemical fertilization to achieve high crop yields, which in turn has resulted in detrimental effects on field crop production, soil quality, and nitrogen cycle [23]. For instance, Lv et al. (2016) reported that the use of chemical fertilizers (urea, phosphorus pentoxide, and potassium superoxide) has increased six to eight times from 1980 to 2014, and the environmental nitrogen from atmospheric dry and wet deposition and irrigation water can yield 33 kg N ha^{-1} and 13 kg N ha^{-1}, respectively, due to the rapid development of agriculture and extensive use of irrigation water in extremely arid oasis cropland regions [23]. Concomitantly, drip irrigation and mulching have become prevalent since 2005 due to reduced water evaporation and improved N utilization efficiency, which has led to a significant improvement in water use efficiency of cropland in arid regions. In this case, the subsequent effects on crop yield, yield stability, N uptake, and migration and distribution of Nmin in the soil profile are not clearly understood. To date, only a few continuous long-term studies have been conducted on the oasis gray desert soil of Western China [24,25], and fewer studies have focused on the impact of chemical fertilizer, straw, and manure application on grain yield, yield stability, and Nmin distribution in the arid zone.

The objectives of this study were (1) to determine the impact of different fertilization treatments on maize–winter wheat–spring wheat production, biomass, and nutrient recovery efficiency based on long-term experiments in an oasis cropping system; (2) to study the concentration and distribution of NO_3-N and NH_4-N in the 300 cm soil profile of different treatments after 24 years of experimentation in the extremely arid oasis cropland areas; (3) to trace the migration dynamics of ^{15}N-labeled urea in the 100 cm soil profile in two different crop seasons; and (4) to determine the urea-N recovery, soil residual, and other losses of different treatments for spring wheat and maize cropping seasons. Additionally, the long-term experiment reported herein began in 1990 and is one of the longest-running annual cropping system experiments in the arid regions of China. This research is based on supporting data gathered from 1990 to 2013, which represent the early and mid-term status of croplands in the extremely arid region of Northwestern China. Therefore, the research on soil mineral nitrogen's effect on crop yield stability under fertilization management is typical and representative. Due to the long sequence, multiple parameters, a huge amount of data, and heavy workload of data collation, the experimental data collected from 2014 to 2021 are being analyzed, and follow-up research will progressively realize the complete time-series data for the research. All in all, this study systematically explored the effects of typical fertilization treatments on crop yield, crop yield stability, and nitrogen migration and distribution based on long time scales in extremely arid areas, which will provide a scientific basis for the future food security and efficient nitrogen use of cropland in arid or extremely arid areas.

2. Materials and Methods

2.1. Study Site

This study was conducted at Anningqu, a long-term experimental site located near Urumqi, capital of Xinjiang, Northwest China (43°56′ N, 87°28′ E). The site is a typical oasis farmland with a maize–winter wheat–spring wheat rotation system. The mean annual precipitation is 310 mm, 70% of which falls in winter and summer. Evaporation is approximately 2570 mm, and mean annual temperature is 7.7 °C (Figure 1). Sunlight amounts to an average of 2594 h per year. The annual frost-free period is approximately 156 d long [7]. According to the Chinese Soil Taxonomy Classification, the soil is gray desert soil, with topsoil layer (27 cm) clay, silt, and sand fractions of 30.3%, 52.5%, and 17.2%, respectively [24].

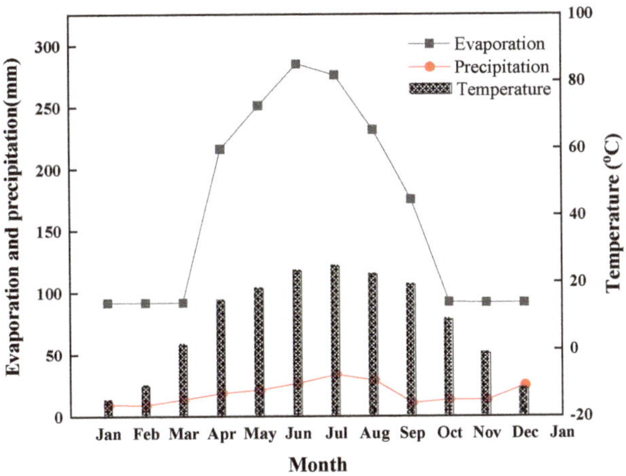

Figure 1. Average evaporation, precipitation, and ambient temperature during a long-term experiment from 1990 to 2014 in a gray desert soil.

2.2. Long-Term Experiment

The experimental field was a wasteland before the start of the long-term experiment in April 1990, and thus the soil had retained its original quality. The 0–20 cm topsoil layer had an organic matter content of 16.7 g kg^{-1}, total N, P, and K contents of 1.00, 0.35, and 18.18 g kg^{-1} and available P (Olsen-P) and K contents of 9 and 505 mg kg^{-1}. Soil pH (1:1 for the ratio of water and soil) was 8.1 (Table 1).

Table 1. Initial soil physicochemical characteristics at the experimental site in 1990.

Soil Layer	Organic Matter	Total N	Total P	Total K	Available P	Available K	pH	$CaCO_3$
(cm)	(g kg^{-1})	(g kg^{-1})	(g kg^{-1})	(g kg^{-1})	(mg kg^{-1})	(mg kg^{-1})	(H_2O)	(g kg^{-1})
0–2	17.4	1.03	0.36	18.10	11	495	7.87	65.5
2–16	16.7	1.00	0.35	18.18	9	505	8.03	64.2
16–31	17	1.03	0.35	17.85	7	383	8.13	99.4
31–43	12.3	0.77	0.30	18.35	1	335	7.92	93.0
43–68	9.33	0.55	0.28	18.43	2	243	8.07	77.6
68–90	6.9	0.35	0.16	17.93	1	73	8.22	27.6
90–110	6.46	0.36	0.25	17.19	1	73	8.39	62.2

The experiments included the following eight fertilization treatments:

1. N: 240 kg N ha^{-1} yr^{-1} as urea.
2. NP: 240 kg N ha^{-1} yr^{-1} as urea and 60.3 kg P ha^{-1} yr^{-1} as calcium superphosphate

3. NK: 240 kg N ha^{-1} yr^{-1} as urea and 48.1 kg K ha^{-1} yr^{-1} as potassium sulfate.
4. NPK: urea, calcium superphosphate, and potassium sulfate at 240 kg N, 60.3 kg P, and 48.1 kg K ha^{-1} yr^{-1}.
5. NPKM: 168 kg N ha^{-1} yr^{-1} as urea, 20.1 kg P ha^{-1} yr^{-1} as calcium superphosphate, and 16.3 kg K ha^{-1} yr^{-1} as potassium sulfate as above and an additional application of farmyard manure (30 t ha^{-1}) containing 2.9 t C ha^{-1} yr^{-1} and 72 kg organic N ha^{-1} yr^{-1}.
6. NPKM+: 216 kg N ha^{-1} yr^{-1} as urea, 30.5 kg P ha^{-1} yr^{-1} as calcium superphosphate, and 24.7 kg K ha^{-1} yr^{-1} as potassium sulfate and an additional application of farmyard manure (60 t ha^{-1}) that contained 5.9 t C ha^{-1} yr^{-1} and 144 kg organic N ha^{-1} yr^{-1}.
7. NPKS: 197 kg N ha^{-1} yr^{-1} as urea, 51.1 kg P ha^{-1} yr^{-1} as calcium superphosphate, and 42.3 kg K ha^{-1} yr^{-1} as potassium sulfate as above and the straw return (4.5–7 t ha^{-1}) containing on average 1.5 t C ha^{-1} yr^{-1} and 43 kg organic N ha^{-1} yr^{-1}).
8. CK: without any fertilization.

Prior to 1995, the application rates of N, P, and K were 100.9, 35.8, and 18.8 kg ha^{-1}, respectively, and 245, 60.3, and 48.1 kg ha^{-1}, respectively, after 1995 to date (Table 2). Nitrogen fertilizer was split into two applications: 60% as basal fertilizer and 40% as topdressing in the N, NK, NP, NPK, NPKM, and NPKM+ treatment plots. Phosphorus, potassium, and organic fertilizers were applied as basal fertilizers. Organic fertilizer was obtained from sheep manure (see Table 2 for details). The experiment was laid in a randomized blocks design. Each 466.5 m^2 plot was isolated by cement banks buried 70 cm deep and raised 10 cm above the soil surface to prevent leaching.

Table 2. Quantity of fertilizers from the different fertilizer treatments in the long-term experiment in a gray desert soil.

		N Fertilizer [1]		P Fertilizer [2]	K Fertilizer
Treatment	Basal	Topdressing	Total	Total	Total
	kg N ha^{-1}	kg N ha^{-1}	Kg ha^{-1}	kg P_2O_5 ha^{-1}	kg K_2O ha^{-1}
CK	0	0	0	0	0
N	144	96	240	0	0
NK	144	96	240		58.5
NP	144	96	240	138	
NPK	144	96	240	138	58.5
NPKS	163	77	240	138	58.5
NPKM [3]	173	67	240	138	58.5
NPKM+	274	86	360	184	78

[1] N fertilizer was obtained from chemical sources, straw, and manure for NPKS, NPKM and NPKM+ treatments. [2] P_2O_5 represents phosphorus; K_2O represents potassium. [3] Manure was fresh sheep.

Spring wheat was sown in late April and harvested in August. Winter wheat was sown in October and harvested in early June the following year. Maize was sown in late April and harvested in August. Drip irrigation was applied both in the wheat and maize seasons. Plastic mulching film was used only in the maize season. For the specific content, see Lv et al. (2016) [12]. The field was irrigated five to six times during the wheat season and four to five times during the maize season, depending on precipitation. The volume of water used for irrigation was 2850 m^3 ha^{-1} yr^{-1}. Herbicides and pesticides were used to control weeds and insect pests, respectively. Wheat and maize were harvested up to the level of the soil surface; thus, the stubble left in the field was negligible, but roots were left in the soil. All straw was removed from the field. Grain and straw were weighed separately after air-drying.

2.3. ^{15}N-Labeled Urea Experiment

The ^{15}N experiment was conducted in mid-April 2011 and 2013 to trace the fate of urea-N in winter wheat and summer maize. ^{15}N-labeled treatments were as follows: N, NP,

NK, NPK, NPKM, NPKS, and NPKM+ in the wheat cropping season, and NPK, NPKM, NPKS, and NPKM+ in the maize cropping season. Four microplots (0.7 × 0.6 m) were established on the northeastern side of each plot. A syringe was used to inject the urea solution into the microplots to ensure that the crops would receive even nutrition. Metal squares (0.35 m high) were driven 0.30 m deep into the soil to prevent surface runoff and lateral contamination. ^{15}N-labeled urea (abundance: 5.2%, produced by the Institute of Chemical Industry in Shanghai, China) was applied to the soil at the same N rate. All P and K applications and field management practices in the microplots were the same as those in the corresponding large plots for winter wheat and summer maize. Soil samples from the microplots were air-dried and ground to pass a 150 mm (100-mesh) screen for total N and ^{15}N isotope analysis. Grain, straw, and soil samples were analyzed for total N and ^{15}N abundance using the micro-Kjeldahl procedure and isotope ratio mass spectrometry (Nimmo et al. 2013). Percentage of fertilizer N recovery in the grain and straw from the crops and the soil from all the microplots at harvest was determined using Equations (1)–(3), where all ^{15}N was expressed as the atom% excess corrected for background abundance (0.37%).

$$\text{N (kg N ha}^{-1}\text{) derived from fertilizer treatment (N}_{\text{dff}}\text{) in plant} = \text{N uptake by plant} \times {}^{15}\text{N atom\% excess in plant} / {}^{15}\text{N atom\% excess in fertilizer} \quad (1)$$

$$\text{N}_{\text{dff}} \text{ in the soil (kg N ha}^{-1}\text{)} = \text{Total N in soil} \times {}^{15}\text{N atom\% excess in soil} / {}^{15}\text{N atom\% excess in fertilizer} \quad (2)$$

$$\text{Fertilizer N recovery (\%)} = \text{Total labeled N in the aboveground biomass} / \text{labeled N applied} \times 100 \quad (3)$$

2.4. Other Ancillary Measurements

Soil samples were collected with a 5 cm inner diameter auger tube from 0 to 100 cm depth from all plots and microplots and separated into 20 cm depth increments at the beginning of the experiment and after harvest of winter wheat and summer maize. Soil samples from 0 to 300 cm were collected after maize harvest in 2013 to determine the distribution of nitrate nitrogen (NO_3-N) and ammoniacal nitrogen (NH_4^+-N) as Nmin in the soil profile after 24 years of experimentation. Soil samples were stored in an ice box immediately after sampling and transported to the laboratory for analysis. Within 12 h, all the fresh soil samples were extracted with 0.01 M $CaCl_2$ solution (soil-to-solution ratio of 1:10). Soil extracts were analyzed for Nmin using continuous flow analysis (Bran and Luebbe TRAACS 2000, Hamburg, Germany). Plants were separated into grain and straw after harvest, and biomass, grain, and straw production were calculated. Grain and straw samples were subsequently oven-dried at 60 °C in a forced-air oven and ground to pass through a 150 mm screen.

2.5. Statistical Analysis

Coefficient of variation (*CV*) was used to assess the range of variation and stability. The *CV* of wheat and maize was calculated as follows:

$$CV = \delta_{sd} / \mu_{ave} \times 100\% \quad (4)$$

where δ_{sd} represents the standard deviation of yield, and μ_{ave} represents the average value of yield (kg ha^{-1}).

The sustainable yield index (I_{sy}) is typically used to assess production sustainability (Li et al., 2011).

$$I_{sy} = (Y - S_D) / Y_{max} \quad (5)$$

where Y represents mean yield (kg ha^{-1}), S_D represents the standard deviation of the yield (kg ha^{-1}), and Y_{max} represents maximum yield (kg ha^{-1}).

The N harvest index (NHI) is frequently used to assess N transport from the shoots and leaves to the grain (Li et al., 2011). NHI values for wheat and maize were calculated as follows:

$$NHI = \frac{N_{grain}}{N_{aboveground-biomass}} \quad (6)$$

where N_{grain} and $N_{aboveground\text{-}biomass}$ are the N uptake in the grain and crop biomass in each treatment (CK, NPK, NPKS, NPKM, and NPKM+).

All statistical analyses were performed using SPSS 16. Prior to statistical analysis, the data (three replications) were examined for homogeneity of variance when required; however, untransformed means and standard errors are presented in figures and tables. Comparisons of the effects of the treatments on yield, yield stability, and Nmin were analyzed using one-way Analysis of Variance (ANOVA). Tukey's honestly significant difference (HSD) test for multiple comparisons among means was employed to test for differences among the treatment means at $p < 0.05$.

3. Results

3.1. Crop Yields, CV, I_{SY}, and NHI of Wheat and Maize

Statistical analyses of grain yield of spring wheat (cropped for four years), winter wheat (cropped for eight years), and maize (cropped for nine years) indicated that fertilizer treatments significantly affected grain yield (Table 3). During the 23-year planting process of the three crops, both crop yield and aboveground biomass showed a significant upward trend (except for individual years) due to the continuous improvement of cropland management level (Figure 2). For spring wheat and winter wheat, the combined application of N and P fertilizers (NP, NPK) led to a significant increase ($p < 0.05$) in biomass and yield compared to the treatments without N and P (control, N and NK). A comparison of the NPK and NPKS treatments for all three crops indicated that straw return did not have any significant effect on yield in the long-term experiment ($p > 0.05$). The NPKM and NPKM+ treatments did not show significant differences in yield during the study period (Table 3). Furthermore, a comparison of the CV of the treatments showed that NPKM and NPKM+ did not differ significantly in the wheat season but were significantly lower for the NPKM treatment in the maize season ($p < 0.05$), illustrating overuse of manure amendment is not conducive to the stability of crop yield in arid regions due to excess nutrients. Compared to the NPK treatments, the NPKM treatment showed significantly lower CV values in the winter wheat season. The NPKS treatment showed lower CV values in the maize season, suggesting that under the condition of equal nitrogen application, appropriate manure amendment and straw returning can better ensure the stability of crop yield. The results for sustainable yield index I_{sy} were similar; the NPKM and NPKS treatments resulted in lower I_{sy} values than those of other treatments in the wheat season, but not in the maize season, indicating that the yield stability of the different crops under study responded differently to different treatments in extremely arid cropland systems. A comparison of NHI for balanced fertilizer treatments showed a significantly higher NHI value for the NPKM treatment than those of other treatments in the wheat and maize seasons. The NHI of the NPK or NPKS treatment did not significantly differ in the wheat season, but it did in the maize season ($p > 0.05$), thus confirming that the addition of organic fertilizer is more conducive to the accumulation of N in the grain under the extreme-drought cropland management model.

Table 3. Mean yield, coefficient of variation (CV), sustainable yield index (I_{sy}), and nitrogen harvest index (NHI) of spring wheat, winter wheat, and maize and the ANOVA results showing 'Groups' of treatments from 24 years' data.

	Spring Wheat [1]				Winter Wheat				Maize			
	Yield ($t\ ha^{-1}$)	CV (%)	I_{SY} (%)	NHI (%)	Yield ($t\ ha^{-1}$)	CV (%)	I_{SY} (%)	NHI (%)	Yield ($t\ ha^{-1}$)	CV (%)	I_{SY} (%)	NHI (%)
CK	1.1 (0.6) ^{c} [2]	51.8 ^{a}	28.8 ^{e}	76.3 ^{a}	1.2 (0.5) ^{c}	43.2 ^{a}	31.5 ^{f}	81.4 ^{b}	4.5 (1.5) ^{c}	23.8 ^{c}	43.4 ^{d}	61.8 ^{bc}
N	2.1 (0.4) ^{b}	20.9 ^{e}	60.2 ^{a}	76.9 ^{a}	2.3 (0.6) ^{b}	25.8 ^{cd}	49.0 ^{c}	81.0 ^{b}	6.5 (1.3) ^{b}	35.5 ^{b}	58.9 ^{c}	60.9 ^{c}
NK	2.0 (0.4) ^{b}	17.6 ^{e}	67.8 ^{a}	72.8 ^{b}	2.4 (0.7) ^{b}	27.1 ^{c}	47.5 ^{cd}	82.4 ^{a}	6.3 (1.4) ^{b}	24.0 ^{c}	61.2 ^{bc}	66.8 ^{a}
NP	3.8 (1.2) ^{a}	32.6 ^{d}	45.5 ^{b}	70.0 ^{b}	5.2 (1.0) ^{a}	23.8 ^{d}	51.5 ^{c}	79.9 ^{bc}	7.6 (1.7) ^{ab}	20.4 ^{d}	64.9 ^{b}	63.3 ^{b}
NPK	3.6 (1.1) ^{a}	33.8 ^{d}	44.5 ^{b}	70.3 ^{b}	5.3 (1.3) ^{a}	24.0 ^{d}	51.1 ^{c}	80.6 ^{b}	7.5 (2.0) ^{ab}	25.8 ^{c}	60.3 ^{c}	58.7 ^{d}
NPKS	3.7 (1.6) ^{a}	42.9 ^{b}	35.2 ^{d}	69.2 ^{bc}	4.7 (1.7) ^{a}	35.5 ^{b}	39.6 ^{e}	80.8 ^{b}	8.0 (1.4) ^{a}	19.2 ^{d}	69.1 ^{a}	61.7 ^{bc}
NPKM	4.2 (1.6) ^{a}	37.2 ^{c}	41.3 ^{c}	78.8 ^{a}	5.7 (1.2) ^{a}	20.4 ^{e}	56.8 ^{b}	82.7 ^{a}	8.2 (2.0) ^{a}	27.1 ^{c}	63.6 ^{b}	65.8 ^{a}
NPKM+	3.9 (1.4) ^{a}	35.0 ^{cd}	43.4 ^{b}	68.5 ^{c}	6.2 (1.2) ^{a}	19.2 ^{e}	61.5 ^{a}	79.5 ^{c}	8.2 (1.7) ^{a}	43.2 ^{a}	68.0 ^{a}	59.0 ^{d}

[1] Different letters (a, b, c, and so on) in yield, CV, I_{SY}, and NHI showed significantly different from LSD at the 0.05 level, the same as below. [2] Data within brackets represent the standard deviation of mean yield.

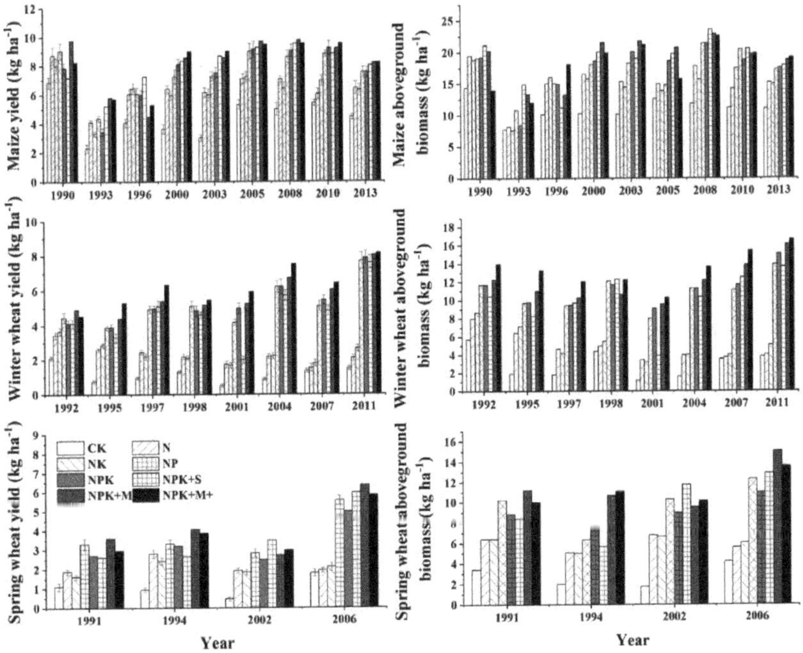

Figure 2. Effect of fertilizer treatments on crop yield and aboveground biomass under long-term fertilization in 1990–2013.

3.2. Residual Mineral N in the Soil Profile under Different Fertilization Treatments

Fertilization treatments resulted in different distribution patterns of NH_4^+-N in the 300 cm soil profile after 24 years of experimentation. The CK treatment had the lowest NH_4^+-N concentration, with values in the range of 0.9–3.4 mg N kg^{-1}; meanwhile, other treatments showed relatively high NH_4^+-N contents, especially for N and NK treatments, with values in the range of 6.8–12.8 and 4.1–15.6 mg N kg^{-1}, respectively, in the 300 cm soil profile, while the NH_4^+-N concentration of other treatments was basically within the range observed for CK, N, and NK treatments. The distribution of NH_4^+-N across all treatments did not show any peaks in the soil profile. The distribution of NO_3^--N in the soil differed from the NH_4^+-N concentration profile. The NPKM+ treatment resulted in the highest NO_3^--N concentration, with a value of 118.9 mg N kg^{-1} in the 20 cm soil profile (Figure 3 shows the distribution after maize harvest), followed by a significant downward trend down to 150 cm into the soil profile. This was followed by the NPKM, NPK, and NPKS treatments, with the peak NO_3^--N concentration in the 20–40 cm soil layer. The distribution pattern in the unbalanced treatments differed and peak values of NO_3^--N concentration are seen between 200–250 cm, suggesting the nutrients that cannot be absorbed by the crop are more likely to migrate deep into the soil.

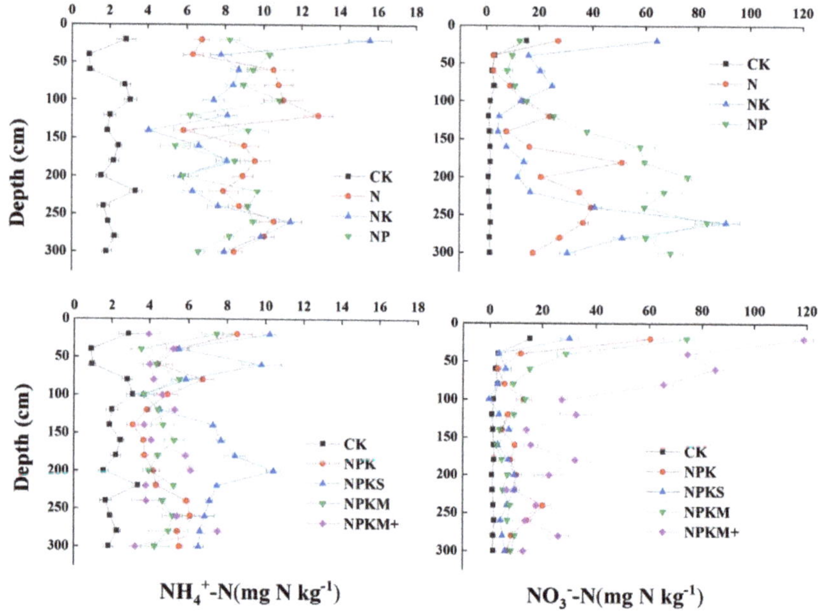

Figure 3. Distribution of NO_3^--N and NH_4^+-N after the maize harvest in 2014 in the 24 years of experimentation in a gray desert soil cropland. Bars indicate error of the mean, n = 3.

Comparing the residual Nmin in different soil layers, we found that the concentration of Nmin in the 300 cm soil profile for the CK treatment was minimum (Figure 4). In contrast, the highest levels of Nmin (mainly in the form of NO_3^--N) were found in the N treatments, concentrated mainly in the 100–300 cm soil profile (Figure 4), followed by the NPKM+ treatment, with 64.5% NO_3^--N accumulated in the 0–100 cm soil profile. NK and NP treatments also resulted in high NO_3^--N, with 56.9% and 47.7% accumulated in the 200–300 cm soil profile, respectively. The NPK and NPKM treatments contributed to relatively low NO_3^--N values of 585 and 617 kg ha^{-1}, respectively, in the 300 cm soil profile, concentrating mainly in the 0–100 cm layer. NPKS showed the lowest NO_3^--N (234 kg ha^{-1}) in the 0–300 cm soil profile among all treatments. Overall, the stock of mineral N decreased in the order N > NPKM+ > NK > NP > NPK, NPKM > NPKS, which was expected in the unbalanced (N, NK, NP) or excessive fertilizer treatments due to P and K limitation or excess fertilizer input.

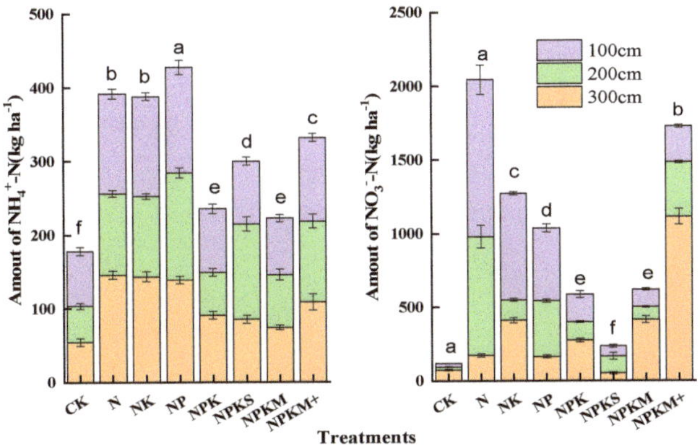

Figure 4. Accumulation of NO_3^--N and NH_4^+-N in the 300 cm soil profile under different fertilization treatments in 2013 (Different letters showed significantly different from LSD at the 0.05).

3.3. N Recovery, Residual N, and Other Losses in Winter Wheat and Summer Maize Seasons

The experiment with ^{15}N-labeled urea to trace the fate of N in terms of crop uptake, soil residual, and other losses in 2011 and 2013 showed significant differences among the different long-term treatments (Table 4, Figure 5) ($p < 0.05$). In the case of winter wheat, ^{15}N-labeled urea was applied only in the topdressing period, having missed the early stage of wheat growth. The results showed that ^{15}N mainly accumulated in the topsoil (0–20 cm), while much less ^{15}N-labeled urea reached the deeper soil profiles. However, in the maize season, although ^{15}N-labeled urea was mainly concentrated in the topsoil, a large amount of ^{15}N-labeled urea still migrated down to a depth of 100 cm, indicating that a significant amount of N is lost by leaching in gray desert soils undercurrent fertilization and irrigation schemes. Furthermore, the NPK and NPKM treatments showed the highest ^{15}N crop uptake rates, with values of 46.4% and 49.4%, respectively, in the wheat season and 28.2% and 32.2%, respectively, in the maize season, followed by the NPKS treatment, in which case the uptake rate was 41.6% and 25.3% in the wheat and maize season, respectively. At the same time, NPKM showed the lowest ^{15}N-labeled urea loss in the wheat season and a relatively low value in the maize season, illustrating that the combination treatment consisting of chemical fertilizer and manure is a relatively good choice for improving NUE while reducing environmental loss of nitrogen in extremely arid croplands.

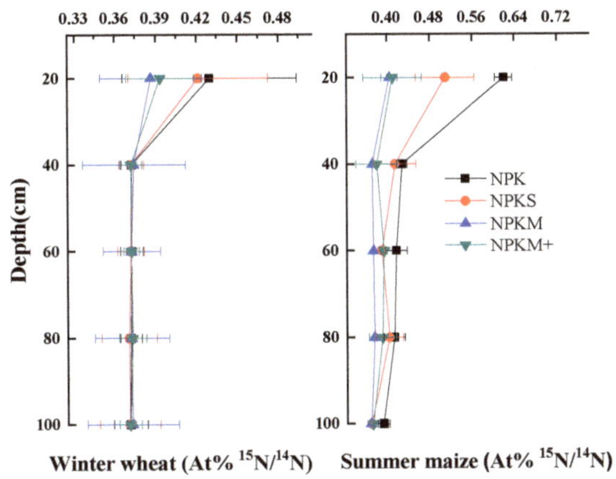

Figure 5. Distribution of at $\%^{15}N/^{14}N$ along the 0–100 cm soil profile under different fertilization treatments in winter wheat and summer maize.

Table 4. The fate of fertilizer N in winter wheat in different long-term treatments using ^{15}N tracer method in 2011–2013.

	Treatment	^{15}N Applied	Crop Uptake		Soil Residual		Unaccounted for	
		kg ha^{-1}	kg ha^{-1}	%	kg ha^{-1}	%	kg ha^{-1}	%
Winter wheat	N	96 [1]	8.7 ± 1.1 [2]	9.0 ^{f}	60.2 ± 5	62.7 ^{a}	29.2 ± 3	30.4 ^{a}
	NK	96	12.8 ± 2.0	13.3 ^{e}	53.6 ± 1.4	55.8 ^{b}	31.6 ± 1.7	32.9 ^{a}
	NP	96	38.9 ± 4.8	40.5 ^{c}	33.4 ± 2.2	34.8 ^{e}	25.7 ± 3.5	26.8 ^{bc}
	NPK	96	44.5 ± 3.3	46.4 ^{ab}	26.1 ± 3.4	27.2 ^{f}	27.4 ± 3.3	28.5 ^{b}
	NPKS	86.4	35.9 ± 1.7	41.6 ^{c}	26.8 ± 0.8	31.0 ^{ef}	25.3 ± 1.3	29.3 ^{ab}
	NPKM	67	33.1 ± 1.6	49.4 ^{a}	31.3 ± 4.1	46.8 ^{c}	3.3 ± 2.6	5.0 ^{d}
	NPKM+	86	26.2 ± 2.4	30.0 ^{d}	36.6 ± 1.9	41.9 ^{cd}	24.5 ± 2.2	28.1 ^{b}
Maize	N	240	/	/	/	/	/	/
	NK	240	/	/	/	/	/	/
	NP	240	/	/	/	/	/	/
	NPK	240	67.8 ± 2.1	28.2 ^{b}	99.9 ± 3.2	41.6 ^{ab}	72.3 ± 2.7	30.1 ^{c}
	NPKS	216	54.7 ± 2.5	25.3 ^{c}	85.8 ± 2.4	39.7 ^{b}	75.6 ± 2.9	35 ^{b}
	NPKM	168	34.3 ± 2.2	32.2 ^{a}	73.1 ± 2.3	43.7 ^{a}	60.1 ± 3.0	35.8 ^{b}
	NPKM+	216	22.6 ± 1.7	10.3 ^{d}	58.4 ± 2.2	26.7 ^{c}	137 ± 4.6	62.9 ^{a}

[1] The amount of fertilizer N was in topdressing period in wheat season. [2] Data are means followed by standard deviation (n = 4).

4. Discussion

4.1. Fertilization and Crop Yield

Crop yield potential is normally defined as the yield of a crop variety grown under optimal management conditions, where water and nutrient supply are optimum, and there is effective control of pests and diseases [26]. Further, yield potential is also affected by the intensity of incident solar radiation and ambient temperature [27,28]. In our long-term study, water was supplied by irrigation because of extreme drought and high evaporation rates; therefore, moisture was not a limiting factor for crop production. However, the various nutrient/fertilizer treatments resulted in different crop yields. Thus, NPK, NPKM, NPKM+, and NPKS provided long-lasting high crop productivity in the 24 years of experimentation in the selected gray desert soil area. Further, except for the first year, the CK treatment consistently rendered the lowest productivity. Balanced fertilizer treatments showed higher production potential than the unbalanced fertilizer application without NP, indicating that oasis gray desert soil was rich in K but deficient in N and P.

Crop yield after balanced fertilizer treatment did not show significant differences ($p > 0.05$), although the production was slightly higher in the NPKM treatment, as seen from the 24-year average. Our results differ from those of research conducted at the Huang-Huai-Hai Plain of China, where Cai et al. (2006) found that the NPK treatment resulted in higher production, compared to all other treatments [29]. This can be explained by the higher proportion of organic fertilizer input and microbial activity due to adequate irrigation, a significant temperature difference between day and night, and alternating freezing and thawing overall, leading to the rapid degradation of soil organic matter [23]. In addition, manure amendment treatments displayed higher yield stability (CV) and NHI than any other treatment in both wheat and maize seasons, which is consistent with the results of other long-term experiments in nonarid areas [1,20,28,30,31]. In addition, the NPKS treatment showed a significantly higher CV than the NPKM and NPK treatments, both in the spring and winter wheat seasons, but not in the summer maize season, as seen from the 24-year data. These findings indicated that straw return is more beneficial for yield stability in the summer maize season than in the winter or spring wheat seasons in extremely arid cropland regions, which might be attributed to the difficulty of decomposing straw under extreme drought and nonirrigated conditions due to its high cellulose content, thus affecting the wheat germination rate [30]. However, it is not clear whether straw return application can ensure crop yield stability.

4.2. Nmin Distribution in the Soil Profile over a 24-Year Nutrient Management Period

Nitrate leaching showed that NPKM+ might lead to a large residual N in the soil layers. This is attributed to the excessive input of manure N, which led to more mineralized N entering the deeper soil profile. However, the results showed that the Nmin after manure amendment treatments (NPKM+ and NPKM) accumulated mainly in the 100 cm soil profile and a low proportion leached deeper than the 100 cm soil profile, while treatments involving the application of chemical fertilizer (N, NK, NP, and NPK), resulted in a large amount of mineral N leaching 100 cm below the soil, reaching even down to a depth of 300 cm due to weak uptake by crops and high-volume irrigation in an extremely arid region. These results further demonstrate that manure amendment is more useful in preventing the downward migration of Nmin under extremely dry environmental conditions. The same phenomenon was observed in our 2-year study using ^{15}N-labeled urea. In addition, the observations of some other studies were similar to our results [27,28,32]. The NPKS treatment showed the lowest N residual and N loss, compared with the other treatments; this finding may be explained by the following two reasons: first, compared with the NPK treatment, the NPKS treatment had less chemical fertilizer input due to the inclusion of N in the straw; second, the high carbon/nitrogen (C/N) ratio of straw holds mineral N more easily in the soil by increasing the microbial abundance and activity [31,33,34]. Comparedwith other nonaridcroplands [13,26,35,36], extremely arid croplands have a higher Nmin residual in the soil profile due to the high volume of irrigation water, which increases N deposition, and to overapplication of N fertilizers. Lv et al. reported that dry and wet N deposition in the arid oasis area of western China can reach up to 33 kg ha^{-1}, while NO_3-N goingintothe groundwater can be 8–17 kg N ha^{-1} yr^{-1} due to heavy irrigation [12]. The resulting input of environmental N in extremely arid areas is not lower than in other regions, such as North China [14]. Overall, although balanced fertilizers combined with manure and straw return can maintain a suitable concentration of mineral N in the 0–100 cm soil profile, excessive nitrogen input still exists in extremely arid region cropland systems due to neglect of the environmental nitrogen input.

4.3. Fertilizer N Recovery and Loss

The combined fertilizer treatments tested herein resulted in different fertilizer N recovery and loss rates. In gray desert croplands, unbalanced fertilization can result in high residual N in the soil due to insufficient absorption, as seen in our study using labeled ^{15}N. The treatment with NPK showed a relatively lower ^{15}N residual in the 100 cm soil profile than either the NP or the NK treatment. This finding clearly indicates that N absorption is higher when the supply of nutrients is balanced. Among the balanced fertilizer treatments, NPKM showed the highest proportion of residual ^{15}N in both wheat and maize seasons, concomitant with other losses being significantly lower, thus illustrating that suitable manure amendment does help to retain mineral N in the soil and reduces other losses, due to N uptake and fixation by the abundant soil microorganisms [15,16,32,37]. Further, the NPK and NPKS treatments did not show significant differences in ^{15}N soil residues or environmental losses, although there were slight variations in the different crop seasons, suggesting that straw return to the field did not significantly improve NUE, which might be attributed to the low efficiency of straw decomposition on account of the extremely arid environmental conditions and high C/N ratio, which in turn may account for the different results reported for nonarid areas [27,30,35,38]. In addition, the NPKM treatment resulted In the highest ^{15}N-urea recovery in both wheat and maize seasons, with NUE of 49.4% and 32.2%, respectively, indicating that manure amendment may promote N absorption from the chemical fertilizer. Our results further demonstrate that the combined application of organic and inorganic fertilizers is an adequate choice to improve NUE in extremely arid cropland areas.

5. Conclusions

High organic fertilizer input (NPKM+) was not conducive to crop yield stability or nitrogen harvest index from the perspective of long-term application in extremely arid cropland regions. Crop yield under the NPKM treatment did not significantly differ from that of the NPK or NPKS treatments, but the coefficient of variation and NHI were lower under conditions of equal nitrogen input. Compared to the NPK treatment, the NPKM and NPKS treatments seemed to be more conducive to the accumulation of soil mineral nitrogen in the 0–100 cm soil profile and reduce nitrogen losses due to leaching under heavy irrigation in extremely arid regions. The highest recovery rate and the lowest unaccounted losses were observed in the NPKM treatment, indicating that the combined application of inorganic and organic fertilizers or straw return, such as treatments NPKM and NPKS, was commendable to ensure the yield security and environmental safety in extremely arid regions.

Author Contributions: Conceptualization, S.L., J.L.; Data curation: S.L., X.W.; Formal analysis: S.L.; Funding acquisition: J.L.; Investigation: S.L.; Methodology: S.L.; Project administration: J.L., J.G.; Resources: X.W.; Software: S.L., J.L.; Supervision: J.L., J.G., C.K.; Validation: J.L., J.G.; Visualization: S.L.; Writing—review and editing: S.L., J.L., J.G. All authors have read and agreed to the published version of the manuscript.

Funding: This work was funded by the Natural Science Foundation of China (Grant no. 41807098, 2019), and National Gray Desert Soil Fertility and Fertilizer Efficiency Monitoring Station of China.

Institutional Review Board Statement: Not applicable.

Informed Consent Statement: Not applicable.

Data Availability Statement: The data presented in this study are available on request from the corresponding authors.

Acknowledgments: We are thankful to all mumbers of staff from National Gray Desert Soil Fertility and Fertilizer Efficiency Monitoring Station of China for their contribution during the field experiments.

Conflicts of Interest: The authors declare no conflict of interest.

References

1. Srinivasarao, C.; Kundu, S.; Venkateswarlu, B.; Lal, R.; Singh, A.K.; Balaguravaiah, G.; Vijayasankarbabu, M.; Vittal, K.P.R.; Reddy, S.; Rupendra Manideep, V. Long-term effects of fertilization and manuring on groundnut yield and nutrient balance of alfisols under rainfed farming in india. *Nutr. Cycl. Agroecosyst.* **2013**, *96*, 29–46. [CrossRef]
2. Sainju, U.M.; Ghimire, R.; Mishra, U.; Jagadamma, S. Reducing nitrous oxide emissions and optimizing nitrogen use efficiency in dryland crop rotations with different nitrogen rates. *Nutr. Cycl. Agroecosyst.* **2020**, *116*, 381–395. [CrossRef]
3. Jiang, C.M.; Yu, W.T.; Ma, Q.; Xu, Y.G.; Zou, H. Alleviating global warming potential by soil carbon sequestration: A multi-level straw incorporation experiment from a maize cropping system in northeast china. *Soil Tillage Res.* **2017**, *170*, 77–84. [CrossRef]
4. Yang, B.; Xiong, Z.; Wang, J.; Xu, X.; Huang, Q.; Shen, Q. Mitigating net global warming potential and greenhouse gas intensities by substituting chemical nitrogen fertilizers with organic fertilization strategies in rice–wheat annual rotation systems in china: A 3-year field experiment. *Ecol. Eng.* **2015**, *81*, 289–297. [CrossRef]
5. Sainju, U.M.; Stevens, W.B.; Caesar-TonThat, T.; Liebig, M.A.; Wang, J. Net global warming potential and greenhouse gas intensity influenced by irrigation, tillage, crop rotation, and nitrogen fertilization. *J. Environ. Qual.* **2014**, *43*, 777–788. [CrossRef] [PubMed]
6. Kundu, S.; Bhattacharyya, R.; Prakash, V.; Gupta, H.S.; Pathak, H.; Ladha, J.K. Long-term yield trend and sustainability of rainfed soybean–wheat system through farmyard manure application in a sandy loam soil of the indian himalayas. *Biol. Fertil. Soils* **2006**, *43*, 271–280. [CrossRef]
7. Lv, J.; Yin, X.; Dorich, C.; Olave, R.; Wang, X.; Kou, C.; Song, X. Net field global warming potential and greenhouse gas intensity in typical arid cropping systems of china: A 3-year field measurement from long-term fertilizer experiments. *Soil Tillage Res.* **2021**, *212*, 105053. [CrossRef]
8. Voss, M.; Hietanen, S. Biogeochemistry the depths of nitrogen cycling. *Nature* **2013**, *493*, 616–618. [CrossRef]
9. Liu, X.J.; Duan, L.; Mo, J.M.; Du, E.Z.; Shen, J.L.; Lu, X.K.; Zhang, Y.; Zhou, X.B.; He, C.N.; Zhang, F.S. Nitrogen deposition and its ecological impact in china: An overview. *Environ. Pollut.* **2011**, *159*, 2251–2264. [CrossRef]
10. Nosengo, N. Fertilized to death. *Nature* **2003**, *425*, 894–895. [CrossRef]
11. Luo, X.S.; Wen, X.; Liu, X.J. An evaluation of atmospheric nr pollution and deposition in north china after the beijing olympics. *Atmos. Environ.* **2014**, *74*, 209–216. [CrossRef]

12. Lv, J.; Liu, H.; Wang, X.; Li, K.; Tian, C.; Liu, X. Highly arid oasis yield, soil mineral n accumulation and n balance in a wheat-cotton rotation with drip irrigation and mulching film management. *PLoS ONE* **2016**, *11*, e0165404. [CrossRef]
13. Li, K.H.; Liu, X.J.; Song, W.; Chang, Y.H.; Hu, Y.K.; Tian, C.Y. Atmospheric nitrogen deposition at two sites in an arid environment of central asia. *PLoS ONE* **2013**, *8*, e67018. [CrossRef] [PubMed]
14. Fan, M.S.; Lu, S.H.; Jiang, R.F.; Liu, X.J.; Zeng, X.Z.; Goulding, K.W.T.; Zhang, F.S. Nitrogen input, n-15 balance and mineral n dynamics in a rice-wheat rotation in southwest china. *Nutr. Cycl. Agroecosyst.* **2007**, *79*, 255–265. [CrossRef]
15. Bah, H.; Ren, X.; Wang, Y.; Tang, J.; Zhu, B. Characterizing greenhouse gas emissions and global warming potential of wheat-maize cropping systems in response to organic amendments in eutric regosols, China. *Atmosphere* **2020**, *11*, 614. [CrossRef]
16. Oladipo, D.G.; Wei, K.; Hu, L.; Medaiyese, A.; Bah, H.; Gbadegesin, L.A.; Zhu, B. Short-term assessment of nitrous oxide and methane emissions on a crop yield basis in response to different organic amendment types in sichuan basin. *Atmosphere* **2021**, *12*, 1104. [CrossRef]
17. Zhang, X.; Guo, J.; Vogt, R.D.; Mulder, J.; Wang, Y.; Qian, C.; Wang, J.; Zhang, X. Soil acidification as an additional driver to organic carbon accumulation in major chinese croplands. *Geoderma* **2020**, *366*, 114234. [CrossRef]
18. Hao, T.; Liu, X.; Zhu, Q.; Zeng, M.; Chen, X.; Yang, L.; Shen, J.; Shi, X.; Zhang, F.; de Vries, W. Quantifying drivers of soil acidification in three chinese cropping systems. *Soil Tillage Res.* **2022**, *215*, 105230. [CrossRef]
19. Wu, Z.; Sun, X.; Sun, Y.; Yan, J.; Zhao, Y.; Chen, J. Soil acidification and factors controlling topsoil ph shift of cropland in central china from 2008 to 2018. *Geoderma* **2022**, *408*, 115586. [CrossRef]
20. Steiner, C.; Teixeira, W.G.; Lehmann, J.; Nehls, T.; de Macêdo, J.L.V.; Blum, W.E.H.; Zech, W. Long term effects of manure, charcoal and mineral fertilization on crop production and fertility on a highly weathered central amazonian upland soil. *Plant Soil* **2007**, *291*, 275–290. [CrossRef]
21. Ning, S.; Zhou, B.; Shi, J.; Wang, Q. Soil water/salt balance and water productivity of typical irrigation schedules for cotton under film mulched drip irrigation in northern xinjiang. *Agric. Water Manag.* **2021**, *245*, 106651. [CrossRef]
22. Wu, L.; Zhou, X.; Zhao, D.; Feng, T.; Zhou, J.; Sun, T.; Wang, J.; Wang, C. Seasonal variation and exposure risk assessment of pesticide residues in vegetables from xinjiang uygur autonomous region of china during 2010–2014. *J. Food Compos. Anal.* **2017**, *58*, 1–9. [CrossRef]
23. Lyu, J.; Liu, H.; Wang, X.; Olave, R.; Tian, C.; Liu, X. Crop yields and soil organic carbon dynamics in a long-term fertilization experiment in an extremely arid region of northern xinjiang, china. *J. Arid Land* **2017**, *9*, 345–354. [CrossRef]
24. Lv, J.; Liu, X.; Liu, H.; Wang, X.; Li, K.; Tian, C.; Christie, P. Greenhouse gas intensity and net annual global warming potential of cotton cropping systems in an extremely arid region. *Nutr. Cycl. Agroecosyst.* **2014**, *98*, 15–26. [CrossRef]
25. Kuang, W.; Gao, X.; Tenuta, M.; Gui, D.; Zeng, F. Relationship between soil profile accumulation and surface emission of N_2O: Effects of soil moisture and fertilizer nitrogen. *Biol. Fertil. Soils* **2019**, *55*, 97–107. [CrossRef]
26. Andruschkewitsch, R.; Koch, H.-J.; Ludwig, B. Effect of long-term tillage treatments on the temporal dynamics of water-stable aggregates and on macro-aggregate turnover at three german sites. *Geoderma* **2014**, *217–218*, 57–64. [CrossRef]
27. Zheng, J.; Chen, J.; Pan, G.; Wang, G.; Liu, X.; Zhang, X.; Li, L.; Bian, R.; Cheng, K.; Zheng, J. A long-term hybrid poplar plantation on cropland reduces soil organic carbon mineralization and shifts microbial community abundance and composition. *Appl. Soil Ecol.* **2017**, *111*, 94–104. [CrossRef]
28. van der Bom, F.; Magid, J.; Jensen, L.S. Long-term fertilisation strategies and form affect nutrient budgets and soil test values, soil carbon retention and crop yield resilience. *Plant Soil* **2018**, *434*, 47–64. [CrossRef]
29. Cai, Z.C.; Qin, S.W. Dynamics of crop yields and soil organic carbon in a long-term fertilization experiment in the huang-huai hai plain of china. *Geoderma* **2006**, *136*, 708–715. [CrossRef]
30. Malhi, S.S.; Nyborg, M.; Goddard, T.; Puurveen, D. Long-term tillage, straw management and n fertilization effects on quantity and quality of organic c and n in a black chernozem soil. *Nutr. Cycl. Agroecosyst.* **2011**, *90*, 227–241. [CrossRef]
31. Li, X.; Wen, Q.; Zhang, S.; Li, N.; Yang, J.; Romanyà, J.; Han, X. Long-term changes in organic and inorganic phosphorus compounds as affected by long-term synthetic fertilisers and pig manure in arable soils. *Plant Soil* **2022**, *472*, 239–255. [CrossRef]
32. Vuaille, J.; Gravert, T.K.O.; Magid, J.; Hansen, M.; Cedergreen, N. Long-term fertilization with urban and animal wastes enhances soil quality but introduces pharmaceuticals and personal care products. *Agron. Sustain. Dev.* **2021**, *42*, 1. [CrossRef]
33. Guo, J.; Wang, B.; Wang, G.; Myo, S.T.Z.; Cao, F. Effects of three cropland afforestation practices on the vertical distribution of soil organic carbon pools and nutrients in eastern china. *Glob. Ecol. Conserv.* **2020**, *22*, e00913. [CrossRef]
34. Jiang, M.; Wang, X.; Liusui, Y.; Han, C.; Zhao, C.; Liu, H. Variation of soil aggregation and intra-aggregate carbon by long-term fertilization with aggregate formation in a grey desert soil. *Catena* **2017**, *149*, 437–445. [CrossRef]
35. Congreves, K.A.; Hooker, D.C.; Hayes, A.; Verhallen, E.A.; Van Eerd, L.L. Interaction of long-term nitrogen fertilizer application, crop rotation, and tillage system on soil carbon and nitrogen dynamics. *Plant Soil* **2016**, *410*, 113–127. [CrossRef]
36. Andruschkewitsch, R.; Geisseler, D.; Koch, H.J.; Ludwig, B. Effects of tillage on contents of organic carbon, nitrogen, water-stable aggregates and light fraction for four different long-term trials. *Geoderma* **2013**, *192*, 368–377. [CrossRef]
37. Yi, B.; Zhang, Q.; Gu, C.; Li, J.; Abbas, T.; Di, H. Effects of different fertilization regimes on nitrogen and phosphorus losses by surface runoff and bacterial community in a vegetable soil. *J. Soils Sediments* **2018**, *18*, 3186–3196. [CrossRef]
38. Begum, K.; Kuhnert, M.; Yeluripati, J.; Glendining, M.; Smith, P. Simulating soil carbon sequestration from long term fertilizer and manure additions under continuous wheat using the daily daycent model. *Nutr. Cycl. Agroecosyst.* **2017**, *109*, 291–302. [CrossRef]

MDPI
St. Alban-Anlage 66
4052 Basel
Switzerland
www.mdpi.com

Atmosphere Editorial Office
E-mail: atmosphere@mdpi.com
www.mdpi.com/journal/atmosphere